P9-DMD-571

Diana Mosley

By the same author

1939: THE LAST SEASON

CIRCE: THE LIFE OF EDITH,
MARCHIONESS OF LONDONDERRY

THE VICEROY'S DAUGHTERS:
THE LIVES OF THE CURZON SISTERS

ANNE DE COURCY

DIANA MOSLEY

Mitford Beauty,

British Fascist,

Hitler's Angel

wm

William Morrow

An Imprint of HarperCollins*Publishers*

Published in Great Britain in 2003 by Chatto & Windus, Random House, London.

DIANA MOSLEY. Copyright © 2003 by Anne de Courcy. All rights reserved. Printed in the United States of America. No part of this book may be used or reproduced in any manner whatsoever without written permission except in the case of brief quotations embodied in critical articles and reviews. For information address HarperCollins Publishers Inc., 10 East 53rd Street, New York, NY 10022.

HarperCollins books may be purchased for educational, business, or sales promotional use. For information please write: Special Markets Department, HarperCollins Publishers Inc., 10 East 53rd Street, New York, NY 10022.

FIRST U.S. EDITION

Printed on acid-free paper

Library of Congress Cataloging-in-Publication Data has been applied for.

ISBN 0-06-056532-2

03 04 05 06 07 ❖/ RRD 10 9 8 7 6 5 4 3 2 1

Contents

List of Illustrations

Third section:

Unless otherwise stated, all the photographs are from sources in private hands.

The author and publishers have made every effort to trace the holders of text and illustrative copyright. If any inadvertent errors or omissions remain, they can be rectified in future editions.

For Robert

Preface

Although I came to love Diana Mosley personally, I abhorred her politics. The hundreds of hours I spent with her, the personal papers and correspondence she let me read, her diaries and the many letters we exchanged over the years helped illumine the complex personality and extraordinary life of the woman who became a focus of controversy and hatred, as her chosen course ran counter first to the social conventions of her time and then, devastatingly, to the beliefs of her country as it fought for its life. As I believe objectivity is one of the first duties of a biographer, I have tried to tell her story as it happened, without benefit of hindsight, so that readers can come to their own conclusions.

What first struck everyone who met her was her beauty. Throughout her long life it cast an enduring spell. Even her mosaic image on the floor of the National Gallery, with its blind blue gaze, refulgent blondeness and curved Grecian limbs, expresses this goddess-like quality. Yet the romantic adoration she evoked was, strangely, almost free of sexual content – like her sister Nancy, she was often surrounded by a homosexual coterie.

The cleverest of the six Mitford sisters, Diana was brought up in a household where extremism and prejudice were the norm, overlaid by exquisite, old-fashioned manners. Later, her love of music and the arts, her good looks and legendary charm often blinded both admirers and critics to the obnoxiousness of her views, while her high intelligence and the heritage of her privileged yet eccentric background endowed her with remarkable self-confidence and a disdain for the opinion of others.

As a young woman her elegance, warmth, gaiety and oblique wit ensured that she became an iconic figure, with intimates ranging from social figures such as Lady Cunard and Daisy Fellowes to writers and artists such as Evelyn Waugh, Lytton Strachey, Cecil Beaton, John Betjeman, Harold Acton and Augustus John.

After her meeting with Mosley and her complete enthralment by both the man and his ideas, her political beliefs and attitudes changed and hardened, geared only to what 'The Leader', as she called Mosley, thought and did. It was during his absences, speaking round the country to raise support for his newly formed fascist party, that she first went to Germany, followed by her meeting with Hitler – the man who carried The Leader's political beliefs to their ultimate.

Within a few years, she had become one of the Führer's intimates, closer to him than any other woman except Magda Goebbels and Winifred Wagner; and of necessity on familiar terms with most of his most sinister henchmen. Then, and later, she made no secret of her great affection for and admiration of the man she and her sister Unity called 'sweet Uncle Wolf'.

'If the proper study of mankind is man, then having had the luck to know such an unusual man the only thing to do was to see as much of him as possible,' she declared.

Although quite clear that fascism would never have worked after the war – after all, by that time Mosley had already said so – she remained equally convinced it would have been a valid form of government in the 1930s. Her view was that major reforms could be enacted only with a small number – five, three or preferably one, and that one her husband – at the top, with the power to carry through decisions.

Internment in prison (during the war, and under Regulation 18B) did nothing to modify these convictions. When, eventually, after several years of refusing to believe it, she was forced to admit the Holocaust had actually happened (though in true Diana vein she claimed it had been exaggerated) she was horrified. 'I thought it appalling, wicked, monstrous,' she said.

Hitler, however, escaped her condemnation. To the end, she maintained her affection for this man, who so nearly demolished everything the western world stands for. 'Knowing about the Holocaust absolutely did not change my perspective of Hitler,' she told me. 'I don't think of him as the man who did that, I think of him as the man I knew, who wouldn't have been capable of that. If he became so capable, it was really a form of madness, which in a way is understandable if everything you've worked for is being ruined.'

When asked what would have happened to the Jews if there had, as her husband Sir Oswald Mosley had advocated, been a negotiated peace, she would respond with the same wilfully blind disingenuousness that Hitler 'might have given them Morocco or Tunis or some other emptyish place' (presumably forcibly acquired from its inhabitants in this same 'negotiated peace'). 'And then they would have been told "either go, or stay at your own risk",' a draconian solution that appeared to cause her no qualms at all.

Friends learnt either to avoid the whole question of politics or to appear to submit to the selective truths with which she justified her view of history. To interviewers, she would respond with practised ease. Ask about the Holocaust, and she would reply: 'What about the Gulag? Stalin killed many more than Hitler. Or the thirty million Chinese killed in the so-called Cultural Revolution? What about Pol Pot? The fact is, we live in a cruel world.'

How *could* a woman so intelligent, perceptive and intuitive not only embrace the tenets of National Socialism so wholeheartedly but remain faithful to the memory of a man who led the world into a terrible war and attempted the genocide of an entire race?

The answer – which caused a lifelong conflict at the core of her being – lay in her own powerful, passionate and singlehearted nature.

There is no doubt that Mosley would have been a fascist whether or not he had met Diana – indeed, he had already begun to contemplate it before their meeting – but it is highly unlikely Diana would have embraced such an unEnglish creed if she had never met Mosley.

Instead, her politics might have settled somewhere left of centre. She had been deeply affected by the miseries of the General Strike in 1926, and her political sympathies lay with the underdog; on the other hand, she thought socialism dreary in the extreme. 'I suppose I would have described myself as a Lloyd George liberal,' she told me.

Then came Mosley, a man for whom politics and sexual domination were inextricably intertwined (later, he far preferred Mussolini to Hitler because he regarded the Italian leader as more virile), met at the moment when his genuine idealism for banishing the terrible unemployment of the early 1930s was at its height. His conviction, personal magnetism and well-honed sexual expertise quickly bound her to him. From then on she followed him unswervingly, clinging to many of his beliefs even after he himself had abandoned them.

Or as she said herself when I asked her how, at the age of twenty-two, in love with a married man who had told her he had no intention of leaving his wife, she had seen the future, she replied simply: 'My idea of the future was him.' And for her, that is how it remained.

Anne de Courcy,
London,
August 2003.

ACKNOWLEDGEMENTS

My first and greatest debt is of course to Diana Mosley herself, for the many hours of interviews she gave me and for permission to reproduce photographs from her private albums, and extracts from her letters, diaries and writings, as well as for her generous hospitality. For talking to me so freely and for giving me permission to quote relevant material, I am grateful to Lady Mosley's immediate family: her sisters the Duchess of Devonshire, the Hon. Mrs Pamela Jackson and the Hon. Mrs Robert Treuhaft; her sons Lord Moyne, the Hon. Desmond Guinness and Alexander Mosley and her daughter-in-law Charlotte Mosley; and her stepchildren Lord Ravensdale (Nicholas Mosley), Mrs Forbes-Adam and Michael Mosley.

I owe much to Jane Lady Abdy, Erica Betts, Michael Bloch, Simon Blow, Lady Butler, Shirley Conran, Elsie Corrigan, Quentin Crewe, June Ducas, Lord Gladwyn, Derek Hill, Louise Irvine, Joy and Richard Law, the Rt. Hon. Sir Frederick Lawton, Walter Lees, James Lees-Milne, Patrick Leigh Fermor, Lord Longford, Lady Longford, Lady Cecilia McKenna, Lady Alexandra Metcalfe, David Metcalfe, Lord Monckton of Brenchley, Frances Partridge, Lord Skidelsky, Hugo Vickers and John Warburton for their recollections and insights.

I am also most grateful to the Duchess of Devonshire for allowing me to reproduce photographs from her private albums and to quote writings and letters; to Rosaleen Mulji for permission to quote the letters of her father Bryan Guinness and extracts from his books and poetry; to Winston Churchill for permission to quote the letters of his father Randolph Churchill; to Lord Moyne and Desmond Guinness for allowing me to quote from letters and writings; to Nicholas Mosley for allowing me to quote from writings and letters, and for his kind permission to quote from his parents' letters and from his aunt Irene Ravensdale's diary, and for reading the chapters concerned with his biography of his father; to Francis Sitwell for permission to quote from his mother's diaries; to Francis Beckett for permission to quote from his father's unpublished manuscript; to Auberon Waugh for permission to quote extracts from his father's diary and letters; to Mr Frank Cakebread for letting me see his *History and Description of Savehay Farm*; to the Beineke Library at Yale for the letters of James Lees-Milne; and to Steve Torrington and the staff of the *Daily Mail* library for their help with archival material.

Quotations from *Carrington: Letters and extracts from her diaries*: chosen with an introduction by David Garnett, published by Jonathan Cape are reprinted by permission of The Random House Group Ltd. Quotations from *John Betjeman Letters*: edited by Candida Lycett Green are reprinted by permission of The Random House Group Ltd.

Chapter 1

Exceptional beauty is an attribute which defines its possessor's life. When Diana Mitford was only ten, her sixteen-year-old cousin Michael Bowles fell violently in love with her. Her looks, charm and gaiety, enhanced by the setting of a close and vivid family life, had an irresistible appeal to this rather lonely boy. From Marlborough – then one of the most rigorous and least enjoyable of public schools – he wrote daily screeds of devotion, often concluding wistfully, 'I suppose we must wait six years and then you will be old enough to marry.' Although this devotion was entirely proper – Diana was far more interested in the family ponies, dogs and chickens – he was uneasily conscious that it might be misinterpreted, so persuaded Mabel, head parlourmaid and friend of all the Mitford children, to give his letters to Diana privately instead of leaving them on the hall table with the others.

Discovery was inevitable. Lord Redesdale reacted as if his ten-year-old child had planned to elope, rushing straight to Marlborough and storming unannounced into the study Bowles shared with his friend Mitchell.

'Is this Michael Bowles's room?' he shouted furiously. 'My name is Redesdale and I want to talk to him.' His rage was so terrifying that Mitchell, alarmed, set off to find his friend.

'Somebody called Redesdale has come to see you,' he said. 'You've got to hide for a couple of hours, until he goes away. Otherwise I think he'll kill you.'

Eventually, after stumping up and down muttering for some time, the angry father departed. Although Lord Redesdale forgot fairly quickly, Michael Bowles was so terrified by the incident that he lost touch with his cousins for almost forty years.

The episode illustrates not only Lord Redesdale's almost oriental paranoia about the chastity of his daughters but the effects of a personality so powerful that it was stamped on his children indelibly.

The second son and third of the second Lord Redesdale's nine children, David Freeman-Mitford had impressively good looks, with blue eyes under light, brown-blond hair and a smooth skin tanned by a life spent as far as he could manage it out of doors. He was a man of strong and irrational prejudices. He loathed Roman Catholics, Jews and foreigners – especially Germans – and took the occasional instant dislike to some harmless individual. His special fury was reserved for men whom he suspected of wishing to woo his daughters.

His charm, when he exerted it, was formidable; he was loving, affectionate, even sentimental, and immensely funny, with an original, oblique and intelligent but uneducated mind. His rages were terrifying, though actual punishment was seldom worse than being sent out of the room or, occasionally, early to bed. What made them so devastating was their unpredictability: dazzling good humour could give way without warning to ferocious temper. Sometimes his children could get away with cheekiness and uproarious wildness, the next moment they would be sent upstairs in tears for the same behaviour that had made him laugh only minutes before.

To the children's friends this emotional quicksand was petrifying – one shy boy was threatened with a horsewhip for putting his feet on a sofa, another referred to as 'that hog Watson' to his face. James Lees-Milne has described an evening when he was sent away from the house at nine-thirty p.m. in a downpour, for attacking an anti-German film about Edith Cavell. Equally true to form, a couple of hours later, drenched and miserable, he was welcomed back affectionately by his host, who appeared to have forgotten all about the incident.

But these extremes merely toughened the Mitford children's psyches. Grouped on a sunlit lawn in muslin frocks and picture hats they may, as one dazzled visitor remarked, have resembled a Winterhalter portrait but under the ribbons they were of the same steely breed. 'Mitfords are a savage tribe,' wrote Goronwy Rees.

David Mitford was the sun around which his daughters revolved. Forever simultaneously seeking his approval and seeing how far they could go, they led an emotional life that was a switchback between hysterical tears ('floods') and gales of laughter ('shrieks'). Their often outrageous behaviour, designed to attract his attention, found echoes in adult life. Jessica spoke of the 'strong streak of delinquency' in her husband Esmond

Romilly that struck such a responsive chord in her; Unity found a positive joy in shocking people and in extremes of behaviour; and most of the sisters later conceived passions for men with notably strong personalities or convictions. As Nancy wrote of Uncle Matthew, the character founded on her father in her novel *The Pursuit of Love*: 'Much as we feared, much as we disapproved of, passionately as we sometimes hated Uncle Matthew, he still remained for us a sort of criterion of English manhood; there seemed something not quite right about any man who greatly differed from him.'

To Diana, for years of her childhood David's favourite daughter, he handed on his honesty, his funniness, a singleness of purpose so undeviating as to be at times both ruthless and blinkered, the brilliant Mitford blue eyes and an independence of mind that cared nothing for what others might think, say or do. Perhaps significantly, in view of their own later political beliefs, she and Unity described him as 'one of Nature's fascists'.

There was nothing obvious in David Mitford's parentage to produce such an extreme, eccentric personality. His father, 'Bertie' Redesdale, was a well-read, cultivated man, a cousin and contemporary of the poet Algernon Charles Swinburne, with the necessary gravity and cleverness to serve with distinction as a diplomat, first in Russia and then as First Secretary at the British Embassy in Tokyo. Here he became fascinated by all things Japanese, learning the language and being one of the first Europeans to be presented to the emperor. He was a great gardener and his influence can still be seen at Sandringham, where he assisted in laying out the gardens – his speciality was bamboo. He was the author of a number of books, from the best-selling *Tales of Old Japan* (written on his return from the Far East) to his memoirs. David's mother, born Lady Clementine Ogilvy, was a first cousin of Bertrand Russell.

Bertie Redesdale was also intensely musical, admiring especially the German composers and Wagner in particular. He visited Bayreuth frequently, not only for the festival of music and opera but also to see the Wagner family – through his friend, the Wagner expert Houston Stewart Chamberlain (whose works evinced a deep belief in the superiority of the Teutonic race), he had become such an intimate of the Wagners that his photograph was on Siegfried Wagner's writing desk. His passion for Wagner's work, with its mythic themes and stirring martial passages, was a cultural bonding which would prove significant in the empathy later felt by both Diana and her brother Tom for Nazi Germany.

David was born on 13 March 1878. Even as a child his temper was appalling and his behaviour often so bad that his father would not let him follow his older brother Clem to Eton, fearing that his disruptive presence might make life there difficult for Clem. Instead, he was educated at Radley, which he hated. Team spirit, for which Radley was famous, was low on his list of priorities and he detested all academic work. He could, though, speak excellent French, learned in early days in the schoolroom from a M. Cuvalier, who went on to become a French master at Eton.

His career began with tea planting in Ceylon. When the Boer War broke out in 1899, he enlisted as a private in the Northumberland Fusiliers and was so badly wounded that one lung had to be removed. In February 1904 he married twenty-four-year-old Sydney Gibson Bowles, whom he had first met when she was a child of fourteen. She was tall, reserved and attractive, with light brown hair, beautiful eyes and a manner of speaking – with drawling, idiosyncratic emphases, ejaculations and sudden swoops up or down the scale – that later became known as the Mitford voice.

Sydney's father was one of the cleverest and most amusing men of his age. Thomas Gibson Bowles was the illegitimate son of an early Victorian Cabinet minister, Milner Gibson, by his mistress Susan Bowles. Milner Gibson brought up his 'by-blow' in his own large family, though Thomas was sent to France to be educated.

Thomas went into journalism and, with his perfect French, became the Paris correspondent for the influential *Morning Post* – during the siege of Paris, he was mistaken for a German spy because of his flaxen hair and blue eyes. When he was twenty-five he founded *Vanity Fair* which, with its cartoons by Spy of politicians and notables, quickly became the rage. He entered parliament, sitting as the Conservative member for King's Lynn for many years – though he never hesitated to criticise his own party in another of his publications, the *Candid Quarterly*.

He was an original in every sense and inculcated principles of health in his daughter Sydney from which she never deviated. They were an early form of preventive medicine, resting on the belief that, given the right food and healthy living conditions including fresh air and exercise, the body – the 'Good Body', as Sydney called it – would look after itself. He also advocated showers which, in the days when having a bath was often an adventure in itself, were unheard of. 'The sort of bath we all take is a great mistake,' he once told a schoolfriend of his grandson Tom. 'You wash the dirt off yourself, then you sponge it back on.'

All her life Sydney baked her own bread and the food at her table, though plain, was always first class. Like her father, she followed the Mosaic Law – she had a theory that Orthodox Jews never got cancer. Sydney's children were forbidden shellfish or crustaceans – Diana did not taste lobster until she was married – and allowed meat only from animals that had cloven hooves or chewed the cud. Simple dishes like rabbit pie were therefore unknown. Bacon and sausages were, however, a regular breakfast dish for David, who refused to take any notice of his wife's dietary quirks; often, after their parents had left the dining room but before breakfast had been cleared away, the children would sneak in and thrust what remained of these forbidden delicacies up the legs of their elasticated knickers to be enjoyed later. Tom used to make his sisters' mouths water by his descriptions of school breakfasts ('sausages AND bacon!').

Thomas Bowles owned another publication, *The Lady*, which he had founded in order to provide a worthwhile job for his mistress, whom he made its editor. It was to prove a lifeline for his son. When David and Sydney married in 1904, by the standards of their class and kind they were poor: their total income was £600 a year. *The Lady* provided an allowance for Sydney and a job for David, who was employed there for ten years, although he disliked the work almost as much as he loathed London life. Equally typically, Thomas Bowles lent the bride and groom his yacht, the *Hoyden*, for their honeymoon.

Thomas's passion was the sea: he was a master mariner, he put his two sons into the navy, and he spent all the time he could on the *Hoyden*. His family accompanied him on his annual summer cruise, usually to the smart Channel resorts of Deauville and Trouville. Here they met the fashionable portrait painter Helleu, a fellow sailor – his yacht, the *Etoile*, which was moored at Deauville, became his summer studio. Helleu was later a formative influence in Diana's life.

When Thomas Bowles's wife died at the age of thirty-five, there was even less to keep him tethered to dry land. Taking his four children, he would put to sea for months at a time. This curious, isolated upbringing, remote from the ordinary stimuli of everyday life, coupled with the precision, order and discipline necessary in shipboard life, may have accounted for much in Sydney's character. She ran her houses with efficiency, keeping meticulous accounts. To her children she was detached, undemonstrative – she seldom cuddled or physically petted them – and strict. If she disapproved, she presented a stonily implacable face to the malefactor, though as her brood increased in numbers she became noticeably more mellow. All her daughters except the youngest, Deborah, who consistently adored her, went through long periods of disliking and resenting her, though all, except the eldest, Nancy, came later to love and respect her.

It was from Sydney that Diana inherited her love of beautiful houses and gift for furnishing them exquisitely, together with the ability to run them smoothly and efficiently. For all her vagueness, Sydney's life was extremely well ordered. By eight-thirty every morning, even after a ball to which she had chaperoned one or other of her daughters until three, she was at her writing desk, telephoning orders to be delivered that day, organising the menu and setting down every penny spent. She was an excellent manager: for years she kept chickens, which paid for her daughters' education. Twice a week, eggs from these 150-odd hens went up to Kettner's restaurant and her husband's club, the Marlborough. They were despatched from the station at Shipton-under-Wychwood in large wooden boxes lined with felt. These boxes, which held six dozen eggs each, were locked by Sydney at the start of their journey and unlocked with duplicate keys by her customers in London.

When David and Sydney were first married they lived in London at 1 Graham Street (now Graham Terrace), a pretty little house Sydney papered in white, with wreaths of green leaves round the cornices. Here were born Sydney's first four children: Nancy on 28 November 1904, Pamela on 27 November 1907, then the longed-for son, Thomas, on 2 January 1909. When Diana was born, at two o'clock on 17 June 1910, her mother's first reaction was to burst into tears at the sight of another girl.

The children's first nanny was the daughter of the captain of their grandfather's yacht; their second, quickly dismissed, became known as Bad Nanny. Then came Laura Dicks, known as Nanny Blor, who was adored by them all and whose influence was immense – she was warm, loving, clever and uncompromisingly fair, providing a much-needed emotional balance.

When small, the children lived an entirely nursery life, emerging from it only in the early evening when, in clean clothes and with hair neatly brushed, they were taken down to the drawing room to see their parents. At six o'clock they were collected and taken upstairs for bath and bed.

All of them, including the servants, managed to cram into 1 Graham Street – on one occasion it also stabled a pony. On his way home from his office at *The Lady*, David spotted a Shetland which he thought would be an excellent first pony for his young family. He bought it there and then, took it back to the house in a taxi and stabled it on the half-landing amid straw until they could take it down to the Old Mill Cottage, which

Sydney had rented near High Wycombe. As he was not allowed to put it in the guard's van – 'It isn't a dog,' the guard remarked irrefutably – he booked a complete third-class compartment, but even then they were such a tight fit that Pam had to travel in the luggage rack.

Sydney's fifth pregnancy made a move to a larger house essential. By the time war broke out on 4 August 1914, the family was installed in 49 Victoria Road, conveniently close to Kensington Gardens for their twice-daily walks. Here, though still believing themselves poor on £1,000 a year, they had a cook, kitchenmaid, parlourmaid, housemaid, nanny and nurserymaid.

Four days after the outbreak of war, Unity Valkyrie, generally known as Bobo, was born. As a war baby, her second name, after the warrior maidens of the 'Ring', was a tribute to her grandfather Redesdale's passion for Wagner.

David rejoined his regiment, the Northumberland Fusiliers, though because of his Boer War wounds he was not allowed to go to the Front and served instead as a dispatch rider. His army pay was less than he was earning from *The Lady*, and Sydney's father's response to these hard times was to cut his daughter's allowance. Sydney accordingly let both their houses and took her family to live rent-free at Malcolm House, on the Gloucestershire estate of Batsford, which belonged to the children's grandfather, Lord Redesdale. It was their first experience of real country.

In 1915 David's elder brother Clement was killed. David became his father's heir, succeeding to the title the following year. Bertie Redesdale was buried in a flower-lined grave at Batsford church – his little granddaughters wore cotton dresses in the half-mourning colours of grey and mauve – and two months later David and Sydney moved into Batsford Park; David's mother, now the Dowager Lady Redesdale, went to live at Redesdale in Northumberland, where the family owned more land.

At Batsford David led the life of a country gentleman. Although he did not hunt – he had broken his pelvis in a terrible fall and never got on a horse again – riding was an integral part of his children's lives and he himself bred Shire horses. It was a quiet life, as both Sydney and David actively disliked society, seldom went out and, if possible, saw only their relations and a few friends. One Batsford highlight was the removal of Diana's appendix on a table in one of the visitors' rooms when she was eight; another was the – to her – revelation that men too could read books. Their London neighbours the Normans had come to stay to escape the bombs and Zeppelins of wartime London and one evening she saw Mr Norman sitting in an armchair reading a book, an activity she had hitherto believed was confined to women and children. 'Farve abominated the printed word,' she said later. When Batsford was sold, ten-year-old Tom was left to decide which of the books in its huge, well-stocked library should be kept.

This sale, the first of what his children regarded as their father's many attempts to divest himself of property and furniture, took place in 1919. David could not afford to keep up Batsford, although he had closed many of the rooms, but it was a bad time to sell and prices were disappointingly low.

His plan was to build a house designed specifically for his own requirements on a wooded hill above the village of Swinbrook in Oxfordshire, where he had inherited land. Meanwhile, with some of the proceeds of the Batsford sale, he had added to the Swinbrook estate by buying a thousand acres adjoining it, land that included most of the small village of Asthall, two miles from Swinbrook, and its big house, Asthall Manor. Here the Mitfords lived for six years.

Asthall, the house most loved by the children and the most closely identified with their childhood, is an exquisite Elizabethan manor of gold-grey Cotswold stone, its steep roofs dappled like tortoiseshell, lying just above the church and a small churchyard full of the casket-shaped tombs of rich wool merchants. It is set in green lawns that fall away, past a huge copper beech and groves of silver birch, to the valley of the River Windrush below. Along the green verge below its high stone outer wall is a row of lilacs planted by Sydney; the three may trees on the tiny village green were planted by David. The road they flank is called Akeman Street: the village was once a Roman settlement. A little way along from the house is a square of old stone farm buildings and stables where the children kept their ponies and horses. Beyond the pretty old cottages, which then housed the Asthall farm workers, stretches rolling empty farmland.

The children were encouraged to 'farm' in miniature, to teach them to look after animals and to earn pocket money. Each was given a smallholding, for which a small rent was paid to their father. Pam, then as later keenly interested in farming, got herself asked to the annual dinner for tenant farmers on the incontrovertible grounds that she too was renting land from David. Here she discovered that she was paying more than the going rate and negotiated a reduction for them all. Diana kept chickens, pigs and sometimes calves.

Sydney furnished Asthall with her favourite French furniture and family portraits, including one of herself painted by de Laszlo at the height of her beauty. In the drawing room were Louis XVI commodes, gilt-bronze clocks, a secretaire and barometer (later bought by Diana when they came up at one of her father's numerous sales) and white chairs covered with antique needlework. The long hall with its old oak linenfold panelling had a fireplace at each end and deep windows on either side of the door. To the right of this door, which faced the church, was David's ground-floor business room, with his dressing room above and, on the third floor, the bedroom later inhabited by Pam.

As luck would have it, Pam and her father were the two family members most sensitive to the supernatural, and this, the north end of the house, appeared to be haunted. Both were disturbed by voices, inexplicable noises and a general atmosphere of terror, though neither spoke of these phenomena to each other until many years later – David because he did not wish to alarm his family and Pam because few children confided in their parents then, least of all the Mitfords. In any case none of David's children saw much of their parents; Jessica (always known as Decca, and born on 11 September 1917) used to have bets with herself as to whether she could go a whole day without seeing either of them.

Pam first experienced these manifestations when, wanting a bedroom to herself, she was moved up to the top floor. Night after night, she felt that some malevolent entity was watching her. Going to the bathroom meant passing through an empty attic used only for storing luggage. At first she thought the voices she heard talking nearby were the servants, whose bedrooms were also on the third floor; soon she discovered they were not. Her father's experiences were more specific. In addition to the voices and the inexplicable noises which he and Pam heard when sitting alone in the hall and which he dismissed as 'rats' or 'water in the pipes' – although nobody ever saw a rat and the pipes were all on the other side of the house and a floor higher – he regularly saw a female ghost. As time passed, she showed herself to him with increasing frequency. The bedrooms on the first floor, most of which were on the side of the house nearest the road, were fortunately unaffected.

Soon after the family moved to Asthall, Sydney went up to London to have her seventh and last child, a daughter (more tears), born on 31 March 1920, christened Deborah but always known as Debo.

As the children grew, shifting alliances and ambivalent feelings characterised their relationships with each other. Tom, the only son, occupied a special position, adored and spoilt by both parents, worshipped and envied by his sisters – treatment which had no effect upon his happy, self-possessed, confident nature. Diana, closest to him in age, regarded him almost as her twin and for many years he was the most important person in her life. They had the same interests, mainly literature and music – Tom was the most musical member of the family. At Eton, where he was allowed a piano in his room, he played the flute in the school orchestra and won the Music Prize in 1926. Inspired by the teaching of the Precentor of Eton, the organist Dr Ley, he thought for a time of becoming a professional musician.

Much of the sisters' childhood was spent in fear of Nancy. Pam and Diana were her particular targets, as displacers of the love that had been exclusively hers until Pam was born. The jealousy that would manifest itself throughout her life was expressed in the nursery through peculiarly vicious teasing, her unerring eye and sharp wit mercilessly probing the weak spot of her victim. The only defence against it was laughter. Nothing was sacred, any outward display of feelings was to be mocked, everything must be turned into a joke, a tease ('We shrieked!'), a refusal to take anything seriously which persisted through future years. Tom, away at school much of the time, suffered far less than the others. Nancy was particularly cruel to Pam, the new baby who had drawn away the attention of her parents and, even more importantly, of Nanny. Pam was also the most easily victimised because of an early attack of polio which slowed down her development (from which she later recovered fully). For years Nancy made Pam's childhood a terrifying burden, but her whimperings and misery only provoked her tormentor to worse attacks.

Diana got away more lightly. As a small child she was spoilt by Sydney who, greatly influenced by physical beauty – the housemaids were hired largely for their looks – invariably favoured the child or children going through a pretty phase. On Diana's birthday, Sydney would crown her lovely little daughter with a wreath of the Mrs Simpkins pinks that grew under the rosebushes. During the holidays, when the children had more free time, Tom was there as protector and advocate, yet all through Diana's childhood she feared Nancy for her imaginative unpleasantnesses.

The younger ones, Unity, Jessica and Debo, respectively ten, thirteen and fourteen years younger than Nancy, were more or less immune: by the time they were old enough for serious mockery, Nancy was out in the world and her heart was no longer in teasing. Instead, they attached themselves like small pageboys to particular older sisters. Debo followed

Nancy about like a puppy, never happier than when sent on some errand; Diana was worshipped by Jessica (whom she played with, taught to ride, and encouraged to learn French and speak up in front of older people) with a deep devotion that was to turn into a correspondingly icy enmity.

Being a member of a large family meant in many ways an enclosed, cabalistic existence. The sisters saw few people outside the family, and those were chiefly relations and a few chosen friends. There were tribal games – visitors would suddenly see half a dozen blonde children processing in a circle among the Asthall trees holding the family cat in front of them as they chanted, 'The Queen is coming, the Queen is coming!' – family jokes and family teases, most with the obligatory touch of cruelty. 'Who would you rather throw off a tower, Muv or Bobo?'

For Diana and Tom, Asthall's most important feature was a large barn close to the house which had been turned into a library. Neither Sydney nor David visited it – that their parents were lowbrows was one of their children's chief complaints. Here Tom and Diana would retreat for hours, to read, or for Tom to play the piano and Diana to listen, or for Tom to impart to her some of the more interesting things he had learned at Eton, encouraging her to read seriously. When Tom's friends came to stay, he would immediately ask them, 'Would you like to hear me play?' and off they would go to the library, followed by Diana. Inevitably, many of them fell for Tom's beautiful sister.

One of these boys was James Lees-Milne. A friend of Tom's since their private-school days at Lockers Park, he had first come to the house as a ten-year-old. Jim, whose own parents believed that any child found reading was wasting time better spent out of doors, was fascinated by the clever, original Mitford children and the freedom to read and talk afforded by the Asthall library. He taught Diana to read Byron, Shelley, Keats and Coleridge, and fell so deeply in love with her that for the rest of his life no other woman ever evoked in him the same depth of feeling. To him, in the flush of adolescent awakening to beauty of all kinds, she seemed a goddess figure from an earlier world, her blue, caressing eyes, peach skin and long golden hair like those of a Botticelli Venus, her cleverness and beguiling, sympathetic manner hopelessly ensnaring. Diana, he realised even then, was not for him, though they were to become lifelong friends. 'May I treat you as a very cherished sister to whom I can say everything?' he wrote on his return home. 'You don't realise how essential they are to boys. Why are you so amazingly sympathique as well as charming? Very best love . . .'

Another admirer was Winston Churchill's son Randolph, whom Diana had known since he first came to stay at Asthall when he was twelve. He was her second cousin: her grandmother Redesdale and Randolph's grandmother Lady Blanche Hozier, Clementine Churchill's mother, were sisters. Of the five Churchill and seven Mitford children, the cousins closest in age were Diana and Tom Mitford and Randolph and Diana Churchill – the two Dianas were the same age. As they grew older, the four cousins spent more and more time together. Randolph and Diana Churchill came to stay at Asthall; Tom and Diana Mitford were constantly asked to Chartwell. Tom and Randolph became each other's closest friends, while Randolph, precocious, brilliant, surging with an irrepressible vitality and confidence that often expressed itself in appalling behaviour, became ever more besotted with his beautiful cousin. Diana Churchill, red-haired, garrulous from nervousness and eagerness to please, hovered disconsolately on the edge of this tripartite intimacy – often, cruelly, they would run away and hide from her.

Tom and Diana loved these visits to Chartwell. One of the charms of staying with the Churchills was that the rule that children should be seen and not heard did not apply. Randolph was encouraged by his father to take part in intelligent conversation; Tom and Diana were treated as thinking beings. Sometimes Cousin Winston would sing 'Soldiers of the Queen' at the end of dinner, before going off to play a game of six-pack Bezique followed by hours of work on a book; sometimes there would be discussions which lasted for hours. Here Diana first realised the fascination of politics. Here, as a sixteen-year-old, she heard discussion after discussion about the General Strike and the long-drawn-out stand by the miners which followed; it became, in consequence, the first major political event to impinge on her consciousness. Her sympathies were all with the miners.

Here, too, Diana met Eddie Marsh, the first person to treat her on absolutely equal terms. He was quickly followed by one of Cousin Winston's favourite and most frequent guests. Professor Lindemann (the Prof) was the first person Diana had heard abusing someone for being a Jew – his target was Brian Howard, of whom it was his chief complaint. The Prof was dazzlingly erudite and always prepared to answer questions without any condescension. As Diana grew older he became extremely taken with this clever and beautiful girl. He wrote to her often and she was allowed to go to luncheon with him – suitably chaperoned by Tom – in his rooms at Christ Church, where the excellence of the food, drink and conversation contrasted with the ugly furniture and unpleasing reproductions on the walls. He gave her a watch made of three different golds. None of this did her hypersensitive parents mind, although his innocuous suggestion that she learn German, as Tom was already doing, in order to read Schopenhauer was forbidden – the idea of an 'over-educated' female was anathema to David.

Talking to brilliant older men, listening to Tom's friends, reading what was recommended to her, formed the major part of Diana's education. As in many upper-class families, the Mitford daughters were educated at home – David violently disapproved of school for girls. They had, however, a more solid grounding than many. Of the fifteen governesses, including French teachers who came during the holidays, the longest-serving and the sisters' favourite was Miss Hussey, a superb teacher. From her, through the well-regarded PNEU system with its twice-yearly examinations marked at the PNEU College in Ambleside, they learned subjects such as ancient history and geology, as well as basics like English and history.

Nevertheless, not being sent to school was a recurrent grievance with Nancy, Unity and Jessica. They furiously resented what they saw as their parents' deliberate failure to give them a proper education, although Nancy was allowed to go briefly to Hatherop Castle, where a dozen girls did lessons together. Diana, who hated the idea of dormitories, compulsory games, school food, uniforms, lack of privacy and saying goodbye to her pony, guinea pigs and dog, dreaded the possibility of going away. Nancy, of course, played on this fear. 'Farve and Muv say you're not wanted here,' she would begin. 'They were saying last night that you might not be so stupid and babyish if you went to school.' Later, eyes gleaming, she would shoot triumphant glances at Diana, silent and miserable across the dining-room table.

Chapter II

In October 1926 Asthall was sold. David, who by now could hardly bear the haunting, was delighted, though they had no new home to go to as Swinbrook was not yet ready. Twenty-six Rutland Gate, bought by the Redesdales as a London house from which to bring out Diana and her three younger sisters, was let until Christmas.

To economise while waiting to move to Swinbrook, Sydney and the girls went to Paris (the exchange rate was extremely favourable). They stayed in a family hotel in the avenue Victor Hugo, which was round the corner from the Helleus' apartment in the rue Emil Menier and near the Cours Fenelon, an excellent small school at which Diana was enrolled a few days after their arrival. Tom, who had just left Eton, was in Vienna, deciding whether to take up music professionally or read for the Bar, and perfecting his German.

Paris was an awakening for Diana. She became aware for the first time of the world to which she then and later aspired: an essentially metropolitan milieu, aesthetic and cultured, where concerts, paintings, books, and people who could discuss them intelligently and amusingly, were a daily staple. She also became conscious of the uninhibited male French reaction to a beautiful woman. Helleu had always admired Sydney and his painter's eye was fascinated by her blonde brood, in particular Diana, whom he frequently likened to a Greek goddess.

For her part, Diana loved the old Paris houses built around courtyards; she loved the Helleus' drawing room, with its Louis XVI chairs covered in white and pale grey silk, its carved and gilded eighteenth-century wooden frames hanging empty on the white walls; she loved the aesthetic yet practical approach of Helleu himself – and she loved the dazzled admiration which he did not hesitate to express. 'Tu es la femme la plus voluptueuse que j'ai jamais connu,' he would tell her. His everyday name for her was 'beauté divine'.

She was also absorbed by what she was learning at the Cours Fenelon. She would sit in the classroom facing the rue de la Pompe, its windows hermetically sealed against the intrusion of any possible *courant d'air*, with a dozen or so of her thirty-two fellow pupils to listen to lectures of the highest quality, many given by professors at the Sorbonne. After school she walked alone, 100 yards round the corner, to tea with Nanny at the hotel. It was the first time she had been allowed to walk alone on a city street.

At Christmas, the Mitfords came home to 26 Rutland Gate, which had been furnished by Sydney with all the best pieces from Asthall – the French furniture, a Peronneau portrait of a young man in a powdered wig, his blue velvet coat enhanced by the blue taffeta drawing-room curtains. The Cours Fenelon Christmas holidays were brief, so after only a fortnight Diana returned to Paris.

She was not, of course, allowed to travel alone. No young, unmarried girl who was not even officially 'out' could make a journey without a chaperone. In London, the sisters were only allowed to walk the short distance from Rutland Gate to Harrods in pairs, and not until she was twenty-two and had been married for four years did Diana travel on a train by herself. Fortunately, Winston Churchill, who was going to Rome to visit Mussolini, agreed to drop Diana in Paris on the way. With him was Randolph, delighted to be in Diana's company again, although he did not cut quite the figure he had hoped: a devastatingly rough Channel crossing laid him flat with seasickness.

At the Gare de Lyon, Diana was handed over to one of the two elderly sisters with whom Sydney had arranged that she should lodge. She had a ground-floor bedroom with tightly clamped shutters and was allowed a skimpy bath in a tin tub twice a week. Her new home, 135 avenue Victor Hugo, on the corner of the rue de la Pompe and the avenue Victor Hugo, was considered close enough to the Cours Fenelon, her violin lessons near the Lycée Janson and the Helleus' apartment for her to be allowed to walk to all three places unchaperoned. It was her first, intoxicating taste of freedom.

She used part of this new-found independence to take a radical step that would never have been allowed at Swinbrook: she had her hair shingled. David, though forced to accept this, did so with jokes which did not hide his disapproval. 'I shall be very glad when you finish with Frogland and come home,' he wrote from Rutland Gate. 'Remember me to M. Helleu when you see him. It seems there is danger of a "small" dance here at the end of the month. Have you or have you not recovered your hair which was cut off. You must not leave it in Paris. That was one of the stipulations. Much love, Redesdale.' Three weeks later he was writing gloomily, 'I wish you were coming to the dance, though it will never be much joy taking you about now you have no hair.'

She visited the Helleus constantly. Hungry after an afternoon's lectures, she would fall upon Mme Helleu's tea and rich chocolate cake; equally avidly, she would listen as Helleu expounded on paintings, sculpture and fine art on the visits they made together to the Louvre, Versailles and museums. For by now – though his behaviour was always entirely proper – he was infatuated with the lovely young visitor, whom he drew and painted whenever possible. The shower of compliments, the requests to sit for him, the constant references to her sublime beauty from a man of such charm and forceful personality through whose studio the most handsome women of Parisian society had passed, were a heady diet for a girl accustomed only to Nancy's snubs, Sydney's discounting of compliments to her children, and Nanny's endlessly reiterated 'Nobody's going to look at *you*!'

Letters from admirers confirmed what she was slowly beginning to believe. 'How I would adore to have a picture of you by M. Helleu,' wrote Jim Lees-Milne:

> You must be like Emma Lady Hamilton sitting to Romsey. I dare say you are very vain, and indeed you have cause to be. Haven't you even a snapshot of yourself? your Parisian self, of course – that you can send me, as I would love to see for myself whether you have still got that Rafael face . . . I am convinced one can never love a friend too much. Though I will confess it now, I always think of you as something even higher than a friend.

In Paris, Diana had begun to meet a few young men. Some took her out to tea after lectures, others invited her to the cinema. Newly conscious of her power, she flirted, as she wrote to her friend Jim Lees-Milne, 'outrageously. Round the Bois de Boulogne; in a taxi alone with Charlie after dark – you can guess. Don't feel angry with me – I know it isn't lovely to be so sensual but it is exciting and wonderful.' Such expeditions were strictly forbidden, so she would pretend to her two old ladies that she had an extra-long violin lesson or that Helleu wanted her to give him another sitting. Because of their reverence for art and the famous painter, she was never questioned. Only Diana – and, unfortunately, her diary – knew the truth.

At home Sydney was too preoccupied with the forthcoming dance to worry about a child whose letters home were regular and dutiful. Neither of the Redesdales really liked entertaining and, on the rare occasions they had to, made heavy weather of it. '[The dance] is turning into an immense bore, and we have 170 acceptances and quite 100 not answered yet but we are rather in hopes most of these won't come. I think the house would dance 80 couples or so and of course a good many are older people.' Nevertheless, she found time to send Diana a registered parcel containing a pair of evening knickers and a dark blue silk dress with white spots that would need taking in.

The outstanding tuition at the Cours Fenelon had its effect. Diana's report for her spring term there, written on 31 March 1927, described her as 'excellente élève dont nous garderons le meilleur souvenir', and her conduct and punctuality were both considered 'parfait'.

There was one bitter blow that spring. In March 1927 Helleu fell desperately ill. Diana was devastated. 'Nobody I have loved has ever died except perhaps an adorable animal,' she wrote to Jim. 'And now a man whom I have almost worshipped, and who has worshipped me for three months, is going to die. I shall never see him again, never hear his voice saying, "Sweetheart, comme tu es belle," never ring at his door and hear him come to open it with a happy step. How can I bear it?' Helleu died at the end of the month. For Sydney, who had known him since she was a young girl, it was sad indeed; for Diana, it was the loss not only of a beloved father-figure but of the first man who had loved her as a woman.

When she returned for the Easter holidays, the family had moved into the new house, South Lawn, or, as it was always known, Swinbrook.

A large, square, yellowish building two miles from the village of Swinbrook, South Lawn is half-hidden by trees near the top of a hill. Its sides and façade are dotted with windows – all the children wanted their own bedrooms. On Sundays the family would walk to the beautiful twelfth-century church in Swinbrook, at which David had worshipped even when living at Asthall. They sat in a pew at the back, their appetites sharpened by the delicious smell of freshly baked bread a few inches from their noses – a seventeenth-century benefactor had instituted the custom of giving the poor of the village six large loaves after matins.

Soon after the move David, who had always wanted to put new pews in Swinbrook church, was able to do so and to have them made, as he was determined, of oak. His cousin, Lord Airlie, owned a horse called Master Robert (which had once drawn a plough in Ireland) which he had entered in the Grand National. David put on £10 at 40–1, Master Robert duly came in and the pews were presented to a grateful church. But when David asked the Bishop's permission to finish each one with a carved horse's head in honour of Master Robert, this was refused on the ground that it would commemorate 'ill-gotten gains'.

David was delighted with the move to Swinbrook, but everyone else hated the new house. Nancy called it Swinebrook; Jessica described it as having aspects of a medieval fortress. Inside, though large, it was uncomfortable, cold and without the beauty that had characterised Asthall. It was predominantly yellow-toned: the doors, woodwork and panelling were all of elm, not all of it properly seasoned – as his children pointed out, the bathroom door was so warped that even when locked you could peer in – and the resulting draughts made the whole house icy.

Despite the immense trouble David had taken over its design and building, the house had one major planning fault, which became apparent almost immediately. The children had very little privacy.

There was nowhere like the library in the Asthall barn, with its comfortable sofas, walls lined with books and grand piano, where Tom could play to his heart's content and Diana could read, both in the sure knowledge that they would not be disturbed. At Swinbrook, the piano was in the drawing room. Tom, who hated people coming in and out while he was playing, would at once stop. Reading was equally circumscribed: the library was David's study and he would not countenance any of his children sitting there with him. The bedrooms, small, uncomfortable and icy cold, were the last places to which to retire for hours with a book. The only alternative was the large drawing room, with four or five other sisters there, chatting, shrieking or quarrelling.

David, sensitive and intuitive under his fierce exterior, was miserable at his family's reaction to his longed-for new house. He hated their outspoken dislike of it and he, too, was affected by the pressure of too many personalities – most, in embryo, as strong as his own – crowded into the drawing room. Apart from Tom's sporadic presence, David was a lone man in a sea of women; neither of the Redesdales had ever considered having menservants, believing them to be dirty, troublesome and given to drink. At one point David could scarcely bear to be in the same room as Unity and Jessica, already veering towards different ends of the political spectrum and constantly arguing at the tops of their voices. Nancy, green eyes glittering, would begin to tease their father, reducing everyone to shrieks, until she went too far and he roared at her.

It was at Swinbrook that his habit of taking a sudden, inexplicable dislike to one of his daughters began. Perhaps because there were so many of them he would, often for no apparent reason, appear to loathe one for days or weeks until, equally mysteriously, she was restored to favour. Only Debo, the youngest, busily trotting about her own concerns and hunting whenever she could, was exempt, even when she developed one of her inexplicable rages, slowly sliding off her chair to roar and scream under the table. As reasoning or remonstrating with her only caused her to redouble the yells, Sydney sensibly issued an edict that everyone was to ignore her. When the rage had run its course Debo would emerge and climb on to her chair again as if nothing had happened.

The move to Swinbrook marked the end of Diana's childhood. If it had come a year earlier, she would have been heartbroken at leaving Asthall; as it was, she was absorbed in the exciting possibilities that lay ahead. She loved the Cours Fenelon and her life in Paris, she now knew many of Tom's friends and some of Nancy's. Even hunting had lost its charm: Diana's horse was a 'roarer', and at sixteen she found the noise it made embarrassing in front of the smart, well-mounted members of the Heythrop hunt. But nothing had prepared her for the reality of Swinbrook.

She found the lack of privacy oppressive; in particular she resented furiously the interruption in the reading which now filled most of her days and had become the major plank in the long process of self-education. Though she never had any desire to go to school, she was determined to learn and Tom and his friends helped, encouraged and suggested. Eventually, she took refuge in the large linen cupboard, sitting on the warm sheets with book and torch. As the younger children grew, they joined her, and called it the Hons Cupboard (the H aspirated as in hen).

Shortly after Diana came home for the Easter holidays there was a family row so cataclysmic that it became the stuff of legend. One day, after writing her diary at her mother's desk, Diana went out for a walk with Pam. Suddenly, with a shock of horror, she realised that she had left the diary open. It contained not only details of her sittings for Helleu but of her doings with the young men such as Charlie who had taken her out. She and Pam ran back, arriving panting, but it was too late. The storm broke, with accusation of immorality and untrustworthiness. Calling Diana 'wanton', Sydney told her: 'Nobody would ask you to their houses if they knew

HALF of what you had done!' before burning the fated diary. There was no question of returning to the Cours Fenelon, meals were taken in silence, her parents glowering at her furiously, and she was despatched with Nanny and the little ones to a cottage in Devon belonging to a great-aunt. With no money, no books and only her little sisters for company after the delights of Paris, there followed three months of desperate boredom.

One result was to make her determined not to suffer such incarceration again; her letters are spattered with references to 'reformation'. Another was that, after Devon, even Swinbrook seemed reasonably agreeable, especially now that Nancy treated her more as an equal. They both chafed against parental restraints, they shared the same sense of humour and love of books, and now that Nancy's acid wit was not so often directed against her, Diana found her sister excellent company – besides, she was fascinated by the friends Nancy was beginning to bring home. All were clever, most were aesthetes and many homosexual – a combination both sisters were to find irresistible all their lives.

They included the Acton brothers, John Sutro, Robert Byron and John Betjeman. Her brother Tom did not entirely approve. 'I gave the Prof. your various messages,' he wrote after he had been to stay at Chartwell during Diana's exile in Devon. 'He would like you to keep your respectable Oxford friends, and forsake entirely all aesthetes at Oxford and elsewhere.'

Randolph – being Randolph – was blunter. 'I am afraid the Prof. does not share your views on the desirability of homosexualism. In fact, he said that it was "the very negation of all race survival". If this is true, as I feel sure it is, I do not see how you can continue to have such a very lenient view on the subject.' But perhaps the very fact that homosexuals were 'safe' was what drew Diana; her father's frequently expressed prejudice against 'pansies' was as nothing compared to what his reaction would have been at finding his young daughter indulging in a clandestine love affair. All the Mitfords had been brought up with the twin concepts of rigid chastity and the idea of one deep love that led, inevitably, to marriage. (When love did not, as in Nancy's case, lead to marriage, it led to a lifelong fidelity.)

Much of the holidays that summer was spent at Chartwell, partly thanks to the ardent pleading of Randolph, who begged to see her whenever possible. 'My darling Diana,' runs a typical letter:

I am so longing to see you before I go back to Eton. I came out of quarantine on Wednesday but do not return until Monday. All my family, including Papa, Mummy and Diana, are deserting me on Saturday and are going to Trent for the weekend. I wonder whether you would be allowed to come here on Saturday for the weekend? I know that the family is broke, as a result of the dear ones' libel action, and that you are not allowed to travel, but I will come and pick you up, and deposit you again on the Monday. Your Mother may of course think it inconceivable but against this you can urge 1) that you came this time last year under similar conditions, b) that we are cousins, c) that Cousin Moppet is an ideal chaperone, d) that darling Randolph is going back to Eton on Monday and will otherwise be all alone. Of course you may have other plans for the weekend or alternatively be bored to extinction but if you think it at all likely that you will be allowed to come will you ring me up as soon as possible and I will get my mother to write or telephone to your mother? I do so hope you will be allowed to come as I am so SO lonely and have so much I want to tell you and consult you about. Therefore, if you have any affection for poor 'Randyov' do your best to come and earn his undying gratitude.

All my love, darling Diana,

Your loving

'Randyov'

Remember, telephone Westerham 93 as soon as possible.

She came – but she did not respond in the way Randolph wished.

Indulged by his father and treated with coldness and reserve by his mother – Diana believed Clementine was jealous of Winston's adoration of him – Randolph craved affection, particularly from women. But as Diana was to write to a friend many years later:

Winston's spoiling . . . made him unbearably cocky. Poor Randolph never grew up. He always behaved badly – you could rely on that. One dreaded him. But there was something touching and affectionate about him and I was very fond of him. He was so full of life, so full of chat, he was very very fond of me and adored my brother. Really, I looked on him almost as another brother.

So she remained sweet, affectionate and sympathetic, but any advances were met with a gentle, amused rebuff or, as Randolph put it, 'your extraordinarily cruel and callous behaviour'. He was, he wailed, shortly going back to school and would not see her for another four months:

> you might accordingly have humoured me, if you have any affection for me at all. I have been so melancholy about it that I have not been able to bring myself to write to you. Please do write and tell me why you were so unkind. Was it because you wished to destroy my love for you? If so you were unsuccessful, for I love you as much as I ever have.

Other letters held out the bait of the Prof, whom they both adored, being present; most ended reproachfully, 'Goodbye, darling Diana, Your loving but unloved Randy'. So accepted was his devotion that when Diana Churchill wrote to her from Venice, where she was staying with her parents and the Prof, she concluded:

> The Prof. is returning to England next week and says that the only thing that reconciles him to this sad but inevitable departure is the fact that he hopes to see the gloomy skies of England lit up by your radiant smile. He would have written to you himself only that he fears this might rupture his friendship with Randy.

Back at Swinbrook in the autumn, the stupor of boredom after the cleverness, fun and male admiration at Chartwell became almost too much to bear. Hours were spent with Nancy, talking and dreaming of love. When she was allowed to attend her first ball, the Radcliffe Infirmary Ball in Oxford in the autumn of 1927, accompanied by Diana Churchill, it was the subject of feverish speculation. Would Prince Charming appear that night? Would Sydney or Cousin Clementine now allow them to go to other dances?

The answer was no in both cases. It was back to the chickens, the village, the rows, the shrieks, the boring walks, the impossibility of being alone – all the familiar components of a life that had begun to seem claustrophobic, while somewhere out there lay the brilliant, artistic, sparkling world she longed for.

Chapter III

The season of 1928 saw the gates of freedom inch open. Nancy, who had been 'out' for four years, was allowed to bring her friends to Swinbrook more freely; and Diana began the round of balls, receptions and dances that was supposed to introduce well-bred young girls into the society of their peers where, with luck, they would make the suitable marriage that would then become their life's work. It was fortunate for Diana that she was a natural beauty as she had little otherwise to enhance her looks: her dresses were made by her mother's lady's maid rather than by one of the fashionable court dressmakers and her only jewellery was a small pearl ring given to her when she was four by a great-aunt saddened by the prospect of an empty jewel box in the future ('So sad, being a third daughter – she'll never get anything'). She also had the great advantage that she had none of the agonising social diffidence which crippled so many debutantes plunged straight from the schoolroom into adult life, making each evening's 'gaiety' a torture.

For Diana, everything was a pleasure after the confinement of Swinbrook. She was not shy: she already knew a number of young men, from her boy cousins and her brother Tom's friends to the undergraduate circle she had met through Nancy. The visits to Chartwell had accustomed her to holding her own in the company of clever people of an older generation. Though not at all vain, she was by now sufficiently accustomed to male admiration to have plenty of confidence in her appearance. 'The poor boy told me that he was so carried away by the radiance of your beauty (my heart goes out to him) that he could not hold from imbuing his lightly intended words with some of the ardour of his love,' wrote one admirer of another, while Randolph – though as a cousin and her junior by a year she did not take him seriously – was still pouring out his heart from Eton.

'I hope you do not think that the heart of the lover beats cold or rather that a friend's affection has lapsed, as you would prefer me to put it,' he wrote after they had met on the Fourth of June:

> You looked too radiant and beautiful to describe. Even the brief moment I saw you for was worth an age without a name. Please don't think that I am being silly and least of all childish, for I really mean what I say and I think I am still within the limits that I am allowed by you . . . I am afraid that I shall never have such a glorious time as last holidays, when we were all alone together. I don't suppose it will ever happen again. I hope you don't object to this, though I fear my love is one-sided and unreciprocated. But it is genuine, and you must let me say part of what I feel.

Young, happy, glowing and friendly, Diana was an instant success. Surely love, the great, life-transforming, longed-for catalyst, about which she and Nancy had talked so endlessly over the past two years, could not be far off?

She was right. One May evening, at a dinner party in Carlton House Terrace given by Lady Violet Astor for her debutante daughter Margaret Mercer-Nairne, Diana was placed next to a slim, good-looking, fair young man, an inch or two taller than her own five feet ten inches. His name was Bryan Guinness, and he already knew her two older sisters slightly. He had met Pamela at a dance and Nancy – who became a lifelong friend – through mutual Oxford friends. He had also been to Asthall: Nancy had asked him to a fancy-dress dance there the previous year, while he was still an undergraduate. Across the room he had seen Diana, smiling and silent in a dream of delight, dancing with one of his friends, Brian Howard, and had carried away with him the image of this radiant sixteen-year-old beauty. Now, meeting her, he fell instantly in love.

Bryan Guinness was a golden young man, rich, good-looking, delightful and unspoilt. His parents were Colonel Walter and Lady Evelyn Guinness, a couple who could hardly have been a greater contrast to each other. Walter Guinness, the immensely rich Conservative MP for Bury St Edmunds and Minister of Agriculture, was shrewd, ambitious, political to his fingertips, urbane and clear-sighted. His wealth came from the family brewery, founded in Dublin in 1759, but his interests were science and anthropology. He spent much of his time at sea on his magnificent yacht, often sailing to the South Sea islands, the customs of which fascinated him, an interest which conveniently took him away from the sometimes confused atmosphere at home.

Lady Evelyn was the daughter of the fourteenth Earl of Buchan, an extremely short man so handsome that he was known in the family as P.A., short for Pocket Adonis. Lady Evelyn was equally small, pretty in a doll-like way, fey, imaginative, artistic and romantically obsessed by notions of a chivalric medieval world where nature combined with art in a kind of Gothic fantasy land. In pursuit of this ideal, she had persuaded her husband to buy an open stretch of cornfields at Climping between Littlehampton and Bognor, on which stood a dilapidated farmhouse and an ancient dovecote and chapel. From and around these she planned a group of buildings in medieval style, constructed from old materials, a work which took another five years to complete. While this house, Bailiffscourt, was being built, the Guinness family and their friends stayed in wooden huts nearby.

Their two enormous London houses, 10 Grosvenor Place and Heath House, Hampstead, were also decorated and run with this curious blend of the Middle Ages and the countrified. The garden of the Hampstead house was filled with cottage flowers and the maidservants wore sprigged cotton dresses instead of the usual black uniforms. Grosvenor Place, a large and ugly Victorian house, had been converted by Lady Evelyn as nearly as possible into a facsimile of a house built before the invention of chimneys. Smoke-blackened beams had been installed in the ground-floor rooms, and to foster the illusion still further the fires were encouraged to smoke. The fine Victorian plate glass had been replaced with 'old' leaded windows and instead of silver and china there were two-pronged forks and pewter dishes. At dinner parties there were wild flowers on the table instead of the usual roses, stephanotis or smilax, and the maids wore medieval costume.

Bryan, the eldest child of this incongruous couple, was idealistic, unworldly, sentimental and sensitive – he had inherited his mother's love of the arts and, to some extent, her head-in-the-clouds approach to life, while his father's toughness and political bent had passed him by. In any case, as a young man he was extremely shy and the prospect of making even the shortest speech made him physically sick. His salient characteristic was an exceptional sweetness of nature, which was remarked on throughout his life by everyone who knew him. He was lovable, affectionate, gentle and generous. It was impossible not to like him. His faults were the converse of his virtues – a kindness that could sometimes suffocate, a romantic belief in the exclusivity of love that manifested itself as possessiveness, and the sudden obstinacy of someone who usually yielded.

At Eton, where Walter Guinness had been in the Sixth Form, Captain of the Boats and President of Pop, his son led an inconspicuous life in a games-mad house. His housemaster, the great classical scholar C.H. Wells, played cricket for the Gentlemen and coached the Eton XI; the Captain of the House, Frank Pakenham, was also Captain of the Eton XI, and this enthusiasm for cricket permeated the whole house. Bryan, no games player, wisely chose to row instead, like his father before him.

Bryan's Eton career was in any case violently interrupted when he developed polio. Returning after a cross-country run during which he had got his feet wet fording a stream – later he believed this to be the cause of his illness – he felt unwell. His elderly Dame, whose eyesight was too poor to read a thermometer and whose hearing was so bad she used an ear trumpet, thought he had a chill and sent him down the High Street to a nearby doctor who, startled at his high temperature, sent him back with instructions to go straight to bed. After a few days he was allowed up, but that Sunday found himself unable to get to his feet after kneeling during house prayers in the dining room. 'M'Tutor when walking out with one hand held open and perpendicular in front of him as was his custom perceived my ineffectual struggles, and I remember his concern as he helped me to his feet,' he wrote later. Next day a specialist in London diagnosed polio. His right leg was affected but gradually cured by physiotherapy and, after a spell at a tutorial establishment, he returned to Eton. The legacy of weakness, loss of weight and muscular 'tone' that persisted for some time put any thought of athletic distinction still further out of reach and he began to write – chiefly the poetry that became a large part of his literary output. After Eton came Christ Church, Oxford, where he read French, with German as an extra subject. He got a Second.

At first glance Bryan could have been specifically designed to suit Diana. She had inherited Sydney's penchant for physical beauty: Bryan's fair hair, intense blue eyes and handsome, open, friendly face were instantly appealing. He enjoyed conversation and jokes and, like her, his interests were literary, artistic and musical. In addition, he was a connoisseur of the theatre, to which he longed to introduce her.

Then there were his friends. Throughout his school and university life Bryan had been one of a band of aesthetes. Harold Acton, Michael Rosse, Henry Yorke, Robert Byron, Brian Howard and John Drury-Lowe had been Eton contemporaries, joined at Oxford by John Betjeman and John

Sutro. Together they joined the Athanasian Club which, with collars reversed, ran bicycle races down Headington Hill, and Sutro's Railway Club, which dined in the restaurant car of the train to Birmingham, where the members descended and, after wheelbarrow races on the platform, returned in another restaurant car with wine and speeches. Other members of the group were Mark Ogilvie-Grant, the young don Roy Harrod and Maurice Bowra, soon to be Dean of Wadham and their unofficial mentor. Bryan himself edited *Cherwell*, to which Tom Driberg and Louis MacNeice were contributors. His coterie were all clever, talented, witty and artistic; for Diana, who longed for a milieu where she could meet, entertain and be entertained by intelligent, amusing and cultured people, they added an extra, irresistible aura of desirability and glamour to someone she already found delightful.

Much of Bryan's courtship of Diana took place in the unreal world of the ballroom, a place where, surprisingly, he was at his best. Although no athlete, he was a marvellous dancer, a talent he inherited from his mother, who had been considered the best amateur ballet dancer of her day. To watch him spinning his partner effortlessly in a waltz, his coat tails flying and a look of pure enjoyment on his handsome face, was to envy the girl in his arms – who was always, if he could manage it, Diana.

As every day passed, he fell more deeply and desperately in love with her. For any woman, to be adored by a young, charming, good-looking man is a pleasing experience; for Diana, at an age when love engenders love, Bryan's passion was enough to convince her that she, too, must be in love.

But where Bryan's love for Diana was the idealistic, wholehearted adoration of a young man who believed that he had met the great love of his life, her feelings were more complicated. She loved him, yes, but she did not have the same conviction that this love would last for ever – or at least as long as they both lived. His money made him a brilliant match in worldly terms, but for Diana this hardly came into the equation. Though she would never have married a really poor man and though she enjoyed much of what enormous riches can bring, neither then nor later was she mercenary. Provided any potential husband had what she and her parents called 'enough' – for the servants, travel and entertaining that were a recognised part of pre-war upper-class life – she would have been content. Far more important to her than Bryan's potentially limitless bank balance was the world he moved in, the magic circle of clever, fascinating friends of which she longed to be part; and the aspect of his wealth that most appealed to her was the knowledge that she would now be able to invite these people constantly to her own home.

Above all, there was the knowledge that the only escape route from Swinbrook was through marriage – and while she believed that she loved Bryan, she also wanted to leave home. Most of the Mitford sisters found life at home so stifling as to be insupportable. 'Growing up in the English countryside seemed an interminable process,' wrote Jessica of the relentless round of lessons, walks, church and occasional visits from relatives. 'Freezing winter gave way to frosty spring, which in turn merged into chilly summer – but nothing ever, ever happened.' Only Pam, a countrywoman by nature, and Debo, who lived for hunting, were happy at Swinbrook. Unity vented her frustration in behaviour which was becoming steadily more outrageous, while Jessica was saving up to run away.

Diana felt the drawbacks of Swinbrook life even more keenly. The pleasures of books and music on which she was so dependent had been cut off by the design of the house. She loved the country for its beauty but not much else; after a few days at home she was overcome by a boredom so intense that it was almost a physical ache. At an age when three months is like three years, the thought of spending the foreseeable future surrounded by chickens, ponies and turbulent sisters seemed impossible to contemplate.

Like all girls of their class and generation, the Mitford sisters had been brought up in the expectation that they would only leave their father's house to go to that of a husband. Diana was also well aware that neither she nor her sisters were educated or equipped to make an independent life for themselves. On the rare occasions Nancy had tried this, she had always been forced to return home after a few months, largely through sheer inability to look after herself let alone earn enough to support herself. Psychologically, too, Diana had been brought up to believe that marriage was not only the greatest joy in life but also virtually the only career open to a well-born young woman.

Yet Nancy, who at that time probably knew Diana better than anyone else in her family, had misgivings almost from the beginning about her sister's choice. 'The more I see of Bryan the more it surprises me that Diana should be in love with him but I think he's quite amazingly nice,' she wrote to Sydney. And until the die was finally cast, Diana herself had reservations.

In early July the Guinnesses gave a dance. The housekeeper, seeing that two of the young maids looked tired after all the work of preparation, sent them up to bed early. On their way upstairs the two girls, anxious to have a glimpse of the guests in all their finery, climbed over the balustrading round the servants' floor and leaned forward, supporting themselves by their hands, to peer through the glass ceiling at the dancers sitting out in the main hall two floors below. The glass was not strong enough to bear their weight and they crashed through it. One girl was killed outright and the other, a kitchenmaid of fourteen who had partly managed to break her fall by grasping at a chain from which hung a lamp, was badly injured. Bryan, like his parents, was deeply shocked by the tragedy.

The following afternoon they went to a performance of *Hiawatha* at the Albert Hall. On the way their car passed Nancy and Diana just as they were climbing on to a bus. Bryan knew that the Mitford family had permanent seats at the Albert Hall; now, glancing across the auditorium, he thought he saw Diana's gleaming golden head in the front row of the stalls. The music, the aftermath of shock, a sudden sense of the impermanence of happiness, made this the moment at which his feelings for her crystallised into the knowledge that this was the girl he wanted to marry.

What Diana felt about him he did not know. Though always friendly, charming and affectionate, she had given no overt signs of being in love. Inexperienced with women, aware that Diana, in her first season, was not only younger but presumably even more physically innocent than himself, he was inhibited from putting what seemed like a mutual attraction to the test. One evening, when they sat out together during most of a ball given by his Uncle Ernest at his house at 17 Grosvenor Place, hope rose in him like a wellspring. But as they watched the sunrise from a balcony overlooking Chapel Street, Diana's perfect profile and cool self-possession left him in the familiar state of baffled adoration. Later, he summed up his feelings in a poem he called 'Sunrise in Belgrave Square':

> The night stands trembling on the tips of towers
> In waiting for her paramour the day;
> Before his kiss her deep blue body cowers
> And at his touch dissolves in disarray.
> Here at my side, balanced upon a stair,
> You sit politely in your body's box;
> I wonder at the wonder hidden there,
> But, clumsy robber, do not know the locks . . .

Only two months after they had met he could wait no longer. That she was only just eighteen was overborne by his feelings. On one of his brief but frequent expeditions abroad with Oxford friends – this time to Holland – to look at pictures, houses and museums, he wrote:

> in the course of my tossing and turning it occurred to me that you might be thinking I haven't mentioned the future owing to indifference or fickleness. If you understand me at all, though, you must know that this isn't so. It is only because I have been thinking so much about it, because it needs so much thought owing to your being so terribly young that I haven't rushed straight into it. But I have thought about it now till there is nothing else that I can think out, and we'll have to face it together and decide.

On 16 July, a few days after his return, he and Diana found themselves at a ball at Grosvenor House. It was a hot night and, as they were dancing, Bryan suggested taking a turn outside 'for a breath of fresh air'. They strolled along the pavement until, well out of sight of the brilliantly lit doorway, Bryan stammered out his feelings of love and longing, took Diana in his arms and kissed her.

'Do you – could you – love me enough to marry me?' he asked.

Diana's instinctive response, uttered before thought could intervene, was like a douche of cold water. 'I'm very fond of you,' she replied.

'But do you love me?' asked Bryan. 'You *kissed* me.'

'A kiss means nothing,' replied Diana. 'I do it without thinking as I'm used to kissing my family.'

To Bryan, however, a kiss meant everything. On several occasions he had held back from kissing beautiful girls because of his belief that a kiss represented a sacred, intense feeling rather than a flirtatious pleasure. Of one beautiful Italian girl, whom he had been escorting for several weeks during a visit to Germany, he confided to his diary, 'Standing on the terrace in the evening light, it suddenly struck me from some look she gave me that she could have been expecting me to give her a lover's kiss . . . my heart sank since I was determined to give no such demonstration of feeling until true love should some day strike me like a thunderbolt.'

Now the girl for whom he felt true love had shown that to her a kiss was not a pledge but just a kiss. Silently, Bryan took Diana back to the ball and then, sick with misery, suggested to his mother that they go home. Lady Evelyn, however, was enjoying herself, and the wretched Bryan had to spend what seemed hours going round and round the ballroom 'in a speechless daze'.

The following morning his mood of dull misery was abruptly transformed. On the breakfast tray was a note from Diana, written as soon as she had arrived home. It said that she loved him and accepted his proposal. He wrote back at once, 'I still don't know how much you love me, nor really understand what you felt last night. But I am glad. I am glad that you are glad. I am glad that I love you. I am altogether glad again.'

Chapter IV

The decision of Diana and Bryan to marry provoked immediate opposition. Diana's parents said with reason that at just eighteen she was far too young to know her own mind and that she must wait a year before an engagement could even be contemplated. 'You must have one more season in London,' said Sydney, 'so that you can at least meet more people.' This argument carried no weight with Diana, who had inherited her father's characteristic of unswerving adherence to a decision once taken; anyway, she believed that she had already met quite enough young men to make a considered choice.

Sydney's sensible reluctance to allow Diana's engagement on the grounds of her extreme youth was reinforced by her own upbringing and the tenets of her father's generation, namely that sisters should marry in order of age. There would be plenty of time for Diana to find a husband, she felt, when Nancy and Pam were settled. Pam, it was true, was engaged but though Nancy, as nearly always, was in love, the object of her affections, Hamish St Clair Erskine, was entirely unsuitable. He was charming, good-looking, hopelessly irresponsible, several years younger than Nancy and a homosexual – Nancy, innocent if not ignorant, fondly believed that the love of a good woman would 'cure' him. Meanwhile, they tore round London to every party, giggling in corners. ('Hamish *was* funny yesterday. I do wish you'd been there to wink at, he had 5 glasses of brandy & crème de menthe (on top of sherry etc) & then began to analyse himself . . . I do worship that child.')

There was also the question of the immense wealth of the Guinness family, to Sydney a disadvantage; she felt her daughter was far too young to become the possessor of such a fortune. When the season was over Bryan could, of course, come and stay at Swinbrook from time to time; meanwhile Diana would be sent on a round of country-house visits in the cause of meeting other young men – and she was not, on any account, to consider herself engaged.

But even the Redesdales could hardly refuse to let their daughter stay with Bryan at his parents' home. She would be properly chaperoned, in a house of the utmost respectability – Lady Evelyn, though eccentric, was far from 'fast'. Reluctantly, Sydney accepted Lady Evelyn's letter of invitation and, in mid-August, Diana was allowed to travel to Bailiffscourt in Nanny's charge. Here she and Bryan walked on Lady Evelyn's lawn of wild thyme, sat on the beach and went for family picnics on the Downs. Colonel Guinness was not there; following his usual custom during the parliamentary recess, he was away on his yacht.

Bryan's family loved Diana. 'She's enough to turn anyone's head,' said his little sister Grania; and Lady Evelyn, unworldly, anxious only for her son's happiness and now reassured as to Diana's general niceness, began to take a more favourable view of an early marriage.

Not so Diana's parents. Soon after she returned from Sussex, she was briskly despatched to stay in Scotland with the Malcolms of Poltalloch, whose third son, Angus, was in love with her. Lady Malcolm, reputedly the illegitimate daughter of Prince Louis of Battenberg and Lillie Langtry, was a *grande dame* who lived in state on the 480,000-acre Malcolm estate. In the huge sandstone manor house on the Mull of Kintyre, banners hung round the azalea-scented hall and bagpipes skirled round the table at dinner. Lady Malcolm, presiding bejewelled and tea-gowned over a silver teatray, thought that Diana, well bred, well mannered, beautiful and seemingly biddable, would make an excellent wife for her good-looking son and encouraged them to wander on the hills in the long Scottish evenings. Diana faithfully reported these manoeuvrings to Bryan in Vienna, where he had gone with Robert Byron to distract himself while she was away.

'I am green with envy, but not jealousy, I promise you, of Angus,' he wrote back. 'I get really furious when I think of the Clan Malcolm pursuing you, winding smiles around you, stunning you with bagpipe music and then trying to mother and marry you. I suppose one can't blame them for liking you but how dare Lady M be so forward as to propose to you all helpless and alone under her roof.'

But neither Angus's protestations, his mother's blandishments nor the beauty of the Argyllshire hills had any effect on Diana. Pleasantly, firmly and inexorably, she turned Angus down.

Bryan's promised visit to Swinbrook took place on the weekend of 6 September. In heaven at the thought of seeing Diana again, quaking at the prospect of meeting her formidable parents, facing Nancy's teasing and being subjected to the critical gaze of five other Mitfords, he was in a turmoil of emotion. Fortunately his visit was brief, though long enough for his considerable charm to be felt by Diana's younger sisters, so fascinated by the romantic drama in their midst that they trailed him and Diana constantly ('Do you think we will ever be able to see each other alone? or will the seminary follow in a crocodile wherever we go?' he wrote afterwards). But her parents remained adamant.

How to circumvent or erode this rock of opposition was a constant subject of discussion. For Diana was now just as determined to marry Bryan as soon as possible as Bryan was to marry her. Among their wilder schemes was talk of an elopement to Gretna Green. But this crumbled when, after writing to a parson reputed to be sympathetic to lovers who took this road, all they got in return was a lecture on self-discipline. 'Like shouting halt at the Falls of Niagara,' said Bryan in disgust. More realistically, they concentrated on the process known in the Mitford family as 'slowly wearing away'. Logic, pity, the natural affection of parents for their child, all must be pressed into service. 'When you feel most unhappy you must make a point of showing yourself to your Father and then you may convince him of how serious we are,' wrote Bryan the day after he left. 'It depresses me more than I can say to think of waiting any time at all. The thought of it makes me feel absolutely ill.'

The next day saw the first sign of hope. Bryan went for a walk with his mother, during which she declared that she was now unequivocally on the side of an early marriage. 'I can see no point in your waiting, as you are both absolutely sure of your feelings,' said Lady Evelyn. 'When people are in doubt, they ought to wait for ages. But there is certainly no doubt about you.' She promised to do what she could to influence her husband. As a first step, Diana was asked to stay at Heath House. Here she met Colonel Guinness, now returned from his oceanic wanderings, ate delicious food free from all Mosaic taboos and, for aesthetic sustenance, was introduced by Bryan to his passion for the theatre: every night, after an early dinner, they saw a play.

Randolph Churchill had not taken lying down the decision of his beloved to marry another. His sister Diana, who had seen Bryan and Diana together at the Eton and Harrow match, had reported to him that Bryan seemed (as Randolph put it) 'much catched' by Diana; by September he knew of their secret engagement. Still hampered by his schoolboy status, immured in Eton for the autumn half, furious and frantic at the thought of Diana disappearing from his horizon before he had a chance to grow up and court her properly, he attempted to part them by every means he could think of.

He bombarded each of them with letters explaining why the other was unsuitable, arguing ferociously that the marriage would be an appalling mistake ('You'll hate each other in no time!'). He did his best to put Bryan off Diana by explaining what he, Randolph, who had known her for many years, could see so clearly: she was not the right person for Bryan. In case this was not enough, she had been flirting with other men. Diana, he told an astonished Bryan, and anyone else who would listen, was not nearly nice enough or of good enough moral character. To Diana he explained – presciently, as it turned out – that Bryan was not strong enough for her. She would do far better to wait for him, Randyov, to grow up.

Bryan's generosity of nature allowed him to forgive and forget almost immediately. As for Diana, aware of Randolph's feelings but regarding them as a schoolboy crush rather than a first passionate love, she was too fond of him to do more than scold severely.

Randolph's attempt to part the lovers may have been typically bold, resourceful and ill-judged; it was also perceptive. Equally characteristic was the note of injured self-justification:

> You know I never told Bryan anything about you which anyone could possibly resent. The most I ever said was that though you were basically of a good disposition, you have no fundamental moral sense. In other words, though you rarely do wrong, you do not actually see anything WRONG in sin. With all of which you will, I am sure, agree. However, all this was merely a polite way of telling him what I have already told you: that I do not think he is a character capable of retaining your affection . . . However, as you said in your letter, let's forget all about it. I have had two charming letters from Bryan. In fact, the only person who seems annoyed over what I have done is Diana C, who keeps on making the most bloody remarks.

If Bryan had taken any note at all of Randolph's remarks about Diana's 'flightiness', he might have thought that the enquiry she now made was strange in someone who professed herself so deeply in love. She asked him if he would mind if she went out with other men after they were married. 'I had never before thought about your going out, alone, with other men,' he told her. 'But though I know you could perfectly well be trusted to, I shall never allow you to, because you might think I didn't love you so much if I did. I am afraid we shall be terribly exacting towards each other, just because we love each other so frantically much.'

This conviction had begun to impress his most valuable ally. By mid-October Colonel Guinness, too, had come round. He had become very fond of Diana whom he considered an eminently desirable daughter-in-law, he realised that his son's mind was irrevocably made up, and he could see no logical point in waiting. He offered to write to Lord Redesdale to say so, though he pointed out that there was nothing he could do if Lord Redesdale still countered that Diana was too young. Bryan, however, felt that any approach had best come from himself ('your Father might like it better'), so wrote to Lord Redesdale asking if he could talk to him about Diana if he was at any time in London.

David Redesdale's reply was characteristic: 'I never come to London if I can avoid it and as I can avoid it at the moment I am not likely to be there for some time. I understand you are likely to be at Swinbrook within the next fortnight and, if what you want to say will keep till then, well and good.' This delaying tactic threw Diana and Bryan into some confusion. Should they put all their arguments for an early marriage in a letter? Or should Bryan visit Swinbrook and face David's terrifying presence? Eventually, they agreed that a face-to-face discussion had more chance of success. 'It is much more satisfactory to see him, though much more alarming,' said Bryan.

By now Sydney's opposition had more or less crumbled and so David finally agreed to see Bryan, inviting him to lunch at his club, the Marlborough, on 25 October. ('This is the most alarming thing I have ever had to do in my whole life,' wrote Bryan to Diana a quarter of an hour before setting off by bus down Piccadilly from 10 Grosvenor Place. 'The thought of the Marlborough Club alone would justify whole weeks of terror.')

Rumours that they were engaged were already rife among their friends, putting further pressure on David, who detested the idea that his daughters could be the subject of gossip. 'I went to the dance at Londonderry House last night,' wrote Bryan. 'It was very embarrassing because everyone thought each of my partners must be you. Lady Cunard (!) rushed up to Anne Cavendish (!) who was dancing with me and told her she hoped she'd be happy . . .' Pam's engagement to Oliver Watney had just been broken off ('postponed because of illness', said the announcement in the papers); speculation about Diana was not to be borne.

At the club, David led Bryan into the billiard room to ensure privacy and there, after a brief discussion, agreed in principle to an Easter wedding.

At the beginning of November Bryan went to stay with his uncle Lord Iveagh at Elveden, in Suffolk, for a grand house party; Elveden was a famous shoot and the party had been arranged for George V, who was a first-class shot. Though he was only a junior member of the Guinness clan, Bryan's shyness at being even momentarily the focus of royal attention overwhelmed him, manifesting itself in one of his familiar headaches. 'My head is still bursting,' he wrote on 5 November, the day after the King and Queen arrived:

> The Queen was enormously tall and just exactly like her picture, but the King was very much smaller than I had thought, and much younger looking. He said to me that he hadn't seen me since I was so high, so I muttered, 'Yes, sir'. This is so far my total conversation with him. The Queen will keep walking around the house, looking at the furniture and so one doesn't know when she won't come across one.

But all such terrors soon vanished into insignificance. By the end of Bryan's visit, which lasted for ten days, he and Diana were officially engaged and Diana too had been invited to stay at Elveden.

A dejected letter came from Angus Malcolm:

> In spite of the lashing gales that block every step, it is very pleasant. Everything seems to remind me of you when you were here. You seem to have left a trail of pleasantness, not only on roads, hills and woods but among everybody here . . . Oh dear, oh dear, why does everybody when I begin to get fond of them, get engaged to somebody else six months later?

On 28 November the final bastion fell. Diana's father agreed in a telephone conversation with Bryan's father that Diana would be allowed to marry sooner rather than later. 'Don't tell the young people,' David said to Colonel Guinness, 'but between you and me I don't think the end of January or the beginning of February would be altogether impossible.' Bryan, sitting with his mother on the sofa listening to the fateful words 'in terrific trepidation', rushed upstairs on the pretext of an early bed to write to Diana at ten-thirty with the news. 'I think your Father is the most kind and considerate and delightful man with *far* more imagination than any of us, and he hides it under a bushel of ferocity.' He was so glad, he concluded that they would not have to choose between dying ('which is what waiting till April would have meant') and eloping.

He had dismissed from his mind entirely one disquieting episode. Once, caught off guard in the garden at Hampstead, Diana had answered his fervent declaration that their love would last for ever by murmuring, 'Well, for a long time, anyway.' It was, in his own words later, a violent shock. 'I hoped it was a tease.' The Mitfords, as he told himself, were very fond of teases.

Chapter V

Diana Mitford and Bryan Guinness were married on 30 January 1929. At first the Redesdales had raised objections to a church wedding on the grounds that Diana had largely lost her faith. Bryan found this attitude extraordinary, and vehemently opposed it.

> Of all the amazing things I have heard or seen in this very strange world this objection of theirs is the most astonishing. Just as painters embody ideas of spring, summer, autumn or winter allegorically in human shape, so we realise that the idea of beauty, which includes Christian kindness (which is beauty of deed) has been embodied in a personal deity. My worship of this ideal beauty is directed towards its most perfect manifestation, which is yourself. I would therefore like to be married to you in a building dedicated to beauty, which is goodness, which is truth. Beyond this I know that psychologically I am only as good as I am because I was brought up to believe strenuously in God . . . anyway, I insist on being married in a church because to do otherwise might be the death of my whole family.

In the event, one of the most fashionable churches in London, St Margaret's, Westminster, was chosen. Diana changed into her parchment satin wedding dress at the house of family friends in Wilton Crescent (the Redesdales' house in Rutland Gate was let). She was an exquisite bride, her silver-gold beauty emphasised by the pearl and diamanté embroidery on the tight-fitting bodice of her dress and the crystal and silver flower wreath which held in place the Brussels lace veil given to her by Lady Evelyn.

Her bridesmaids wore dresses of gold tissue with overskirts of parchment tulle, with the long pearl and crystal necklaces that were Bryan's presents to them. True to Sydney's principles, Diana's going-away suit had been made at Swinbrook, though Lady Evelyn had wisely given her new daughter-in-law collar and cuffs of mink to trim it with. It was the last time Diana wore anything homemade. The only disappointment was that neither her best friend Celia Keppel (always known as Cela), whose mother had recently died, nor her sisters Decca and Debo, who had contracted whooping cough a few days before the wedding, could be bridesmaids.

The young Guinnesses spent the first part of their honeymoon in Paris, staying in the Guinness flat at 12 rue de Poitiers. On these two floors overlooking the Seine all was luxury, with no trace of Lady Evelyn's obsession with the medieval. Instead of smoke-blackened beams and pewter, her bedroom had grey satin curtains, a grey satin bed canopy wreathed with pink roses, a wood fire that burned brightly and efficiently and a comfortable daybed heaped with lace cushions.

Diana, encouraged by Bryan, spent days shopping, jettisoning the Swinbrook homemades and buying clothes for her new life. Her favourite, in which she invariably caused a sensation, was a short, tight, white faille dress with an enormous pale blue bow at the back which hung down to the ground, from the grand dressmaker Louise Boulanger. When they were visited by Bryan's grandfather Lord Buchan (the Pocket Adonis), who had given them three exquisite Battersea enamel boxes as a wedding present, the statuesque Diana, in her new Paris finery, made a great impression. As he was leaving, the diminutive Lord Buchan drew his grandson to one side. 'Pretty little woman you've married, Bryan,' he said.

After Paris, the Guinnesses went to Sicily, then still primitive and unspoilt. Here, for the first time, Diana saw the Mediterranean. As she later recalled, few more magical introductions can be imagined than to glimpse the sea through a foam of pink almond blossom from the open porch of an ancient Greek temple.

When they returned, it was to their new house in Buckingham Street and, for Bryan, his Bar exam. He was a pupil in the chambers of Stephen Henn-Collins, an expert on copyright (who eventually became a High Court judge). Anything less suitable for the gentle, shy, artistic Bryan would be hard to imagine, but it was the career suggested by his father when Bryan baulked at Colonel Guinness's first choice – following his father into politics and, at a suitable moment, taking over his seat as Conservative Member for Bury St Edmunds.

Bryan had bought 10 Buckingham Street (now Buckingham Place) from Muriel, Countess de la Warr. It had been designed by Lutyens and even then was expensive: its rates bill was £910, which was considerably more than the annual income of most of their friends. They furnished the house together; it was the only time in her life that Diana – still very conscious that her new affluence was due entirely to her new husband, and still only eighteen – failed to put her stamp on a house. Bryan's decorating taste had been greatly influenced by his parents' reaction against Edwardian ostentation and elaboration and the dining room, with its long refectory table lit by tall candles, flowers in a simple pewter jug and Jacobean glasses, was an echo of Lady Evelyn's medieval fantasies. Diana's bedroom had pink walls and a blue brocade bed standing on a dark blue velvet daïs.

There were familiar faces around them: Bryan's favourite parlourmaid from Grosvenor Place, May, came to look after them and their chauffeur was Turner from Swinbrook. Meanwhile, Lord Redesdale sent them one of his dogs, a labrador called Rubbish which he knew Diana loved. Rubbish was gun-shy, which nearly proved his undoing: a few weeks later, when Diana and Bryan returned from Berlin, where they had gone in search of the exotic night clubs recommended by Bryan's friend Brian Howard, they were met by Turner at Harwich – but no Rubbish. Hearing a car backfire, he had bolted. Two days later, the police brought him back, cut and terrified. It was only a respite; a few weeks later Rubbish started to have fits and had to be put down. To console Diana for the loss of this last link with her childhood, Bryan bought her an Irish wolfhound called Pilgrim, which grew enormous, eating pounds of raw meat – which he would usually take only from Diana's hands – and needing constant exercise.

Almost at once, the difference in the bride's and groom's social styles became apparent. Bryan, naturally uxorious and still barely able to believe his luck in capturing his golden goddess, asked for nothing more than evenings alone with Diana after his day's work studying, a shared companionship which he hoped would be broken only by a steadily increasing brood of children. His love for the arts found its outlet in constant theatre-going. Night after weekday night he would take Diana to each new production, from the latest thriller to Shaw, Strindberg, Shakespeare, Ibsen. On Sundays they went to theatre clubs. But he basically disliked urban life, preferring the beauty and solitude of the country and, though he loved his friends, he liked them in carefully regulated doses. At one point he expressed his view of cocktail parties – becoming increasingly fashionable – in one of the poems he was now writing constantly:

Here Friendship founders in a sea of friends
And harsh-lipped Bubbly cannot make amends;
Nor all the poisons shaken in a flask
That hands the insistent host with smiling mask.

Diana's tastes were the opposite. Now she was catapulted suddenly into the life she had always wanted, in which she could entertain clever and amusing people who were prepared to chat endlessly. The Buckingham Street house and its excellent cook quickly became a mecca both for the Oxford coterie, who only had to pick up the telephone to be invited round by Diana, and for new friends such as Lotte Lenya, Peter Quennell and the Sitwells.

I too could be arty, I too could get on
With Sickert, the Guinnesses, Gertler and John

began one of the poems of another habitué, John Betjeman. Much younger than any of them, Diana listened to their sophisticated gossip, shared their reverence for Harold Acton – she and Bryan were constant visitors to the Acton family's large house in Lancaster Gate – read the books and saw the paintings her new friends recommended. She went to concerts with her brother Tom, who was also constantly in and out of the house (like Bryan, with whom he got on extremely well, Tom had chosen the Bar as a profession). In many ways, the taste, aesthetic sensibility and culture of these friends was the nearest thing to the university education she never had.

She flung herself with zest into her new life. There were luncheon parties, dinner parties, tea parties, there were visits to Italy and France – Bryan's suggestion of a bicycling holiday fell on deaf ears – expeditions to museums, concerts and theatres. Always Diana was the central star, the flame round which they all gathered and, increasingly, the shrine at which they worshipped. Impeccably dressed, beautiful, funny, warm and flatteringly attentive, she fascinated them all.

For weekends, Colonel Guinness lent his son and daughter-in-law a small house called Pool Place in the Bailiffscourt 'compound' at Climping. It was very ugly but almost on the beach, with the lovely Sussex countryside behind. Here too friends came constantly, to swim, to sunbathe, to picnic, but above all to chat. At night Bryan would walk Pilgrim, enjoying the summer evenings or, in wind and rain, thinking of Goethe's enjoyment of tempestuous weather.

* * *

In May 1929, just before the general election, the young Guinnesses went to stay with Bryan's parents at yet another of the Guinness homes: Colonel Guinness's house in his constituency, Bury St Edmunds. For each of them it was an ordeal, though in different ways: Bryan, because the prospect of canvassing for his father, with its public speaking, its bearding of strangers in their homes, made him physically ill; Diana, because her strong though as yet unformulated political sympathies were in direct opposition to the politics of her father-in-law's party. In addition, she had taken an instant dislike to Stanley Baldwin, Prime Minister and Leader of the Conservative Party, when she had met him at one of the eve-of-session parties given by the political hostess, Lady Londonderry. It was an aversion shared by her parents, though both remained thoroughgoing Tories.

The general election of 1929 was the first in which women had the vote on a parity with men: the previous year, the Equal Franchise Act had lowered the voting age for women from thirty to, like men, twenty-one, thus enfranchising a further five million women. When the nation went to the polls there were 14.5 million women voters compared to 12.25 million men and many of them were swayed by the romantic looks and compelling platform personality of the Labour Leader, Ramsay MacDonald, a man patently sincere, kind and honest.

The Trades Disputes Act, 1927, which made any repetition of the General Strike of 1926 illegal, had polarised political opinion in the electorate, dramatically increasing support for the still half-fledged Labour Party. Diana, whose first real political awareness was of the long-drawn-out bitterness of the miners' strike and of their miseries thereafter, was heart and soul on the side of these working men and, if she had been old enough to vote, would have supported the Labour candidate.

As it was, Labour emerged for the first time as the strongest party, winning 287 seats to the Conservatives' 261. Colonel Guinness was returned, though with a reduced majority; now out of office, he was able to board his yacht earlier than usual. The Liberals, who dropped to a mere 59 seats, fell for the first time from second place. It was true that they could have combined with the Tories to keep Labour out of office, but Baldwin's dislike and distrust of Lloyd George was far too strong to permit this. Yet it was the Liberal Party's Election Manifesto, *We Can Conquer Unemployment* – based on the brilliant Liberal economist John Maynard Keynes's belief that a planned economy accompanied by massive public investment could provide jobs for all – that was to be the future inspiration for the young Labour politician Oswald Mosley. The rising star of his party and a brilliant orator, Mosley was made Chancellor of the Duchy of Lancaster in the new Labour administration.

After the election, the Guinnesses resumed a social life that was becoming ever more crowded and glittering. When Diana was taken up by Emerald Cunard, it was a notable gesture from one whose parties were famous – Lady Cunard's formula for entertaining was clever or famous men encouraged to greater flights of wit and brilliance by the presence of beautiful women. So impressed was she with Diana's looks and personality that she insisted she was to be her successor as hostess supreme. Another bond between them was their love of music. Diana was an avid concertgoer; Emerald Cunard worked tirelessly to promote not only the interests of her lover, the conductor Thomas Beecham, but opera generally.

It was after an evening at the opera that Diana first met Lytton Strachey. She and Bryan were sitting in the stalls when they saw Lady Cunard waving to them and went up to talk to her in the interval. She invited them to come back after the opera to her house, 7 Grosvenor Square, to join the supper party she was giving there, and at the same time introduced them to Strachey, who was sitting with her in her box. Tall, pale, thin, with an extraordinary fluting voice and a reclusive disposition, he had become a social 'lion' after the publication of *Eminent Victorians* and his presence at a party was considered a triumph. As usual, Emerald Cunard knew exactly who she intended to seat next to whom at the small supper tables arranged in her first-floor drawing room.

'Come along, Mr Strachey,' she said, 'I want you to sit next to Diana Cooper.'

'I'd rather sit next to the other Diana,' said Strachey and, striding across the room with his heron-like gait, he folded his bony frame into a chair by the side of Diana Guinness. Diana was both immensely flattered and slightly apprehensive about her ability to keep this formidably intelligent, much older man adequately entertained, but their rapport was immediate. It was the start of a friendship that was to last until Strachey's death less than three years later.

Most diversions were less intellectual. It was a frivolous era and the society papers, always anxious for a new star, had already realised that this nineteen-year-old girl was someone to watch. When she began to grow her hair it was reported reverently by the *Tatler* ('Mrs Bryan Guinness looking lovelier than ever with her new "Trilby" coiffure'); when she wore her diamond tiara upside down and flat round her throat as a necklace at a dance at which the Prince of Wales was present, the *Bystander* regarded this as news – along with the fact that the yo-yo was the success of the ball. 'It is a new game which fascinates the Prince.' All these doings were observed and noted by Diana's new friend Evelyn Waugh, and glimpses of them can be seen in *Vile Bodies*, his first success.

'At the Tropical Party given that July on board the *Friendship*, one would-be gatecrasher arrived in full Zulu war paint and headdress only to be turned away,' ran one report, 'and Mr Evelyn Waugh sported a solar topee while his wife Evelyn was daringly dressed in flannel trousers, open-necked shirt and a panama.' The party had been given by the Guinnesses and it was the break-up of the Waughs' marriage only a few days later that brought Evelyn Waugh to the centre of Diana's life.

Waugh was already on the fringe of their circle. He had met Bryan at Oxford and his wife Evelyn – they were known as He-Evelyn and She-Evelyn – was a friend of Nancy, who had moved into the Waughs' five-room flat at 17a Canonbury Square the previous month (it was here that Nancy began to write her first novel, *Highland Fling*). Nancy was of course invited to the *Friendship* party, and brought her friends the Waughs. When She-Evelyn left Waugh for another man and the Waugh marriage broke up, convention made it impossible for Nancy to go on living at Canonbury Square. Waugh gave up the flat and returned to his parents' home; Diana offered her sister a room at Buckingham Street and, naturally, Waugh came to see Nancy.

Once in Diana's company, Waugh was dazzled. For Nancy, this repetition of a familiar phenomenon must have been galling. Against the fact of Diana's beauty, Nancy's own wit, sparkle and considerable attractiveness, her years of friendship with Waugh, counted little. Without intention or coquetry – in fact, without even trying – her sister moved to the centre of Waugh's life.

Soon Waugh was spending most of his day at Buckingham Street. In the mornings he would sit on Diana's bed, chatting, while she read or wrote her letters and telephoned. He would accompany her on walks, drives or shopping, lunch with her and often return for dinner in the evening with both Guinnesses before going home to his parents' house.

Diana found him an enchanting companion. His irreverent wit and sophisticated approach fascinated her; he in turn was always at his best with her, so funny and spontaneously merry that she found it difficult to believe his heart had been broken along with his marriage. What she did not realise was that she was his emotional focus, his unattainable ideal. He longed to impress her, but was psychologically shrewd enough to conceal his infatuation behind jokes, fun, entertaining conversation and a teasing affection.

In August Colonel Guinness lent Bryan and Diana the Guinness house Knockmaroon, just beyond Dublin's Phoenix Park. It had a vista of the River Liffey – the quality of the Liffey's water was said to account for the Guinness fortune – with Georgian houses on the opposite bank and cattle grazing in the green meadows beyond.

Diana was now two months pregnant with their first child and Bryan more in love than ever. When, overcome with the drowsiness of early pregnancy, she rested after lunch, Bryan would scatter rose petals round her sleeping head and leave a poem on the pillow. Socially, however, it was an active time. There was the inevitable house party, including Evelyn Waugh, who had come on from the races. They lunched with the MacNeills in Viceregal Lodge and were introduced to W.B. Yeats by Lennox Robinson of the Abbey Theatre, to which Bryan went at least twice a week when in Dublin.

In November, Bryan, Diana, Nancy and Evelyn went to stay in the Guinness apartment in Paris. All of them except Diana were writing books – Bryan was writing *Singing Out of Tune*, Nancy was completing *Highland Fling* and Waugh was working on his travel book *Labels*. Diana, five months pregnant, stayed in bed until lunchtime while the other three worked; Bryan, who could not bear to be parted from her for a moment, wrote in her bedroom, spattering his mother's grey satin curtains with ink when he shook his recalcitrant pen. In the afternoons, the party would go to a cinema, a dress show or a museum before dining out.

Back in London Waugh, struggling with the deadline for *Labels*, spent as much time as he could with Diana. Though the young Guinnesses spent Christmas morning at Grosvenor Place, Waugh joined them for luncheon at Buckingham Street. They gave him a stocking, containing a gold pocket watch; his present to them was the dedication of *Vile Bodies* ('To B.G. and D.G.'), followed on 4 January 1930, by the manuscript, bound in leather with its title stamped in gold, and the inscription:

Dear Bryan and Diana.

I am afraid that this will never be of the smallest value but I thought that, as it is your book, you might be amused to have it (as a very much belated Christmas present).

Best love from
Evelyn

Vile Bodies was published in January and caused an immediate sensation.

Diana and Bryan's son, Jonathan Bryan, was born on 6 March 1930, weighing ten pounds. The birth was uncomplicated, the ecstatic Bryan gave his wife the most sumptuous present he could think of, the huge and wonderful picture 'The Unveiling of Cookham Memorial' by Stanley Spencer, which Diana hung in the Buckingham Street dining room. Jonathan's birth also marked the beginning of the end of Diana's close friendship with Evelyn Waugh.

Chapter VI

Waugh's feeling for Diana was one of the deepest of his life. He had got to know her when he was not yet the success he so quickly became and at a time when he was deeply bruised emotionally. He was also suffering from grievous loss of self-esteem at the break-up of his marriage after so short a time and his wife's humiliating preference for another man. All these factors, coupled with Diana's beauty, charm and flattering, restorative interest in himself and his work, made her for a time the centre of his life. Medical and social opinion both held that pregnant women should lead restful, home-based lives and during her pregnancy she had depended heavily on Waugh's company.

After Jonathan's birth, she was like a prisoner released, flinging herself back into the world of parties, friends and outings. At first there was no apparent change in their relationship and Waugh's diary shows that he still saw both Guinnesses constantly. He played a successful April Fool trick on Diana, ringing her up early in the morning and cajoling her to appear in the role of policewoman at 'a matinée in aid of the Divorce Law Reform League'. He was one of the godfathers when Jonathan was christened on 11 April, when he met Randolph Churchill for the first time – Diana had wanted to deck the baby in long black lace trousers, but had been persuaded to choose an orthodox christening robe instead. And when, after luncheon with Eddie March, Evelyn went with Diana to see a mask done of herself he was delighted when she promised him a white and gold plaster copy (the mask was never completed).

That spring there were still frequent references in his diary to lunching at 'Buck St' and he himself was host at a luncheon at the Ritz for the Guinnesses, Nancy, Georgia and Sacheverell Sitwell and Cecil Beaton, after which he drove down to Pool Place with Diana and a friend. 'Her dog Pilgrim made smells all the way,' he recorded. The party went back up to London by train on Tuesday, and Waugh again lunched with the Guinnesses at Buckingham Street, after which Bryan took him to see 10 Grosvenor Place. He went home to change, then returned to dine at Buckingham Street, after which they went to see Ruth Draper and on to supper at the Savoy Grill.

But it was the last patch of cloudless sky. 'Full of people we knew,' Waugh had written of their supper at the Savoy Grill and 'people' were, in fact, what came between him and Diana. That summer, there was hardly a ball or party to which the Guinnesses did not go. Diana seized every opportunity to go out, when she was not entertaining at home. She lunched with Bryan every day he was working, usually at the Savoy ('So handy for the Temple,' said Bryan). The flock of friends swooped down again on Buckingham Street – the Acton brothers, Osbert Sitwell, the Yorkes, the Lambs, Robert Byron, Cela Keppel and her brother Derek, Diana and Randolph Churchill, Emerald Cunard, the Mitfords' great family friend Mrs Hammersley – all welcomed joyously by Diana.

Waugh found himself no longer her exclusive confidant. His jealousy and moroseness quickly began to colour his attitude: when he gave Diana an elegant Briggs umbrella for her birthday, his growing disenchantment made him write spuriously in his diary that she broke it the next day – in fact, she used it for years until it was eventually stolen. When he came to a supper party a fortnight later he had a fight with Randolph Churchill in the servants' hall.

The friendship unravelled rapidly and, to Diana, mysteriously. On 5 July, Waugh drove with Bryan to Pool Place, where Diana and Nancy joined them. 'Diana and I quarrelled at luncheon,' wrote Waugh in his diary. 'We bathed. Diana and I quarrelled at dinner and after dinner. Next day I decided to leave. Quarrelled with Diana again and left.' Four days later, at Cecil Beaton's cocktail party, he did not speak to her but merely wished her goodbye as he left; on the 17th, at John Sutro's musical party, he avoided her again, sitting in the garden with Bryan and drinking 'a great deal of champagne'. Diana, he recorded, 'was friendly and reproachful-looking'.

That night he sent her a note on his blue writing paper with its curved capital E at the top: 'Dearest Diana, When I got back last night I wrote you two long letters and tore them up. All I tried to say was that I must have seemed unfriendly lately and I am sorry. Please believe it was only because I was puzzled and ill at ease with myself. More later, saying we are all right. Don't bother to answer. Evelyn.'

He did, however, write the catalogue for the Bruno Hat Show, which took place at 'Buck St' on 23 July. This hoax had evolved during one of the Pool Place weekends earlier that summer, its moving spirit Bryan's Oxford friend Brian Howard. Howard, tall, good-looking, witty, malicious, homosexual, alcoholic, later to be the model for the character of Anthony Blanche in *Brideshead Revisited*, spent most of his life going to parties, gossiping and devising elaborate practical jokes that in their planning, wardrobe requirements and execution were akin to a one-night theatre performance. The Bruno Hat show was a prime example.

Bruno Hat was supposedly an unknown artist whose 'very good paintings in the modern French style' had been discovered by Bryan Guinness in the back room of the general store in Pool Place's local village of Climping. The story, as leaked through Lady Eleanor Smith's gossip column in the *Sunday Dispatch*, was that Hat was the son by a German of the woman who owned the store.

For the purposes of the exhibition, Hat was played by Tom Mitford, just back from Germany and virtually bilingual. Though his bright blue Mitford eyes were hidden behind dark glasses and he was smoking a thin, foreign-looking cheroot, his disguise of black wig and long black moustaches sat oddly against his fair skin. Bruno Hat's pictures, painted by Brian Howard on cork bathmats and framed in rope, were 'daring' in subject and treatment – one was a distorted impression of a couple getting out of a bath, another, 'The Adoration of the Magi', showed matchstick figures worshipping a geometric Mary holding a circular Jesus. To the disappointment of Howard, who had secretly been hoping to be 'discovered' as the new Magritte, only one picture was bought – by Lytton Strachey and then only as a kindly gesture towards Diana. The mysterious Hat, sitting in a wheelchair in a corner of the Buckingham Street drawing room, grumbled in a heavy German accent to friends and connoisseurs alike about the unwanted publicity he was getting. Next morning the press was full of the hoax. As a practical joke it was highly successful; on a personal level it was the last time Diana and Waugh spoke to each other on anything like intimate terms.

The final break came immediately afterwards, when Waugh refused to stay at Knockmaroon that August, giving as his reason his dislike of the other people asked – many of whom were his close friends. Diana was bitterly hurt, realising, without knowing why, that he no longer wished to be her friend, let alone her intimate. Thirty-six years later, a month before his death, he explained his behaviour:

> You ask why our friendship petered out. The explanation is very discreditable to me. Pure jealousy. You (and Bryan) were immensely kind to me at a time when I greatly needed kindness, after my desertion by my first wife. I was infatuated with you. Not of course that I aspired to your bed but I wanted you to myself as especial confidante and comrade. After Jonathan's birth you began to enlarge your circle. I felt lower in your affections than Harold Acton and Robert Byron and I couldn't compete or take a humbler place. That is the sad and sordid truth.

One legacy of that intense, disparate friendship was the bestowal of Debo's nickname for Diana on the second beautiful, golden-haired and blue-eyed Diana in Waugh's life, Lady Diana Cooper. Since she was four, Debo had called her older sister Honks, as did some of Diana Guinness's friends. When Diana Cooper got to know Waugh, whom she had met at one of Emerald Cunard's vast luncheon parties, she would ask him about his friend 'Lady Honks'. Because the two Dianas shared the same Christian name, and because Waugh thought the idea of Diana Cooper, this beauty so accustomed to worship, being called Honks so irresistibly funny, he transferred the name to her with enthusiasm.

That spring Colonel Guinness had decided that the young Guinnesses should have a country house of their own. Diana's occasional teasing remark that she and Bryan ought to build a futuristic steel and glass tower 'for the view' beside Pool Place, an idea which horrified Lady Evelyn, may have had something to do with it; more seriously, Walter Guinness believed, as he told his son, 'It is not a good plan for families to live on top of one another.'

One of the first houses they looked at was Biddesden, an exquisite Queen Anne house near Andover. They loved it at first sight and they had friends in Wiltshire: the Byron family lived in Savernake Forest; Henry Lamb and his second wife, Pansy Pakenham, sister of Bryan's school friend Frank, lived at Coombe Bissett; Diana's new friend Lytton Strachey was a few miles away at Ham Spray. It was close enough to London for Bryan's work, but stood solitary in a lovely valley. It was perfect, but the price, which included a 200-acre farm, was almost twice what Colonel Guinness had given them. Fortunately the owner, Mrs Fothergill, was equally convinced that they would be the right owners for Biddesden. When, after finding no other house to measure up to it, they approached her agents again, she halved her price, saying she wanted the Guinnesses to have it.

Anyone seeing Biddesden would understand their delight at securing it. It is a large and beautiful house of faded brick, the numerous windows of its façade decorated by the unknown architect with military trophies and other ornaments. It was built in 1711 for General Webb, one of Marlborough's commanders. Webb did not get on with Marlborough and when, in 1708, he succeeded in routing a French force three times the size of his own, one of Marlborough's favourites, Major-General Cadogan, was given the credit. The Tory Party took up Webb's cause and when, as a result of his action, the beleaguered town of Lille surrendered, Webb was given one of its large bells, cast in 1660. He hung it in a specially built crenellated clock tower beside his new home.

When the Guinnesses first saw it, creepers covered the exquisite wrought-iron gate and honeysuckle, vines and old-fashioned roses clambered up the south front. A flight of stone steps led down on to the sloping lawn, high walls of ancient, rosy brick concealed a ravishing garden.

It was the first house on which Diana placed her stamp. In her bedroom, the five huge windows looking out on to the terrace were curtained in creamy satin, the satin curtains of her four-poster bed with its Prince of Wales feathers were lined with oyster and the long mirrors between the windows reflected eighteenth-century chairs covered in white and oyster damask.

Many of her friends had a hand in the embellishment of Biddesden. Bloomsbury architect George Kennedy designed a gazebo, the niches in its walls filled by portraits of three muses by the mosaicist Boris Anrep – Diana can still be seen there as Erato.* The sculptor Stephen Tomlin, husband of Lytton Strachey's niece Julia (a former flame of Bryan's), made a lead figure as a centrepiece for the garden; Diana had the gates and doors of the farm buildings painted a deep, bright, near-cobalt blue.

Diana had also found her feet in the management of a house. She acquired a good cook, Mrs Mackintosh (always known as Mrs Mack), already with a sizeable repertoire of her own, and introduced the simple, delicious food that Sydney had offered at Asthall and Swinbrook, though without her parents' strictness at mealtimes – the elderly parlourmaid, May, who now came down to Biddesden, often joined in the conversation.

What neither of the Guinnesses was told was that Biddesden was haunted. Locally this was well known and was perhaps a factor in Mrs

* The muse of lyric poetry

Fothergill's decision to drop her price so dramatically. Only four years earlier, Major J.G.W. Clark of the 16/5th Lancers, who had won an MC and bar in the First World War, had rented it while his regiment was stationed at Tidworth – to be driven out after a week by the ghosts. When, in 1944, Bryan met George Clark, by then a Lieutenant-General and former commander of the Cavalry Division, the General was notably reluctant to speak of his Biddesden experiences.

It was true that Mrs Fothergill had carefully explained that the portrait of General Webb on his cavalry charger which hung in the hall should never be moved. It went with the house, she told them; if it was removed, Webb's ghost would ride up and down the stairs until his portrait was returned to its original position. Bryan and Diana duly left the picture *in situ* and hooves were never heard on the stairs. But there were other disquieting manifestations. At night footsteps could be heard pacing up and down the terrace outside Diana's bedroom windows. When Bryan was away in London she would lie terrified and unable to sleep as she listened to the steady sounds of an invisible walker. Eventually she told Lytton, who raised both hands in a gesture of despairing amazement. 'I *had* hoped,' he said in his most swooping tones, 'that the age of reason had dawned.' From then on, she determined, with some success, to treat the footsteps as a harmless acoustic joke.

Some felt the haunting even more strongly. Visitors who had slept in unaffected rooms pooh-poohed the whole thing; others refused to come again. John Betjeman would have been one of these, had he not been in love with Pam – after the tenant of the Biddesden dairy farm had left, Bryan offered its management to Pam, who was still wretched after the breaking of her engagement to Oliver Watney. During the three months it took to prepare a cottage on the estate for her, she lived at Biddesden. Pam was deeply affected by the Biddesden hauntings, as she had been at Asthall.

The bedroom allotted to her was above the two-storey-high hall and overlooked the drive. She experienced the Biddesden ghost the first evening she was alone in the house, Diana and Bryan having gone to London. By ten o'clock the two housemaids had already gone to bed and Pam put the dogs – Diana's Irish wolfhound Pilgrim and Bryan's dachshund – out for a run. Almost immediately, they came rushing in, howling. Scared, but still without thinking of anything supernatural, she left the wolfhound downstairs and took the dachshund up with her to her bedroom for company. She undressed and got into bed, but instead of falling asleep straight away as she usually did after a day's hard work on the farm, she became aware of an unseen, malevolent presence, standing over and behind her at the head of the bed. Sleep was impossible. Too terrified to move or turn round, she heard the clock in the tower on top of the house strike every hour. 'The ghost never left me,' she recalled later. 'I lay rigid, with the dog shivering in its basket. It was the other side of the room, but it shook so much you could feel the vibrations through the floorboards.'

Occasionally, the manifestations took the form of people talking outside Pam's door, so realistically that the first time it happened she said to the maids the following morning, 'I wish you wouldn't talk outside my bedroom,' to be met by their vehement denials. Sympathetically, Bryan and Diana suggested she move to another room and here she had untroubled nights. Only when she had to sleep for a night in Diana's bedroom, and again in a small room on the other side of the house, was she troubled again by the supernatural.

Bryan, who was not affected by the haunting, loved Biddesden, not only for its beauty but for the fact that he had Diana to himself there except for weekend visitors – and she, for her part, absorbed by her new baby and the embellishment of her beautiful new house, was contented.

The Guinnesses' life soon settled into a routine. On the days that Bryan did not go to London he wrote, aiming at 1,000 words before lunch so that he was free for the rest of the day. His chambers specialised in commercial cases of great complexity in which he was neither interested nor expert, but he enjoyed calling in at the nearby Fleet Street offices of the London *Mercury* for a chat with the writer Alan Pryce-Jones or the *Mercury*'s editor, J.C. Squire, who frequently printed his poems.

Bryan was perfectly happy with this life of writing, riding and pottering about. Often, when John Betjeman was staying, Bryan would accompany 'Betj' and Pam on their visits to old churches; often, too, Betj would persuade 'Miss Pam', as he always called her, to drive him about on Salisbury Plain or to Marlborough, where he would show her his classroom, or he would bicycle with her to matins in Appleshaw. 'My thoughts when they are not with you are with Pamela Mitford – I hope I am not a bore,' he wrote to Bryan after one of his many proposals to Pam. In the evenings there would be communal games, Bryan's conjuring tricks and, after dinner, hymn singing, which both Diana and John Betjeman loved (they had to wait until the port was being passed round, as May the parlourmaid refused to remain in the room while they behaved in such a fashion).

> Lovers, absent, choke with sighing:
> Their minds grow dark with groping fears
> Their hearts are wrenched with smothered crying:
> Their eyes burn full of frozen tears

wrote the lovesick Bryan when parted from Diana on a visit to the Guinness hop farm at Bodiam. But this did not exactly describe Diana's feelings.

Once at Biddesden, the friendship with Lytton Strachey was consolidated. Lytton had dined with the Guinnesses several times in London and accepted Diana's invitation to Knockmaroon that August ('Do come and stay here . . . It is icy cold and there is nothing to do but go to the Abbey Theatre and see sickening Irish plays'), arriving in a suit of orange tweed ('Oh dear me! my new tweeds were far too loud'), a visit which induced in him only modified rapture. Along with the other members of the house party, he was taken to a ball at Viceregal Lodge, up a mountain, to the Dublin National Gallery and endlessly to the theatre – until one evening, Diana, who had had enough of these jaunts, swept out in the middle of a performance with the whole party at her heels.

'How, oh how to say how much I enjoyed every moment of my visit and how, how, oh how, to thank you for your angelic kindness. I only hope my occasional vagaries didn't infuriate you,' wrote Lytton afterwards, signing himself 'Your devoué'. He responded to the Irish visit by inviting Diana and Bryan to stay at Ham Spray. As old friends were the only people ever asked to stay there, it was a signal mark of favour and Diana was, for once, acutely nervous. Lytton was rather an alarming person, whose extreme shyness added to the impression of intellectual hauteur felt even by those close to him. More than that, Ham Spray meant Carrington (the painter Dora Carrington, always known as Carrington). And Carrington, as everyone knew, was besottedly in love with Lytton. Diana, far more beautiful and, though very much a disciple, better able to meet Lytton on his own literary ground, might be seen as a rival worshipper at the shrine. In addition, Diana's natural manner with Lytton, teasing, admiring and rather flirtatious, was not calculated to inspire immediate confidence in a woman who lived only for this one man.

Carrington, by contrast, wore simple clothes and her fair hair was cut in a thick bob, with a heavy fringe behind which she hid when shy – and she was often shy. She had many of the mannerisms of a little girl, with her quiet diffidence, her habit of standing with her toes turned in and her head bent on one side, deferring constantly to Lytton in her soft voice. But her charm, though indefinable, was immense; many men had fallen in love with her.

The Ham Spray ménage consisted of four people: Lytton Strachey, Carrington and her husband Ralph Partridge, and Frances Marshall (Ralph's lover and later his wife). They lived by the principles of simplicity and – as it would now be called – 'greenness' which Bloomsbury held in common with the earlier disciples of William Morris. Carrington did most of the cooking at Ham Spray, serving plain country dishes based on homegrown materials. The jellies, the jams, the fresh currant bread and even the wines were homemade; there was farm butter and honey in the comb. One of her specialities was rabbit pie, for which even the rabbits had been home shot. Carrington, who had a touch of the gamekeeper in her, would pot rabbits or pheasants on the Ham Spray lawn from her bedroom window with an old gun of Ralph's. 'Just as I was getting into bed, I looked out for the last time on the moonlighted lawn and there was my enemy the rabbit, who all this week has eaten up my lettuces and cabbages, so I knelt at the open window and shot him,' was how she described one such episode to Gerald Brenan.

Manners were equally simple. No one, for instance, bothered to say goodnight, they simply got up and went to bed. Lytton would tell some witty, acid story or suddenly shoot out some disconcerting remark in his extraordinary voice. Or, rising from lunch, he would play an arpeggio with his long, elegant fingers along the edge of the table, a performance so deft in its physical wit that it reduced most who saw it to shaking uncontrollably with laughter.

Diana and Bryan's first visit to this unconventional ménage nearly ended in disaster. They had been given the best spare room, with its pale coral walls and large fourposter bed, and Carrington had prepared one of her celebrated rabbit pies for supper. That night, Diana developed sudden and acute gastritis. At four a.m. a doctor had to be called out from Hungerford, while Bryan and Carrington hovered anxiously nearby. Her raging temperature took forty-eight hours to abate and after that she had to stay in bed for another two days. As everyone had eaten the same food and no one else was affected, many of the Strachey circle thought that Carrington had tried to poison her. But it was during those two days in bed that Diana became friends with Carrington, who, guiltily conscious that she might have caused her guest's illness and anxious to do what she could to make amends, was constantly at her side.

After this, there were frequent comings and goings between the two houses, often for the walks or picnics that both Lytton and Carrington loved. Sometimes Carrington would ride over on her white pony and go out riding with Bryan on one of the Biddesden horses ('I had a lovely ride on Goldielocks and went some grand gallops'). There was a further link when Frances Marshall's best friend turned out to be Julia Strachey, whom Diana on meeting found to be instantly *sympathique*.

To those at Ham Spray, the Guinnesses seemed a golden couple, beautiful and very much in love. They laughed, they held hands, they made plans for the future. Around them was an aura of youth and happiness in which both Lytton – already flattered that such a handsome and clever young woman should like him so much – and Carrington basked. That Diana was infinitely stronger than Bryan passed them by completely.

That autumn the Guinnesses' travels continued. They made the annual pilgrimage to Venice customary in their circle. ('Nearly all of them are politicians and very boring and nice,' wrote Diana to Lytton. 'They are rowdy in the water and hit one another about.') They went to Greece and to Constantinople, where they lunched with the Ambassador, Sir George Clerk, who as a young man had ridden in Rotten Row with Diana's grandfather, Thomas Gibson Bowles. Life was so enjoyable that when, just after Christmas, Diana realised that she was pregnant again, it was not an altogether welcome discovery. 'Of *course* we'd have had you one day, darling,' she later told her second son Desmond (born on 8 September 1931), 'but not *just* then.'

However, if it had not been 'just then', Desmond's birth might never have happened. Bryan, even more deeply in love with his young wife, was thrilled at the thought – as he saw it – of the second in a large nursery. But the first cracks were already appearing beneath the surface of the marriage and Bryan's cloying devotion was beginning to get on Diana's nerves. By now he was at home much more as his career at the Bar had virtually come to an end: his clerk – no doubt influenced by the daily lunches at the Savoy – gave the few briefs available for junior barristers to hungrier members of the chambers, saying, 'Mr Guinness doesn't need three guineas.' Under the same roof all day long, with fewer distractions in the shape of friends endlessly dropping in, what Diana saw as Bryan's intense possessiveness became more apparent – and more infuriating to someone for whom a degree of privacy had always been essential.

From Bryan's point of view, love meant being with the person you loved – and the more you loved, the more you wanted to be together. His romantic, idealistic nature inclined him to worship and Diana was his goddess. They were, he believed, each other's 'Elective Affinities', though as he wrote later, 'You may always have felt more elected than electing,' an underlying insecurity that caused him to want her with him every moment of the day and, if not, to know exactly where she was.

Diana found his attitude claustrophobic. Though she recognised, and always would, his sweetness and goodness, his misty sentimentality irritated her, his considerable poetic talent did not move her and his whimsicality was antipathetic to her astringent Mitford humour. In the two and a half years since their marriage the strength of her own character had emerged and, in the words of one friend, 'she walked all over him'. Her taste too had developed and her literary, artistic and musical knowledge was infinitely greater and more sophisticated; she regarded her husband's love of what she thought of as the provinciality of Irish life as another proof that in matters aesthetic he, as she put it, liked pottery where she preferred porcelain.

All this, however, would have been bearable but for the stifling constraints on her personal freedom. There were constant quarrels as she attempted to fight her way free from the blanket of his possessiveness, followed by remorseful reconciliations when she was overcome by guilt at being horrible to someone so sweet-natured.

To the rest of the world, they still presented the picture of the perfect couple with close family bonds. Bryan, who loved her sisters, encouraged her to ask them to stay. Carrington described a luncheon party at Biddesden in 1931 in glowing terms. 'There we found three sisters and Mama Redesdale. The little sisters were ravishingly beautiful, and another of 16 [Unity] very marvellous and Grecian.'

When separated by physical distance, Diana became as loving as even Bryan could wish. She spent most of August, the last full month of her pregnancy, in London. She was uncomfortable and unhappy. The heat oppressed her and her marriage had increasingly begun to chafe her, despite her love for Bryan, who was away touring Austria with her brother Tom. She had become very conscious of the country's deepening economic crisis and particularly of the miseries of the distressed areas with their high rate of unemployment. 'You mustn't talk about having no one to love you, because you have got someone who does love you so much, and the few paltry hundreds of miles between us mustn't be allowed to make any difference,' Bryan wrote on 10 August and, two days later, 'I lie awake thinking and worrying about you.' Typically, though, his letter is a screed of forty pages to her crisp two sides of writing paper.

At the end of August Bryan returned. He and Diana settled in Buckingham Street to await the birth, whiling away the time by nightly visits to the theatre. Diana's first labour pains began while they were watching the play *Late Night Final*, but she found it so exciting ('as good as E. Wallace') that she ignored them until the end of the performance. At seven the following morning, Desmond was born. 'I don't know why I have [called him Desmond],' she wrote to Lytton, 'only Bryan wanted such queer names he read in a book of them like for example Diggery. I thought it sounds like the comic man in Shakespeare, perhaps a gravedigger.'

The baby's arrival seemed to solder the weak seams in the marriage and life went on much as before, studded with parties, people and family events. When Colonel Guinness returned from his long voyage to find that Ramsay MacDonald, whom he could not stand, had become Prime Minister he retired, accepting a peerage to become the first Baron Moyne (motto: *Noli Judicare*).

Diana began leaving Biddesdon to go up to London as often as she could. Beneath all her loving gestures, her good resolutions, her knowledge that her marriage could not, as she later wrote, be described in any other terms than perfect, lay one poignant, ineluctable truth. It had been foreshadowed by Bryan, in his poem 'Love's Isolation'. Written two years earlier, at the height of their mutual happiness, it expressed the fear that had always lain beneath his longing.

My touch may not unlock
The movement of your mind
For secret is the clock
Wherein your thoughts unwind . . .
I see into your sight
But oh, I cannot find
A way into the light
Of yourself who lies behind . . .

Chapter VII

1932, the year that shaped Diana's life, opened in unexceptional fashion, with nothing much changed from the previous autumn. Sydney gave a dance for another Redesdale daughter, Unity. John Betjeman's siege of Pam continued ('My thoughts are still with Miss Pam,' he wrote to Diana in February. 'I have been seeing whether a little absence makes the heart grow fonder and my God, it does . . .'). Bryan's longing to keep Diana to himself and her own fierce determination to lead as active a social life as possible still conflicted sharply; she often went to London on her own for parties, concerts and exhibitions. They went on arguing about the size of their family: Bryan wanted a lot of children, whereas Diana told him that two were enough. There was another of the trips abroad that with marriage had become a regular part of Diana's life. This one was at the instigation of her brother Tom, who had written to Bryan suggesting they visit him in Berlin where he was studying law. Tom was full of German politics; talking to him, Diana heard the word 'Nazi' for the first time. 'If I were a German I suppose I would be a Nazi,' said Tom. In the first years of the Thirties, with National Socialism still a minority movement and Hitler a long way from the power he seized in January 1933, few could foresee the horrific future which the adherents of Nazism were to build in Europe; to many of the young, the new movement seemed to hold out hope to a country still starving and economically depressed fourteen years after the First World War had ended.

Almost immediately after the Guinnesses returned from Berlin came the bitter news that Lytton Strachey, whose health had been steadily deteriorating all winter, had died on 21 January. For Diana it was a misery; for Carrington it was the end of the world, and she determined to carry out her declared intention of taking her own life.

Naturally Diana and Bryan knew nothing of this. They were not privy to the inmost secrets of the Ham Spray circle; they were of a younger generation and, though they knew how deeply Carrington had loved Lytton, neither of them had suffered any loss remotely akin to hers. They knew she needed as much help and comfort as possible; the idea that she might kill herself never crossed their minds.

Carrington, once her mind was made up, managed to put on a brave face to most of the outside world. She dreamed of Lytton in his 'old blue dressing gown' and wrote in her commonplace book, 'Every day your white face haunts me. Every hour some habit we did together comes back and I miss you. You were more dear to me every year. What is the use of "adventures" now without you to tell them to? There is no sense in a life without you.' But she was still able to conceal her consuming despair. Even her writing paper – palest pink, with an orange sunburst and an orange lining to the envelope – seemed to exude a kind of brave cheerfulness when she wrote to Diana, while staying with Augustus and Dorelia John.

Monday February 8

Darling Diana,

It was lovely seeing you on Saturday. I can't get over your kindness in coming all the way over here but if you knew how much difference it meant to me and how much everyone loved you and Bryan and the praises for your beauty and charm, you would have felt compensated. Augustus was very impressed and said in confidence to me he hoped to paint you for his own pleasure, and Dorelia saw everything I wanted her to see. Vivian I regret to say talked more of Bryan but I suppose that must be forgiven. I am going back to Ham Spray tomorrow. Ralph is coming to fetch me . . . We went for a long walk yesterday afternoon in the woods. I have finished Miss John's masterpiece today.

Diana, I wanted so much to give you something of Lytton's. He bought an eighteenth century waistcoat years ago and never could think of anyone worthy of it because it was so beautiful. Now it will be yours. Perhaps you could alter it. I'd like to think of you wearing it. For Lytton loved your beauty, and your rare character, and we talked of you *so* often. I'll send it to you tomorrow, when I get back. My love to dear Bryan and to you

Your very loving

C

When she returned to Ham Spray, Carrington continued to come over to Biddesden. Once, with Ralph Partridge and David ('Bunny') Garnett, she came for one of the picnics she had so enjoyed while Lytton was alive –

Bryan drove them to the chosen spot in a ponycart. More often she arrived on her own to ride with Bryan or Pam.

These rides were always rather an adventure. Bryan, who had acquired the Biddesden horses for the benefit of guests, had bought them from a riding school in Littlehampton, thinking they would be well schooled. At the riding establishment, where they were taken out every day, this was so, but at Biddesden, where they were full of oats and often not taken out for several days on end, they were apt to take hold and bolt once they reached open country. Carrington, who had not learned to ride until her late thirties, took a number of falls. Always brave, after Lytton's death she had the recklessness of one who no longer wished to live.

'Yesterday at Biddesden I came face to face with death,' she wrote in her commonplace book on 22 February:

> The horse bolted down the road and I could not pull him up as my wrists were so weak with trying to hold him in. I saw the long road with the bend and the logs on the Bank, and the horse tearing along the tarmac road towards the bend. He swerved round the corner and I came off and just missed the logs and the road and fell on the Bank. I was completely unhurt only winded by the sudden fall, and I think of the irony of fate that I who long for death find it so hard to meet him.

Six days later she again visited Biddesden on her own. This time she asked Bryan if she could borrow a gun – there were about half a dozen in the gun cupboard – explaining that Ralph's old gun had disappeared. Bryan, who knew how unhappy she was but not that she had already attempted suicide, wondered momentarily if he should lend her one. But her demeanour seemed so normal and her habit of shooting the rabbits on the Ham Spray lawn for the pot was so well known that he agreed, lending her the one he thought she would find easiest to handle, a Belgian 12-bore. It would, in any case, have been awkward to refuse someone almost old enough to be his mother (Carrington was nearly thirty-nine).

Back at Ham Spray, a doubtful Bunny Garnett asked her, 'What's all this about rabbits?' Just at that moment, one ran across the lawn. '*There!*' said Carrington, pointing triumphantly. She took the gun up to her room and hid it. Ralph Partridge, constantly on the *qui vive* in case she made another suicide attempt, had no idea it was there.

A few days before she killed herself Carrington came to stay at Biddesden. Diana was in London, and Bryan welcomed Carrington's company, hoping also that this meant she was gradually beginning to recover from Lytton's death. She arrived on 5 March, accompanied by Ralph Partridge; the following day, they were joined by Bunny Garnett and Frances Marshall. The weather was superb, with a clear blue March sky; Bryan, who had been walking in the woods with Pilgrim, took them on a picnic the following day. On her return home, Carrington wrote:

> Darling Bryan,
> If 'Collins' are to be written, it is *I* who should have written to you not you to me. You never realise how much I love my visits to Biddesden, and those gallops, even if they do end by falling off. I must admit I thought our picnic was an *especially* good one this time and our kitchen party delightful. David Garnett and Ralph and Frances enjoyed themselves enormously and loved you.
> I hope you have a lovely time tonight. I can hardly bear not to see you both in your dresses . . . but I have enjoyed so many parties at the Johns' that I couldn't face sitting out watching or feeling gloomy.
> The country is so lovely. Yesterday was a beautiful drizzling spring day. Very sympathetic. Give darling Diana all my love and a hundred kisses and all my love to you and one kiss.
> Your loving
> C

Apart from her reference to 'feeling gloomy' it was a letter that could have been written by any affectionate guest – and certainly not one to cause alarm bells to ring. But Carrington had only three days to live. 'What a relief just to finish all the jobs that are undone and then "go to sleep",' she scrawled in her private notebook.

On 10 March, Virginia and Leonard Woolf came to see her. They found her alone, which would have been unthinkable if anyone at Ham Spray had felt apprehensive about her. Indeed, so reassured had her husband Ralph become that early the following morning he went up to London for the day. After he had gone, just before eight o'clock, Carrington put on Lytton's yellow silk dressing gown, placed the butt of Bryan's gun on the floor and its muzzle against her side and, facing a mirror so that she could see what she was doing, pressed the trigger with her toe. The first time she forgot to release the safety catch; the second time the gun fired. She did not manage to kill herself outright, but took six hours to die.

It was a dreadful time. For Bryan, who was deeply fond of Carrington, there was the added torture of the guilt of having been the one to pass her the means of her death, of wishing, as he said to Diana, that he had thought quickly enough to say to Carrington, 'I'm sorry, but I never lend my guns.' Diana consoled him as best she could . . . but was soon to inflict a devastating blow on him herself.

Two and a half weeks before Carrington's death, Diana had met the man who would henceforth rule her life. On 21 February a friend with whom she had come out a few years earlier, Barbara St John Hutchinson, invited the Guinnesses to her twenty-first birthday party. 'I'm putting you next to Tom Mosley,' said Barbara. 'I want to see what effect you have on each other.'

Chapter VIII

The man Diana met in February 1932 was fourteen years older than herself, one of the best-known politicians of the day and a notorious philanderer. Drive, vitality, intelligence of a high order, vision, energy and a delightful gaiety were immediately apparent on meeting him. His physical presence was magnetic: his striking good looks, which today seem those of a handsome cad, were cast in the Clark Gable mould. Indeed this film-star charisma was one reason for his success both as an orator and with women. Arrogant, impatient, ambitious, conscious of his physical appeal and exuding a rather brutal masculinity, he gave the overwhelming impression of a personality of immense strength who would stop at nothing to achieve what he wanted or to conquer a woman he desired. A greater contrast to Bryan Guinness could hardly be imagined.

In his earlier political career the lust for personal power that was later to be Mosley's motivating force was not yet to the fore. An admirable idealism, combined with political talents, was bent towards bettering life for the poor. As for so many, his experiences in the First World War had left an indelible impression and he was determined to do everything he could to ensure that slaughter and devastation on such a scale never happened again. In the General Strike he (like Diana) had been on the side of the miners, championing their cause so fervently that he and his wife Cimmie became enormously popular speakers at miners' galas. The Depression of the 1930s increased his determination to help the unemployed.

Like most born orators, Mosley was not afraid to show emotion. He had been biting and sarcastic in parliament, he could demolish a heckler with a brilliant riposte, but his audience was left in no doubt that he cared passionately – he could move himself to tears with his own speeches. But possibly his greatest gift, and the one which accounted for much of his appeal as a leader, was the ability to create in his hearers the sense that anything was possible. His best speeches produced an intoxicating euphoria that lifted his listeners with him and made his followers feel that a glorious future was there to be won.

Physically, he was magnificent, six feet two inches tall, upright, hard and fit. All his life he had taken strenuous exercise at least twice a week, chiefly fencing. His style of oratory was so physical that it demanded fitness – during the Ashton-under-Lyne by-election of the previous year Harold Nicolson described him as 'an impassioned revivalist speaker, striding up and down the rather frail platform with great panther steps and gesticulating with a pointing, and occasionally a stabbing, index.'

His looks emphasised this dramatic pantherine quality: sleeked-back black hair and pencil moustache against a pale skin. Around him hung an aura of danger which served to challenge rather than frighten the worldly beauties he pursued. Like that earlier seducer, Lord Byron, he had a pronounced limp: his right leg was two inches shorter than the left and he had to wear a built-up shoe. Unless fencing, speech-making on a platform or, later, marching, he carried a walking stick or swordstick.

Most memorable were his eyes, dark, penetrating and seeming to look directly into the soul of the person to whom he was talking. Sometimes he would employ a curious trick of 'flashing' his eyes at a woman or an audience he was seeking to fascinate, suddenly dilating and then contracting the pupils ('rather like car headlights being turned on and off' was one description). His voice was sometimes melodic, sometimes harsh, but always hypnotic, able to whip up a platform audience with a frenzy of demagoguery or seduce a woman with its blend of unabashed desire and old-fashioned compliments.

Such a powerful personality stirred up extremes of feeling. In 1924 Beatrice Webb called him 'the most brilliant man in the House of Commons . . . the perfect politician and the perfect gentleman . . . a Disraelian gentleman-democrat', though by the time he was in government she was remarking more sharply, 'That young man has too much aristocratic insolence in his make-up.' His looks, his arrogance, his physical toughness and his uncompromising masculinity appealed strongly to many women, though Harold Nicolson's wife Vita Sackville-West said that he gave her 'the creeps'. The hot-tempered Duff Cooper described him as 'an adulterous, ranting, slimy, slobbering Bolshie'.

Mosley was born on 16 November 1896, eldest of the three sons of the fifth baronet, Sir Oswald Mosley. Though christened Oswald Ernald, he was invariably known as Tom. His father was a wealthy landed squire; the Mosleys owned Ancoats in Lancashire, but from the late eighteenth century Rolleston, near Burton-upon-Trent in Staffordshire, had been the family seat. When Tom Mosley was five, his parents separated and he was brought up by his mother, who moved to Shropshire, and by his grandfather, the fourth baronet, at Rolleston.

Educated at West Downs preparatory school and Winchester, he was not particularly interested in studies, preferring games, or rather sport; although athletic, he hated all team games. One strong – perhaps the strongest – thread of his personality was the need to dominate, to be the one in control, and he enjoyed only those sports which allowed individual combat or expertise. Fencing, at which he became junior champion while still at Winchester, and boxing fulfilled these criteria perfectly; in the holidays, there was riding, and hunting with the Meynell with his mother and brothers.

At sixteen Mosley left Winchester to cram privately for Sandhurst. For the first time, he put his formidable intelligence to work, so successfully that he passed in top of the cavalry list. He was commissioned into the 16th Lancers, then stationed at the Curragh. Mosley, who longed to see action in the war that had just begun (on 3 September 1914), was attracted by the idea of the fledgling Royal Flying Corps, and managed to get permission to transfer, with the aim of training to become a pilot. By Christmas 1914 he was flying as an observer over the German lines. Then, in 1915, he crashed while landing, breaking his leg. It was this crash, he always maintained, that gave him his lifelong limp, though another story has it that he was so unpopular at Sandhurst that he was once chased by a group of fellow cadets, all baying for his blood, and to escape them jumped out of a first-floor window, landing awkwardly and breaking his leg.

He spent a year in England on sick leave – an additional (Army) version of the story being that he had injured his right ankle in an escapade at home – then went to the Curragh in June 1917, only to be eventually invalided out because he could not march. He began work at the Ministry of Munitions, followed by the Foreign Office.

His spell in hospital had given him a chance to think of the future. Politics seemed the obvious choice for someone determined to influence events, and as a wounded ex-officer he was a popular figure. A month after the Armistice, aged just twenty-two, he offered himself as a candidate in the December 1918 election. To a young man in a hurry, getting into parliament was more important than party: Mosley stood as a Conservative because it was the accepted thing for people of his class and background rather than from any deep-rooted conviction, and because he had been offered the safe Conservative seat of Harrow. Once elected, he abandoned the idea of life as a country gentleman and devoted himself wholeheartedly to politics.

He soon began to make a mark. Within months he had become Joint Secretary of the New Members Group, a cross-party association dedicated to promoting a centrist party. When he married Lady Cynthia Curzon (always known as Cimmie), the second of Lord Curzon's three daughters by his immensely rich first wife, the seal seemed set on a glowing future. Cimmie's father's position as Foreign Secretary and former Viceroy of India, Mosley's reputation as a gallant young officer and their combined good looks made their wedding, attended by King George V, Queen Mary and the King and Queen of the Belgians, one of the social events of 1920.

Soon, though, Mosley began to find himself out of sympathy with his party. His marriage to Curzon's daughter did not stop him attacking the coalition government for its use of the Black and Tans in Ireland (an attitude later to stand him in good stead). 'They confused the right of men to defend themselves with the right to wander round the countryside destroying the houses and property of innocent persons.' He found the lack of discipline in this irregular force particularly repugnant.

In the election of 1922 he renounced the Conservative whip and stood as an Independent Conservative. Now a political outlaw, he was ignored by the party and its new leader, Baldwin, and was rapidly losing favour with his constituents. He realised that if he wished to remain in parliament, the path to the power he craved, he had to do something drastic. In 1924, after a brief flirtation with the Asquithian Liberals, he crossed the floor of the House to sit with the Labour Party.

He was welcomed delightedly, the Labour leader, Ramsay MacDonald, calling him 'one of the greatest and most hopeful figures in politics'. For the party to have acquired this brilliant young man, already known as an orator, from the other side of what was then an almost impassable class barrier, was indeed a coup, though many of the rank and file were mistrustful. This youthful heir to a baronetcy and his rich, titled, beautiful wife, daughter of the arch-Conservative Curzon, who constantly figured in the glossy pages of society magazines – were they really serious in their commitment to the cloth-cap values of a movement founded to right the wrongs of an oppressed working class?

Then there was the question of Mosley's women. Though only a small, but increasing, number of his own circle knew that he had been unfaithful to Cimmie from the first weeks of their marriage, the aroma of sexual success hung heavily about him. The Labour stalwarts sniffed it and were alarmed. To a party in which the moral climate had been formed by the crusading zeal of Keir Hardie, the earnest uprightness and abnegation of self-indulgence of the Webbs and the saintliness of George Lansbury, a puritanical approach to matters of the flesh was to be preferred to Mosley's carefree hedonism.

Among many Conservatives, Mosley's morals were considered more or less his own business, especially as he confined his amours to his own circle. 'I may vote Labour but I sleep Tory,' he would remark. More serious for the Tories was what they regarded as his treachery to everything he and his class stood for. Indignation rose so high that at one point there was a move to inflict the final sanction: expulsion from White's Club. Curzon, of course, was furious at his son-in-law's defection, especially when Cimmie, loyally changing her political allegiance to that of her husband, discovered a real talent for public speaking, and decided to stand as a Labour candidate herself.

Knowing that the voters of Harrow would not tolerate a further turn of their member's coat, Mosley had to find another seat to fight in the 1924 general election. Characteristically, he decided to battle for one which, if won, would cause the greatest sensation – and bring him the most personal glory: Neville Chamberlain's constituency of Ladywood, in Birmingham, held by the Chamberlains, the great Birmingham political dynasty, for fifty years. In an election which saw the first, short-lived Labour government swept from office Mosley came within a whisker of victory. The first count gave Chamberlain a majority of 20, a recount reduced this to 7, a third count went to Mosley, and a fourth finally put Chamberlain on top with a majority of 77, a victory so narrow that for the next election Chamberlain abandoned Ladywood for Edgbaston.

Two years later, Mosley was back in the House of Commons, winning a brilliant by-election victory in 1926 at Smethwick, where he raised the Labour majority from 1,253 to 6,582. By 1927 he had reached the five-strong National Executive of the Labour Party – an extraordinary feat for a man from the other side of the class barrier who had been a party member for only three years.

Cimmie, despite being pregnant, was hardly less active in the 1929 general election. She was standing for Stoke-on-Trent, but she campaigned vigorously wherever she was needed, often taking the chair for Ramsay

MacDonald and making one of her effective speeches in his place. She was victorious in the election, achieving a majority of 7,850 over her Conservative opponent, Colonel Ward. Mosley, admiring and delighted, gave his wife a diamond brooch in the shape of the Palace of Westminster with the figures of her majority across it in rubies (one of the few occasions on which he bought jewellery for a wife). But the weeks of unremitting effort had taken so much out of Cimmie that she was exhausted. Two days after the poll, she miscarried. She recovered in time to take her seat on 27 June and made her maiden speech – on Widows' Pensions – in October.

Mosley soon established a national reputation. He was energetic, ambitious, quick, witty, an excellent speaker who was just as effective on the floor of the House as when answering hecklers on the hustings, and he campaigned tirelessly for the party. He led as highly social a life as was consistent with the demands of party politics; against the drabness of much of the Labour movement Mr Oswald and Lady Cynthia Mosley stood out as a glamorous and exciting young couple. When Mosley succeeded his father in 1928 ('a baronetcy', he declared airily, 'is not worth renouncing') his title, like Cimmie's fur coats, only added to the Mosleys' aura of prestige and allure; already rich through Cimmie, he now inherited family estates worth £250,000.

The same year, he was re-elected to the National Executive. Largely through Cimmie, the Mosleys were now close to the Party's leader, Ramsay MacDonald, whose penchant for pretty, charming, aristocratic women who made much of him was already drawing scorn from the serious-minded Webbs. Mosley himself was politically closer to Ernest Bevin.

On the party's return to government in 1929 it was inevitable that Mosley, its rising star, would be given office and he was made Chancellor of the Duchy of Lancaster. With George Lansbury, the first Commissioner of Works, and under the aegis of J.H. Thomas, the Lord Privy Seal and former railway leader, he was entrusted with the preparation of schemes for 'national reconstruction' and, in particular, for the reduction of the demoralising unemployment gripping much of the country.

The plan he produced in January 1930 became known as the Mosley Memorandum. In the previous year's general election, the Liberal Party had largely campaigned on the promise 'We can conquer unemployment', a manifesto based on the economic theories of John Maynard Keynes. Keynes believed that, if there was sufficient determination, unemployment could be cured by a planned economy that combined the low interest rates necessary to stimulate demand with a programme of massive investment in public works and building. When the Liberal Party was virtually extinguished in the 1929 election – it won only 59 seats – Keynes and many of his supporters left it. The Mosley Memorandum was broadly Keynesian in its appeal.

It was a fifteen-page document summing up many of the ideas Mosley had been promulgating for the past four years. The first, and most controversial, section dealt with the executive powers that he believed necessary to put the second and third parts into operation; these, authoritative and similar to a wartime 'inner Cabinet', seemed to suggest dictatorial possibilities to his critics. The second and third sections dealt with long-term and short-term unemployment. To counter the first, industry had to become competitive, which meant rationalisation, up-to-date machinery, redeployment of manpower – and a probable increase in short-term unemployment for several years while this process was taking place. Short-term unemployment (even more urgent) would be dealt with by a programme of (badly-needed) road-building. To make both or either of these work, everything depended on effective administrative machinery.

Mosley's proposals were supported by Lansbury, the Grand Old Man of the Labour Party. But between Mosley and Thomas there was both personal and intellectual antipathy, aggravated when Mosley sent his Memorandum direct to the Prime Minister rather than through Thomas – who learned of it through the Premier himself. Thomas opposed it; his view prevailed with the Cabinet, who rejected it; and on 20 May Mosley resigned the Chancellorship of the Duchy of Lancaster. Clement Attlee was appointed in his place.

Mosley knew, however, that many in the Labour Party were strongly sympathetic to his Memorandum, so at a meeting of the Parliamentary Labour Party at the end of May he insisted that it be put to the vote despite its earlier rejection by the Cabinet.

Strategically, this was a grave mistake. A defeat for the Cabinet so soon after taking office would signal serious internal dissension and effectively transmit the message that the government, already shuddering in the economic blizzard, had neither cohesion nor a coherent policy. The government Whips told members that if Ramsay MacDonald had thirty votes cast against him on the Mosley motion he would resign. They did not need to spell out the unspoken corollary: the resignation of the Prime Minister would throw the fledgling Labour government into total disarray within weeks of achieving office for only the second time. The party swung behind MacDonald and Mosley was defeated by 202 votes to 29.

He was still convinced of the validity of his ideas and he was determined, tenacious and ambitious. If his Memorandum went through he would be catapulted at once into one of the most significant political positions in the country. He determined on a further trial of strength. At the Party Conference at Llandudno in October he put the Memorandum forward yet again. The result, though a defeat – he won 1,046,000 votes against 1,251,000 – was so close that he was encouraged to believe that he could take followers with him. Impatient for power, he made the fatal mistake of resigning from the Labour Party.

Had he waited, continuing to work within the party, he might have become the next Labour leader. By June 1930 unemployment had already reached two million and many of the younger, more left-wing politicians had begun to think that the economic situation was so serious it might lead to a breakdown of the party system and the emergence of a radical new government. A year later MacDonald's 'betrayal', when he accepted the King's commission to form a national government, caused such unrelenting bitterness towards him and the few senior members who followed him that they were forced to stand as independent Labour candidates, leaving the vast mass of the party leaderless. As Harold Macmillan later wrote, 'Here in Mosley the whole party would have found its leader: the man of ideas, the man of courage who alone had faced the realities and tried to bring precise and constructive solutions to the pressing problem of unemployment.' But Mosley was not a man who liked waiting.

Chapter IX

The idea of a new party had been mooted during discussions between Mosley and a group of like-minded fellow MPs. One of the most enthusiastic was John Strachey, a disciple of Mosley's since 1923 when, shortly after leaving Oxford, he had fought Aston as Labour candidate in the general election; he was also a great friend of Cimmie's. Others were W.E.D. (Bill) Allen, MP for a Belfast constituency and head of a large advertising firm, and W.J. Brown, Labour MP for East Wolverhampton and Secretary of the Civil Servants Association. It was to Brown, a man who threw off ideas like a catherine wheel, that the MP John Beckett, later to join the New Party, attributed its genesis. 'Brown had the idea of a great New Party, backed by Mosley, Allen and their wealthy friends,' wrote Beckett. 'Mosley would be the necessary picturesque public figure, and Brown would act as mentor.' Whoever had the original idea, it was Mosley who took the initiative, grasping the concept with the same speed and certainty with which he had embraced Keynesian economics for his Memorandum; from the start, it was *his* party.

By November 1930 Mosley had come to a firm decision that such a party should be launched but was not certain when to make his move. 'If I strike now I may be premature. If I wait I may be too late. If I could have £250,000 and a press I should sweep the country.'

None of this uncertainty was visible in his demeanour. That November, at a Cliveden house party at which the thirty-two guests included Duff and Lady Diana Cooper, Harold and Dorothy Macmillan, Oliver and Lady Maureen Stanley, Brendan Bracken, Robert Boothby, Malcolm Bullock, Frank Pakenham, J.L. Garvin (the editor of the *Observer*), Harold Nicolson and a crowd of clever young men and pretty girls, Mosley was the undoubted star. He was fit, handsome and arrogant; at only thirty-three, he had resigned from government with great éclat on a point of principal while announcing that he alone knew the way forward. He was able to command the most beautiful women as his mistresses. The female members of the house party whispered excitedly to each other that some esoteric drug must account fot his sexual prowess; most of the men were fasinated by his powers of persuasion, his effect on women and his buccaneering courage. He told Harold Nicolson that he wanted to launch a new National Party and hoped to get the industrialist William Morris (the future Lord Nuffield) to finance him. At dinner he held forth to the younger generation. 'After Peel comes Disraeli,' he intoned to Frank Pakenham, recently down from Oxford. 'After Baldwin and MacDonald comes . . . ?' He looked expectantly at Frank, who asked innocently, 'Who?' There was a long and awkward pause, until Mosley answered harshly, 'Well . . . someone very different.'

By the beginning of February 1931, the plans for a new party were well advanced. On 4 February Cimmie, lunching with Harold Nicolson at Boulestin, secured his promise to join; neither of them saw anything incongruous in discussing the miseries of the 'working man' in such an elegant and expensive setting. The allegiance of Nicolson, then working for Lord Beaverbrook's *Evening Standard*, was especially satisfactory, as through him Mosley hoped to gain the support of Beaverbrook – since the press was then the only real medium of publicity, a newspaper platform was invaluable.

Events moved swiftly. On the 15th a party of potentially sympathetic MPs came to Savehay Farm, the Mosleys' home near Denham in Buckinghamshire. They included David Margesson, who became Conservative Chief Whip nine months later. Policies were discussed between energetic games of rounders, and the suggestion was made at dinner that night at the Stanleys that they should send Baldwin a parrot in the hope of giving him psittacosis.

A fortnight later, on 28 February, the New Party was launched. It had a brisk programme of reform – including, of course, the 'abolition' of unemployment by spending public money on road building, slum clearance and the building of new houses, which would simultaneously put money in the workers' pockets and improve their living conditions.

At first the omens were good. Macmillan and Margesson hinted that they might become Mosley supporters, John Strachey followed his friend over from the Labour Party, as did Stanley Baldwin's son Oliver, Allan Young, Robert Forgan and a dozen or so others. There was one disappointing dissenter: W.J. Brown, the fertility of whose ideas was equalled only by his ability to take up the opposite position the following day (he later became editor of the *Daily Express*). He now declared himself unable to join the New Party because his union disapproved and he could not afford to forfeit their support. When Mosley offered to guarantee him against financial loss and obtained his union's permission for him to join, Brown's only recourse was to retire to bed ill and remain there incommunicado.

Harold Nicolson was convinced that Mosley would some day become Prime Minister but Lady Oxford (Margot Asquith), who believed in Mosley and had been exceptionally kind to him, was so furious at his defection that she pinched him sharply with a long clawlike hand. Cimmie, unlike most of Mosley's early adherents, required considerable persuasion. A committed socialist, she was distressed and agitated at the idea of leaving the Labour Party, but gradually came round to her husband's view that only the New Party could save Britain.

It was fortunate she did. Almost at once Mosley went down with a virulent bout of influenza (dangerous and debilitating in those days before antibiotics). As he sat white and exhausted in their Great George Street flat, hoarsely dictating the New Party Manifesto to a secretary and a rota of typists, Cimmie, supportive as ever, stepped into the breach and took meeting after meeting.

Mosley recovered in time to speak on behalf of Allan Young on 27 April at the Ashton-under-Lyne by-election. His 7,000 hearers included friends and supporters such as the Stanleys, the playwright Freddie Lonsdale and Professor Joad (later to become famous in the BBC's 'Brains Trust'). But despite the enthusiasm of the audience, swayed as always by Mosley's oratory, Young's Conservative opponent was elected in May with 12,420 votes. Labour polled 11,005 and Young got a mere 1,172.

It was a bad beginning. But the New Party, with its calls for 'action', was nevertheless attracting members – especially the young Maynard Keynes, sympathetic to a movement dedicated to his economic principles, offered encouragement, and it was learned that the Prince of Wales looked on the New Party with favour. Mosley and Nicolson – who was now deeply involved – decided that what was needed was a weekly paper to disseminate the New Party credo and serve as rallying point for the faithful. Nicolson, who was to edit it, bravely left his job at the *Evening Standard*. ('Go to hell with ye,' said Lord Beaverbrook, who felt Nicolson was making a profound mistake, 'and God bless ye.') Offices were found, at 5 Gordon Square, furnished by Nicolson ('Carpets from Maples and blue curtains throughout'), a dummy prepared and a printer commissioned. Mosley expressed himself as willing to lose £15,000 over two years. The paper was called, predictably, *Action*; and the message 'It's time for action!' was hammered home constantly, amid book reviews, articles on modern architecture and advice on rose-growing from Vita Sackville-West.

As the weeks passed the New Party began to veer further to the right. This unexpected change of political orientation was profoundly disturbing to its original adherents. On 23 July John Strachey and Allan Young both resigned, giving as their reasons their deep suspicion of tendencies they had begun to perceive as fascist. The publicity this attracted and the photographs of their unhappy faces, wretched at the prospect of breaking with the man they admired and liked, damaged the New Party.

Eight days later, their fears were confirmed. Mosley publicly demonstrated his switch to the right by crossing the floor of the House again. He now sat behind the Conservatives.

The real test of the New Party's hold on the public imagination was soon to come. In August, while Mosley was holidaying in Cap d'Antibes, Ramsay MacDonald resigned. As leader of a Labour government, he was unable to countenance the cut in the dole demanded by foreign bankers before they would lend the money essential to resolve Britain's financial crisis. He set off for Buckingham Palace with the resignations of the entire Cabinet in his pocket. But within hours, to cries of 'Traitor!' from his party, he was reinstated as Prime Minister of a National Government, promising to call an election as soon as the immediate situation was restored.

New political alignments sprang up overnight. The Labour Party, headless, splintered its forces: at Peckham alone, the election was to see three Labour hopefuls – one a MacDonald Labour candidate, another the rebel left-wing MP John Beckett for the Independent Labour Party and a third representing the main Labour Party. Randolph Churchill, charged by his father to find out if Mosley would join Churchill and the 'Tory toughs' in opposition, rushed over from Biarritz, but Mosley had no hesitation in refusing. The crowd of photographers waiting to greet Mosley on his return to London with Harold Nicolson (Nicolson had gone to Dover to meet him) demonstrated his importance on the political scene.

Despite the fact that no date had yet been set for the general election, the parties all began campaigning at once. Mosley's, from the first, attracted the kind of opposition that spilled over all too readily into violence. On a wet September day he appeared in Trafalgar Square in front of a crowd of 300, escorted by the burliest members of his newly formed 'Youth Movement' and guarded by the East End boxer 'Kid' Lewis – there had been rumours that the communists intended to break up the meeting. Just over a week later, on 20 September, razors, stones and fists were used against him by a small group of communists at a meeting of 20,000 in Glasgow.

Mosley's reaction was to say that this was a sign that the New Party needed a trained, disciplined and uniformed force of stewards to keep the peace at future meetings, rather than supporters identifiable only by the marigolds in their buttonholes. On 12 October a headquarters for the Youth Movement was established at 122 King's Road, Chelsea, where classes for fencing, boxing, jujitsu and gymnastics were organised.

Deciding on a uniform was more difficult. Harold Nicolson's suggestion of grey flannel shirts and trousers – grey flannel trousers, then worn by nearly everyone, had the great merit of cheapness – was considered too unadventurous and the question of uniform was put on one side to be considered later.

The general election took place on 27 October. New Party candidates stood in twenty-four constituencies. Theoretically, they had a sporting chance: the New Party appealed to the fair number who viewed MacDonald as dangerously left wing but who also believed that only a radical solution would end the misery of the poor.

Mosley himself stood for Cimmie's seat of Stoke-on-Trent (she was pregnant again), campaigning in favour of a tariff barrier against imports. By now his view of leadership was not the traditional *primus inter pares* of British Prime Ministers since Walpole but that of an autocrat – even of an aspiring dictator. In the issue of *Action* which appeared a week before the election he had advocated draconian reform of the legislature and executive. 'Legislation must be passed, or rejected, within a short time-limit,' he proclaimed on *Action*'s front page. 'An inner Cabinet, analogous to the War Cabinet, should be constituted and consist of five Ministers only without portfolio. These proposals are so simple as to seem obvious.'

Within the New Party there were similar disturbing signs. Mosley believed, 'For a new party to achieve power, its ideology must be based on emotion', and emotion of a raw and alien kind was what he increasingly played on in his speeches. With his physical magnetism and oratorical skill he could always invoke wild enthusiasm. He would stand on platform or daïs, chin raised as he inhaled the applause, while behind him stood the Praetorian Guard of the Youth Movement. 'Members of Party Branches may be of either sex,' said the rules, 'but the Control Group must consist of young, fit and proven male members of the party.'

Mosley's personal style, too, had become that of the full-blown demagogue. James Lees-Milne, who spent a fortnight canvassing for him, became increasingly uneasy:

> it became clear that Mosley was a man of overweening egotism. He did not know the meaning of humility. He brooked no argument, would accept no advice. He was overbearing, and over-confident. He had in him the stuff of which zealots are made. His eyes flashed fire, dilated and contracted like a mesmerist's. His voice rose and fell in hypnotic cadences. He was madly in love with his own words. It could be a terrible day, I fancied, when they ran away with him and took the wrong turning. The posturing, the grimacing, the switching on and off of those gleaming teeth and the overall swashbuckling so purposeful and so calculated, were more likely to appeal to Mayfair flappers than to sway indigent workers in the Potteries.

The election may have been a disaster for Labour, which mustered only 52 seats against the Conservatives' 473 (other National candidates brought the government's total strength up to 521), but it was a catastrophe for the New Party. Few of its 24 candidates had any political experience whatsoever – 'Kid' Lewis was one, another was Lees-Milne's Oxford friend Christopher Hobhouse, known as the 'Babies' Candidate' because he was only twenty-five – and all were defeated. Twenty-two, including Harold Nicolson, lost their deposits.* At Stoke, Mosley, with 10,500 votes, came bottom of the poll.

The party's paper, *Action*, was now the only hope. There was still £14,000 left to run it, but it was haemorrhaging money. Despite distinguished contributors such as Christopher Isherwood, Peter Quennell, Osbert Sitwell, L.A.G. Strong and Nicolson himself, its circulation had fallen from 160,000 to 16,000 within ten weeks of its launch.

By November Nicolson had realised that Mosley's ideas were diverging irrevocably from his own. 'He believes in fascism. I don't. I loathe it.' Nevertheless, he agreed to accompany Mosley on a visit to the two fascist leaders, Hitler and Mussolini, in early January. In the last issue of *Action*, published on 31 December 1931, they explained that they were going in order to study 'new political forces born of crisis, conducted by youth and inspired by completely new ideas of economic and political organisation'.

* Then £150, forfeited if the candidate achieved less than 12.5% of the poll.

Though they went on to say that they did not intend to import Italian or German practices into Britain, this did not allay unease at home. In the event, illness prevented Mosley from travelling to Berlin, so he only visited Mussolini.

Il Duce and his fascist government had been in power for ten years. In 1932, fascism did not have the iniquitous and bloody reputation it later acquired – especially as the Italian brand of fascism did not at this stage include anti-semitism. The average Briton, inasmuch as he thought about Mussolini at all, regarded him as a vaguely good thing albeit a trifle comic with his strutting pomposity and general flamboyance (the large carbuncle on his skull was carefully erased from official photographs). Il Duce had done much to quell the Mafia in Sicily, had drained the Pontine Marshes and – of prime importance to the English traveller – made the trains run on time.

Mosley and Nicolson arrived in Rome on the evening of 1 January 1932, having spent New Year's Eve in Paris – Mosley celebrating it with friends and staying up until eight a.m. They were met by Christopher Hobhouse, who drove them to the Excelsior. Here, in the vast drawing room of a luxurious suite with opulent Edwardian décor – Mosley invariably did himself well – Hobhouse reported that Hitler had been dismissive of their political stance. 'He says that "We British Hitlerites are trying to do things like gentlemen. That will never do. We must be harsh, violent and provocative."'

Harshness, provocativeness and violence were to become part of British fascism, but at that time Mosley seemed more impressed by the milder Italian model.

Mosley was full of these ideas when he met Diana at Barbara Hutchinson's party at the end of February. She had spent most of the evening talking to the man on her other side, Victor Rothschild (later to marry Barbara) and was unaware that Mosley had already marked her down. The summer before, he had seen her at a ball given by the Duchess of Rutland at Sir Philip Sassoon's house in Park Lane, framed by the rose-entwined pillars flanking the ballroom as she danced dreamily past and, like Bryan before him, he never forgot that first glimpse. Later, he wrote in his autobiography, 'Her starry blue eyes, golden hair and ineffable expression of a Gothic madonna seemed remote from the occasion but strangely enough not entirely inappropriate.'

But at that first meeting no spark was struck between them. What conversation they had was largely confined to politics and when, at the end of dinner, Mosley made a short and jokey speech in honour of Barbara, most of it was taken up with ragging her father, 'Hutchie', of whom he was an old friend. Diana, who knew that Mosley was supposed to be a brilliant speaker, was a trifle disappointed. After dinner she was one of a sceptical group listening to Cimmie vehemently defend her husband's views on fascism.

Nor did Diana find him particularly attractive. Bryan, always conscious of what she was doing and whether or not she appeared to find a man appealing, was so untroubled by her reaction that he did not even mention Mosley afterwards. For his part, Mosley thought Diana was argumentative – she described herself then as a 'Lloyd George Liberal'.

Diana left the Hutchinsons' party with no idea that she would ever see Mosley again, and no particular wish to do so.

But now that he was no longer in parliament, he and Cimmie once more took up their life in the small, tight nucleus of London society of which Diana was now a part. From then on, his path and Diana's crossed almost daily. Invariably, their conversation turned to politics and, as Diana listened to Mosley expounding his ideas with all the fire, brilliance and persuasiveness of which he was capable, she became increasingly convinced that here was the only man who could save the country. He himself had no doubt of the rightness of his approach.

The leaders of both major parties felt that now he had learned where impetuousness led, this young, ardent and brilliantly gifted man might be wooed back into their respective folds. Early in March, he was approached by David Margesson, now Tory Chief Whip, who, realising that straightforward Conservatism would hold no appeal for Mosley, tactfully suggested that he stood as a National Independent. At about the same time, Joseph Kenworthy offered him the leadership of the now-minute Labour Party – something Mosley would have jumped at only a year earlier. He turned it down flat, as he had the less enticing Conservative proposal. 'He is still obsessed by Mussolini ideas,' noted Harold Nicolson gloomily in his diary.

On 5 April 1932 the Executive Committee of the New Party decided to dissolve the Party but to keep on the Youth Movement. By now Mosley was a fascist in all but name, and close to the dangerous principle that violence 'in certain circumstances' might be necessary. It was a crossroads, and Nicolson severed his connection with him. His friends remarked on symptoms of the developing political creed that filled his mind. 'Lunched downstairs with Tom,' noted Georgia Sitwell that spring, 'he was a trifle absurd as usual about Superman, Nietzsche, Schiller, Napoleon etc.'

For Mosley, fascism blended the Utopian ideal with the Nietzschean concept of the Superman – the individual striving to raise himself above the common clay of humanity while working for the good of the community. He believed that the goal of a better society could be achieved by that supreme instrument of power, the will, expressed through action, backed up where necessary by a cadre of tough, muscular, disciplined and dedicated young men. As for who set the course of that action, there could be only one possible answer: the Leader. It was the name by which he would shortly become known to Diana.

Chapter x

No shred of compunction for his wife – beautiful, sweet-natured and helplessly devoted to him – had ever prevented Mosley from pursuing whatever woman took his fancy. 'Tom Mosley had a blonde siren in tow at the party and had no eyes for anyone else – not even that eye trick which he obviously thinks very fascinating and which indeed has a great effect on some people,' wrote one prospective mistress jealously of another when she saw him up to 'his usual tricks' at one of Syrie Maugham's parties.

Most of these women were friends of Cimmie's, the wives of couples with whom the Mosleys dined, danced or went abroad. Mosley's tastes were catholic, ranging from sophisticated older beauties such as Sylvia Ashley, Katherine d'Erlanger and Paula Casa Maury, with whom he had a long-standing affair, to Georgia Sitwell, the dark and pretty young wife of Sacheverell Sitwell. All would be asked by the unsuspecting Cimmie to stay at Savehay Farm, near the village of Denham, in Buckinghamshire.

Savehay Farm was an old half-timbered red-brick manor farmhouse; its lawns, on which Mosley used to practise pistol shooting, ran down to the River Colne, the county boundary. The Mosleys had bought it in 1926 for £5,000, a price which included the surrounding 123 acres of land. A long, winding drive led past red-brick barns and a cottage or two to a high red-brick wall which sheltered the house. Downstairs there was plenty of room; upstairs, what with their own bedroom, their two small adjacent bathrooms, Mosley's dressing room, day and night nurseries for their children and Nanny Hyslop's room, there were few spare rooms. To provide for overflow guests and their own servants – butler, footman, cook, kitchenmaid, housemaids and Cimmie's French lady's maid Andrée Eberon – the Mosleys added an extension designed by their friend and fellow socialist, the architect Clough Williams Ellis.

Here, amid Cimmie's trestle tables, wheelback chairs and chintzes, there would be gossip in front of the log fire in the panelled drawing room, or dressing up for charades and amateur theatricals in costumes designed by another great friend, Cecil Beaton (known to the Mosleys as 'Ceckle'). On sunny summer weekends there would be croquet on the lawn between the herbaceous borders, tennis on the two courts beside the drive, or sunbathing and swimming in the River Colne. 'Tom evidently fancies himself very much in bathing shorts and displays with pride a sunburnt, muscular torso,' noted Georgia Sitwell, who spent her twenty-third birthday at Savehay.

So open was Mosley about his penchant for one or other or sometimes several of these women that Cimmie at first did not suspect that he was being unfaithful. Yet, as Georgia later admitted, 'Of course I went to bed with Tom. We all did, and then felt bad about it afterwards.'

Georgia became so deeply involved with Mosley that her obsession was the cause of endless fierce quarrels with her husband Sacheverell, exacerbated by the awkwardness of constant social involvement – the Mosleys often came to Sitwell poetry readings, the Sitwells would join the Mosleys for supper at the Savoy after a play. 'Sash and Tom came with me, a curious trio. I felt alternate waves of hysterical laughter and annoyance at the situation,' wrote Georgia in her diary after one such occasion. At the beginning of September Georgia dragged her husband off to join her lover and his wife in Antibes, where she found a party that included the Casa Maurys, John Strachey, Robert Boothby and Cimmie's sister, Lady Ravensdale – the long-suffering Cimmie lent Georgia a bathing hut, cushions and beach loungers.

Georgia was, however, clear-sighted enough to recognise that, charming as Mosley might be, his own emotions remained unengaged. 'I wonder if Tom felt anything at all, then or ever?'

Ironically, she was also in the process of becoming friends with the woman who would soon supplant her – Diana. She and Sacheverell went to supper at Buckingham Street ('Sashie admires Diana enormously'), they met at parties, Georgia went to tea with Diana and they admired each other's opposite looks. Bryan and Diana, believed Georgia, were utterly devoted to each other.

Mosley's pursuit of other women followed a regular pattern, beginning with lunch at restaurants such as the Ritz, the Devonshire or Boulestin –

both the Mosleys followed the fashionable habit of lunching separately. First, politics would be discussed, then, imperceptibly, the conversation would shade into more personal matters. When, in 1929, Mosley acquired a bachelor flat, 22A Ebury Street, seduction became even easier. Officially a *pied à terre* for his political work in London, in reality it was little more than a *garçonnière*. Entered through a discreet front door past two doric columns, it had one huge, pillared room, the shapeless sofas and chairs done up in browns and beiges, the walls hung with eighteenth-century fencing prints. There was a small kitchen plus a bathroom, and the chauffeur's wife came in and cleaned. *The pièce de résistance* was the gallery bedroom with its huge double bed. Here, at the touch of a concealed button, warm air wafted down over the occupants.

Cimmie, wanting to believe in her husband, was easily deceived. When Mosley, out of parliament after the formation of the National Government in 1931, announced that he was taking up fencing again, she was overjoyed. She was expecting their third child and experiencing a difficult pregnancy. Harold Nicolson, who was devoted to her, often called to see her as she lay on her sofa. One afternoon, as they discussed Mosley's 'incurable boyishness and joie de vivre', Cimmie declared that fencing would serve as a safety valve for her husband's overflowing physical energy. In fact, it was the perfect pretext for slipping away to assignations with mistresses, now that the many excuses open to a Member of Parliament were no longer available.

During the twelve years since his marriage to Cimmie in 1920 Mosley had had up to three dozen affairs. His libido and his singlemindedness played a part in the relentless impulsion towards these conquests, as did his upbringing. His father had been a successful womaniser in the louche convention of the 'fast' Edwardian set whereby young unmarried girls were 'out of bounds' but it was almost obligatory to pay court to an attractive married woman if you found yourself alone with her.

A man of great energy and force, Mosley's desire to control, to dominate, to be the victor, the most powerful in any group or relationship, was so strong as to be a compulsion. Early on he had discovered his power over women and his successes with them had become a drug, feeding his need for adventure, challenge and admiration. His gaiety, handsomeness – in the idiom of the day the *Tatler* described him as having 'John Barrymore looks' – buccaneering attitude and notoriety meant that he usually succeeded and, as with all pleasurable skills, he enjoyed practising them. In the winter there was shooting; in the spring and summer what he called 'flushing the coverts' – going round the balls and parties of the London season to spot the latest crop of attractive young married women. Until he met Diana, he had never fallen in love with any of them.

On the whole, Cimmie did not find out. When she did, she minded bitterly. Mosley would cajole her, make love to her, swear that not one of them was anything more than a diversion and assure her, truthfully, of his love. For she was, in every way except as provider of sexual excitement, the perfect wife and partner for a man so deeply committed to politics. Mosley had few men friends: men tended to dislike his arrogance and mistrust his behaviour. But everyone adored Cimmie. The deep affection between them made many, who would not otherwise have done so, accept Mosley. 'They are really fond of each other for all their squabbles and infidelities,' wrote Harold Nicolson. The distress her husband often caused her was, however, plain for all to see.

Cimmie was peculiarly vulnerable to such emotional blows. Her nature was gentle and straightforward; her upbringing in many ways had been similar to that of an orphan, with no reassuring older figure to offer intimate advice, comfort or understanding. She and her two sisters had lost their mother when they were still in the nursery (Lady Curzon died in 1906) and had been brought up by their widowed father in conditions of terrifying strictness and grandeur. Curzon's natural style was lofty and autocratic, while the meticulousness and passion for detail that marked his Viceroyalty of India meant that he enquired closely and penetratingly into every aspect of his children's lives, from what they were learning to the cost of their underwear.

Inquisitions took place at mealtimes. Luncheon, and later dinner, in a dining room five minutes' walk away from the kitchen and always eaten off silver plates and dishes (part of a service for seventy-two), became a form of *viva voce* that the Curzon daughters dreaded. Their father would first discourse on great events, then put his children through their historical paces by firing a series of questions at them. Unsurprisingly, they were usually so paralysed with fear that they were unable to answer, upon which he would round on the unfortunate governess, who would find herself equally struck dumb with nerves.

No pains were spared to make the girls fit consorts for princes, or at the least dukes or aristocratic conservative statesmen in the Curzon mould. The eldest, Irene, was sent to Dresden for two years to study music, singing and piano before returning to make her debut in 1914;

the theatre and opera, together with hunting, remained her passions all her life. Curzon, who had not allowed her to meet any young men before, now gave a great ball for her at Carlton House Terrace (at which he forbade her to wear make-up). Soon afterwards, he remarried, and in 1917 gave a small coming-of-age dance for her at Trent, Sir Philip Sassoon's house (reported by the press as a huge London ball under the headline 'Curzon dances while London burns'). It was, though, an important festivity: the equivalent of the coming of age of an heir. Curzon, who had no son, had secured that one of his titles, the barony of Ravensdale, should pass to his eldest surviving daughter. It also marked the beginning of the breakdown of what relationship there was between Curzon and his daughters.

The ground was the fortune their mother had left them. Lord Curzon had always treated his daughters' incomes as his own, spending them freely to maintain the way of life to which he felt entitled. When they were small this was, of course, no cause for discord and, in any case, largely justified since Kedleston was their home and no expense was spared on their education or debuts. When they grew up and wished to have the use of their trust funds themselves, it was another matter. The first to argue the point was Irene: she wanted to buy a hunting box in the shires, she wanted to travel, she wanted to patronise artists and musicians. Her father was furious, Irene stuck to her guns and there was an open breach.

When Cimmie told her father she wanted to marry Mosley, his first reaction was to pull down *Debrett* (always known as the Red Book in the Curzon family) from his library shelves and to look up the Mosley family. He could discover no evidence on which to refuse his daughter her choice and reluctantly gave his consent. Even more reluctantly, for by now he had learned that she, too, wished to remove from his grasp the money her mother had left her, he allowed her to be married from Carlton House Terrace.

For Curzon, his worst fears were justified when, shortly after Cimmie and Tom Mosley were married, his new son-in-law carelessly signed an answer to an official invitation, which had been worded and typed by his secretary: 'Dear Lord Curzon, The wife and I will be pleased to accept . . .' Curzon wrote to his daughter castigating her for this solecism, for her slipshod approach, for her lack of education – here pointing out the money he had spent on governesses – and listing a number of ways in which the invitation could have been answered correctly. Curzon's letter covered eight pages.

After this, Cimmie seldom saw her father.

* * *

Curzon's youngest daughter, Alexandra (Baba), devoted to Cimmie, was sixteen and still in the schoolroom when Cimmie married. Thanks to her cloistered upbringing, her sister's glamorous husband was the first man with whom any kind of close friendship was permitted. From the start, Mosley fascinated her. To Baba, just beginning to dream of love and romance, Cimmie and Tom seemed like creatures out of a fairy tale and she found a vicarious pleasure in imagining their love. What she did not realise was that though her feelings for her sister were of passionate admiration and love, those for Mosley were a schoolgirl crush that would later develop into a full-blown infatuation.

By the time Baba came out the pattern had been set: the three sisters, estranged from a father who had never been more than a distant figure, turned to each other for support, advice, comfort and love. When Mosley married Cimmie, he stamped himself as the predominant male figure not only on Cimmie's life but on the lives of Irene, older but single, and the adolescent Baba. Both fell in love with him, first Irene and then, more deeply, Baba. Both suffered unadmitted passion that put a fierce edge of jealousy on their feelings for Cimmie and each other. If Mosley appeared to prefer one, it was torture for the other; and while they hated to see their beloved sister made miserable by his unfaithfulness, they could not help forgiving him.

When Diana, so clearly more important than his other infidelities, appeared, she provided a focus for the sisters' rage and guilt, and she was soon mythologised into a demon figure. Both Irene and Baba hated Diana for the rest of their lives – half a century on, Baba still could not bear to speak of her.

For his part, Mosley had not been worried that Diana had almost ignored him at their first meeting. She was so great a prize that he was prepared to take his time. The most beautiful young woman in London, adored wherever she went but with an impeccable reputation, she presented a challenge more compelling than any of his previous conquests. They had met at the psychological moment: he was out of both government and parliament and therefore had the time to concentrate on winning her; she was finding that the bonds of her marriage were chafing. The luncheon invitations began, the web of words was spun. And soon they were deeply involved in a passionate love affair.

A cause, be it political, patriotic or religious, presented in the person of someone to whom one is greatly attracted sexually, is almost irresistible. Diana, like most of her generation, believed, as she put it, that 'things had gone desperately wrong in England'. She was deeply conscious of the poverty, the hunger marches, the lethargy of Baldwin followed by the vacillations of Ramsay MacDonald, the stock market crash, the flight from the pound, the unemployment that left once proud and self-respecting working men and their families starving and humiliated. She already knew of the Mosley Memorandum; now, as Mosley expounded his ideas, she listened with the growing feeling that here was the answer. 'Even at the beginning there was not only passion, though of course it was very strong, but also politics,' she wrote later of those first months:

> Every time we dined or lunched – restaurants mostly – he talked. Occasionally we argued. But on the whole I was completely converted.
>
> In 1932 we all – everyone with the slightest intelligence – thought about politics. We believed that our parents' generation had made the war, that by *will* plus *cleverness* its horrible legacy could be cancelled out, and the world could be changed . . . He seemed to me to be a prophet of this changed world one longed for and I thought: If I can help him as he seemed to think I could, then nothing else mattered in the very least.

She would come back glowing to Buckingham Street and pour it all out to her friend Cela Keppel, who was staying with the Guinnesses for her second London season. 'Isn't the Leader *wonderful*!' she would exclaim as she recounted their lunchtime conversation, or she would tell Cela how marvellous his lovemaking was compared with Bryan's inexperienced advances. As Cela was by now equally fond of both Guinnesses, she found it difficult to know how to answer these confidences.

It was clear Diana was in the grip of a physical passion so strong that it blinded her to everything else. She knew that Mosley was a philanderer, but she did not care. Gentle-seeming, amusing and affectionate as Diana was, she was also completely ruthless. She took not the slightest notice of Bryan's objections to her lunching with Mosley even when he took the painful step of writing her a letter forbidding it. She was very fond of Bryan, but she was not going to let loyalty to him, the effect on their two children should the marriage be threatened, or the devastation she might cause, stand in her way.

She was perfectly open about these meetings and inevitably her friends soon realised her involvement with Mosley. They were horrified, adding their voices to the growing chorus of disapproval. 'You can't – you *mustn't* see him. Come and lunch with me instead.' But nothing, not even the knowledge that she was jeopardising a marriage to a kind and loving husband and an enviable way of life, and least of all the quarrels and reproaches over her behaviour, deflected her for a minute – especially when Mosley let her know that he had dropped Georgia Sitwell and Paula Casa Maury, the two women he had been sleeping with when he met Diana.

In April 1932 the Guinnesses moved from Buckingham Street to a larger and more beautiful house, 96 Cheyne Walk. It had been built in 1760 and had once belonged to Whistler. Diana redecorated it completely, painting most of the old walls and panelling white, with some rooms washed in pale blues, pinks or golds. One of the chief beauties of the house was its first-floor panelled drawing room, its tall windows at each end overlooking a long, L-shaped garden at the back and the river in front. Diana painted the panelling grey and white and put on the floor a huge Aubusson, which Lord Moyne, who was extremely fond of her, had given them. One of the two smaller Aubussons, which she had acquired for Buckingham Street, was put in the children's nursery. '*So* good for them to see pretty things when they're crawling about,' she would explain. On the same principle, her two small boys, angelic-looking with their blond curls, were often put into dresses, because 'little girls' clothes are so much prettier'.

Once in Cheyne Walk, Diana became more of a social icon than ever. She was offered the part of Perdita by C.B. Cochran in a production of *The Winter's Tale*. The society paper, the *Bystander*'s reporter saw her 'looking lovely and talking to – I should have said being talked to by – Mr Randolph Churchill'; the *Tatler* informed its readers that the Honourable Mrs Bryan Guinness was among the young hostesses 'responsible for the fashion of wearing trousers, not beach pyjamas, in the privacy of their gardens. At Biddesden she wears corduroy slacks in the colder weather, with hand-knitted jumpers, and grey or white flannel trousers with sleeveless shirts in the summer.'

On 7 July the Guinnesses gave a grand party in Cheyne Walk, part housewarming and part coming-out ball for Unity, now eighteen, and a Mitford cousin, Miss Robin Farrar. It was a warm night; in the long, floodlit garden filled with the scent of roses a Russian orchestra played for those of the 300 guests who did not feel like dancing in the panelled drawing room. Supper was served in the two dining rooms, monastic-looking with their white walls, sparse furnishings, refectory tables and arches. In an echo of Lady Evelyn, the maids who served were wearing green and white flowered dresses instead of uniform. The beauties were there in force: Lady Weymouth in black and white cotton, Lady Jersey in pale blue with a blue feather boa, Lady Lavery in white, Penelope Dudley Ward in pale blue and silver lamé, Unity in white satin with a black velvet sash. Winston and Clementine Churchill brought their daughter, Diana, and Augustus John arrived with one of his daughters, Poppet. All three Curzon daughters were there: Irene (now Lady Ravensdale), Baba with her husband Major Edward Metcalfe whom she had married in 1926 – known as 'Fruity' Metcalfe, he was a close friend of the Prince of Wales – and Cimmie with Tom Mosley.

Diana, in love and by now certain that she was loved, looked glorious in a grey tulle and chiffon dress with a skirt that fell in soft flounces to the ground. Round her neck was the diamond necklace given her as a wedding present by her parents-in-law and on her head a small tiara of diamonds and rubies. 'You were dazzling as the presiding goddess, fresh and dewy from Olympus,' wrote Emerald Cunard afterwards.

Others felt the same. 'Now I know the meaning of the old term, "rout" for dance – someone who routs the rest. Thus did you,' wrote Cela's adoring brother Derek.

'Thank you so much for your beautiful party, where I left my hat and the remains of my heart,' volunteered Augustus John, who was painting Diana at his Mallord Street studio. 'When can I get hold of you? You looked so ravishing last night that I am reconsidering the picture.'

'It was the best party ever given, even by you,' said Robert Byron. 'I feel as if I had been raised from the dead by it.'

For Diana, it was something more. She was twenty-two and she had found her fate. That night, some time before a pink and gold dawn glittered on the river outside, she and Mosley had committed themselves irrevocably to each other.

At the same time, Mosley had told her that he would never leave Cimmie.

Next morning, Diana told Cela, 'I'm in love with the Leader, and I want to leave Bryan.'

Chapter XI

Diana felt no guilt at the effect her liaison had on Mosley's marriage. Even if she had, it would have altered her conduct only in detail. Like the whole of London, she knew that Mosley had had affairs virtually from the moment he and Cimmie had emerged from their wedding reception at Carlton House Terrace. She rationalised it: Cimmie would find it easier if Mosley had one steady mistress, rather than a constant succession with all the pain of fresh deceptions and discoveries.

That summer, Diana and Mosley met constantly: at the balls and parties of the London season, at Emerald Cunard's fifty-strong supper parties after the opera – where Diana laid the foundations of an enduring friendship with another Cunard habitué, the witty, cultivated and eccentric Lord Berners – and secretly at Mosley's Ebury Street flat.

They were determined to spend every possible moment with each other. The chic destination for the rich, smart set during the last weeks of August and most of September was then Venice, and the meeting place its smallish and not very pleasant Lido. It was easy to manipulate dates so that the Guinnesses and the Mosleys would be there at the same time. Diana and Bryan had arranged to motor leisurely through the south of France to Italy with Barbara Hutchinson and Victor Rothschild (to whom Barbara was now engaged). Mosley was also motoring, but Cimmie, her kidney trouble aggravating the after-effects of childbirth, was not well enough to drive; she, the two elder Mosley children and her lady's maid Andrée would follow by train. Desperate at the prospect of being parted even for the seven to ten days the car journey would take, Diana and Mosley planned that he would 'run into' the Guinnesses as if by chance in Arles.

But in Avignon, staying at the Jules César, Diana woke with a throat so sore she could not speak, an infection quickly diagnosed as diphtheria. 'So sorry you are ill with pleurisy in s. of France (to my mind something worse than Maidenhead),' wrote John Betjeman sympathetically on 12 August. To Diana it was a desperate matter. She was terrified that a letter from Mosley might arrive at the hotel and, because of her prostrate condition, be opened by Bryan. She took Barbara and Victor into her confidence and they managed to get a message through to Mosley in Arles that she would see him in Venice when she was well. She was young, strong and determined to recover as soon as possible; aided by frequent injections from a doctor at the Institut Pasteur she was able to travel on to Venice eight or nine days later.

That September was eventful in the annals of the Lido. The usual group congregated every day on the small beach, where the cabins were allocated year after year to the same people. Diana, Bryan, Barbara and Victor arrived to find the Mosleys and their friends John Strachey and Robert Boothby, Emerald Cunard and a party including her lover Sir Thomas Beecham, Duff and Diana Cooper, Randolph Churchill, Brendan Bracken, Doris Castlerosse, the Austrian Tilly Losch and Diana's brother Tom. Always there were one or two reigning hostesses, usually American, principessas or duchesses, vastly wealthy heiresses attracted by the glamour of an ancient European title, around whom would group parties of their friends. That year the queen of Venice was the principessa San Faustino, an American known as Princess Jane.

The Lido day would start at about eleven a.m. Every twenty minutes or so the Hotel Danieli's launch – or as Diana pronounced it, the 'larnch' – would disgorge men in short-sleeved, open-necked shirts and linen shorts and women in fashionable beach pyjamas and espadrilles. Mosley, because of his deformed leg, always wore trousers, with a navy blue shirt. Diana, who refused to follow fashion, kept out of the sun as much as possible and wore plain, pale cotton dresses. Most of the huts' occupants knew each other, and there was much sauntering along the beach to chat. Eventually, like smaller balls of mercury drawn into a larger one, these knots of two or four would eddy towards Princess Jane, grouping around her on cushions. In the afternoons, for those who did not want to swim or siesta, there might be a sightseeing expedition. Most evenings began with drinks at the cafés in the Piazza San Marco, the band playing *Lohengrin*, before dinner at the palazzo of a friend or at one of Venice's excellent restaurants. They would return late at night by gondola along some dark canal, silent except for the water lapping its stone sides.

In that rich, rather louche circle many of the men were pugnacious, promiscuous or both. Love affairs and quarrels abounded. What should have been a *fête champêtre* on Torcello was ruined by a furious row, in which Duff Cooper was the prime mover. Another inevitably involved Randolph, often drunk and as usual behaving badly. After a good-humoured lunchtime discussion at the Lido restaurant, he turned to Brendan Bracken, who had suffered all his life from the rumour that he was Winston Churchill's illegitimate son, and asked loudly, 'And what does my dear brother say to that?' While Diana and the others sat horrified and Brendan made an unavailing lunge at him, Randolph ran away down the beach and into the sea. The much older Brendan lumbered after him, whereupon Randolph snatched off Bracken's spectacles and threw them in the water.

Venice, romantic and beautiful, with its tiny squares and great dark churches, was a city made for secrecy and assignations. Randolph Churchill was pursuing Tilly Losch the dancer; Tilly's husband Edward James was pursuing Serge Lifar. Tom Mitford, staying with the Guinnesses at the Danieli, was openly conducting a liaison with the rich, beautiful and stylish Princesse de Faucigny-Lucinge, a woman so chic that wherever she went she was stared at and copied. Those of their circle who had not yet heard of Diana's entanglement with Mosley were soon made aware of it. On sightseeing parties, the two would suddenly disappear. A few steps round a corner or down some narrow alleyway, and they would be at the water's edge; once in one of the gondolas that plied every canal, they were invisible, speeding off to spend passionate afternoons in small hotels.

The rest of the party, uncomfortably aware of the anguish of Bryan and Cimmie, would struggle to pretend that everything was normal. Eventually the lovers would reappear, armed with some excuse, in time to join the others for dinner. Once, when all of them had spent the whole day on the Lido so that one of these stolen afternoons was impossible, Mosley leant across the dining table in the Lido restaurant and said to Boothby, 'Bob, I shall need your room tonight between midnight and 4 a.m.' Boothby, who was there as the Mosleys' guest, was so taken aback by this public request – both Cimmie and the Guinnesses were nearby – that he merely asked jokingly, 'But Tom, where shall I sleep?' 'On the beach,' said Mosley. It was no joke: Boothby had to retire to one of the recliners in the Mosleys' cabin on the Lido.

So miserable was Cimmie, who wept for much of what she had hoped would be a Venetian idyll with her husband, that her already weakened state of health was exacerbated by despair.

For Bryan, forced to confront publicly the fact that his wife had fallen in love with another man, Diana's behaviour was not only a dagger through his heart but a bitter humiliation.

After their Venetian fortnight the Guinnesses returned to Biddesden. The daily proximity of her lover in the world's most romantic city, the aesthetic delight afforded by the buildings, paintings, canals and bridges of Venice, the spice of secrecy, had lifted Diana on to a plateau of happiness. Though she was thrilled to see her two small boys again, once back at Biddesden the familiar sensation of being in a prison returned. Everything irritated her and when she was irritated she became uncaringly cruel. The half-whispering speech that Bryan sometimes affected – especially for the sort of question to which he dreaded the answer – infuriated her, while endless quarrels were provoked over what she saw as his possessiveness. Because of his unhappiness and the realisation that his wife was drifting away from him, Bryan's anxious questioning had taken on stifling proportions: sometimes Diana could hardly leave the room without being asked where she was going and what she was going to do; and she would round on him furiously.

The truth was that they were basically unsuited. Bryan, gentle, whimsical, poetical, idealistic and uxorious, uninterested in the world of politics or the salon, could not hold Diana. Although neither of them had realised it when they married, she was infinitely the stronger and more independent character – and she needed a strong man. In this respect, Mosley was similar to the father whose character and behaviour had so imprinted her childhood. Both had dominant personalities, both had powerful convictions (or, in truth, iron prejudices) which they expressed freely, both were Alpha males in a household or group of women.

Bryan's adoration, once so touching, now served only to irritate. When his eyes followed her around the room she would drawl out some remark so brutal that Pam, who often spent evenings with them, would gasp. 'He worshipped her, and she walked all over him,' said Pam of that unhappy time. Bryan had never been able to join in the Mitford habit of 'teasing', with its vein of savagery concealed beneath irony or wit; against a Diana who had withdrawn her love he was defenceless. For by now Diana believed that her future lay with Mosley. Even in those early days she had, she said, 'a feeling so definite as to be a knowledge that we were made for each other'.

Mosley, on the other hand, was trying to reassure Cimmie that they could have 'such a lovely life together if his little frolicsome ways did not upset her'. As a diversionary tactic this failed hopelessly. Cimmie, who by now believed that Diana was a serious threat to her marriage, was in agony. Her letters breathe the despair of love betrayed.

> If only you would be frank with me, that is what I beg. If when you refused Mereworth you had said it was because you thought you would like to take Diana out for the day Sunday I would have known where I was. I started by thinking it odd; but as you said nothing about another plan I began to think it must be that you wanted to stay at home and just be with us and I was so pleased. Then you tell me about Sunday as if it was vague and only planned last night – and then I realise the whole thing was arranged before and you had been putting off telling me . . . the feeling that you are not telling me, that you do things behind my back, that you are only sweet to me when you want to get away with something, gives me such a feeling of insecurity and anxiety and worry I am more nervy and upset than I ought or need to be.

In a passion of jealous misery she added, 'Oh darling darling don't let it be like that. I will truly understand if you give me a chance, but I am so kept in the dark. That bloody damnable cursed Ebury – how often does she come there?'

The answer was constantly. Diana and Mosley saw each other as often as they could and when they were not able to meet Diana attended his fascist meetings.

The British Union of Fascists rose phoenix-like out of the ashes of the New Party. Cimmie, loyal as ever, helped to design a fascist flag and discussed with her husband how Sousa's 'Stars and Stripes' could be turned into a fascist anthem with words by their friend Osbert Sitwell. Irene, a masterful, forthright figure who moved in the wealthy hunting circles of Melton Mowbray, organised a fund-raising ball and helped to persuade figures such as Lord Rothermere to contribute to the BUF coffers. This was not through any deep belief in fascism *per se*, as she was swayed not so much by her own convictions as by her brother-in-law's personality – although her earlier feelings for him had passed, she was still fascinated by his charisma and sexual magnetism. His personality was such that people were either passionately for or passionately against him; if the admirer was a reasonably attractive woman, it was equally in his nature to try to seduce her. This had happened with Irene: their brief affair had begun after they had been to bed one night after a hunt ball in Melton.

On 1 October 1932 Mosley, tanned and healthy after weeks of sunshine and swimming, officially launched the British Union of Fascists with a flag-unfurling ceremony in its offices (formerly those of the New Party) in Great George Street. This was followed a fortnight later by the first public meeting, in Trafalgar Square. Mosley, in dark suit and tie and white shirt, spoke from the plinth at the bottom of Nelson's column, with his now-customary bodyguard of strong young men. There were eight of them, and one significant difference could be noted in their appearance: all were wearing the black shirts that would soon be infamous. At a later meeting in the Memorial Hall in Farringdon Street, the 'Blackshirts', as they quickly became known, took an active role, ejecting several hecklers. In the march down Fleet Street and the Strand that followed, there were scuffles with communist supporters.

Throughout the autumn Diana went to these BUF meetings, usually with a friend, watching admiringly as Mosley, by now in full fascist rig of black shirt, breeches and boots, surrounded by his bodyguard of tough young Blackshirts, strode theatrically on to platforms. 'All this swagger and vanity to Mrs Bryan Guinness and Doris Castlerosse,' wrote Irene angrily in her diary – for Cimmie, too, had a bodyguard, in the shape of her two sisters who came to most of her husband's meetings with her.

At the end of October there were two fancy-dress balls on consecutive nights. The first was given by the Guinnesses at Biddesden and the second by Cecil Beaton, who lived nearby at Ashbrooke. Cimmie and Tom Mosley came to both. For the ball at Biddesden on Friday – gatecrashed by Prince Henry (the Duke of Gloucester) – Diana wore a white dress and a silver wig, cleverly constructed out of silver string by Robert Byron. Mosley was a demon-like figure in black; Cimmie was a shepherdess, a symbolic representation of innocence lost on Diana, who spent much of the early part of the evening telling her friends of Mosley's brilliance and the inspiration of his ideas. 'It was rather like having a crush on a film star,' said Lady Pansy Pakenham. On the lawn outside the long drawing room was a bonfire, its flames flickering beyond the tall sash windows as couples danced. By the end of the evening, it was obvious what was happening: as soon as possible after dinner Diana and Mosley had vanished upstairs, only reappearing when it was time to say goodbye. For once, Cimmie was unable to hide her anguish in public; her devastated face as she left Biddesden that night remained fixed in the minds of all who saw it.

It was then that Diana began to think of abandoning Bryan and leading an independent life in London.

She seemed calm, almost insouciant, as she discussed it with Cela, appearing unconcerned at the thought of the difficulties and the social problems facing a young separated woman. Georgia Sitwell, after having tea with Diana on 11 November, described her as 'youthfully arrogant'.

When Diana told Mosley she was thinking of leaving Bryan, he encouraged her to do so. At the same time he made it perfectly clear that he had no intention of leaving Cimmie. To those who loved Diana his behaviour appeared unspeakable: he was persuading a young and inexperienced girl to leave her husband and home simply to make herself more available for him. In the parlance of those days, it was the action of a cad.

At the end of November Diana finally told Bryan that she wanted to be free. It took her only a few minutes. She said first that she knew she was making him suffer, then, her voice calm and reasonable, continued, 'I think the only thing is for us to part.'

It was a momentous decision for a girl of twenty-two, married to one of the nicest men in England and with two small children. On the material side, she would be throwing away exactly the sort of life she enjoyed – the beautiful houses with the money to furnish them exquisitely, the means to travel whenever and wherever she wanted, the clothes, the jewels, the ability to entertain anyone she liked on a grand scale. Socially, it was disastrous: while a blind eye was turned to discreet affairs, it was an unwritten rule that there should be no public scandal. Diana was leaving someone who had treated her with love and generosity in order to set herself up as a mistress to a known philanderer who had declared that he had no intention of marrying her.

It was also an extraordinary effort of will on her part, in the teeth of fierce and justified opposition. What carried her through was the Mitford confidence, the intransigence that was later to show itself in her younger sister Jessica's behaviour – indeed, Diana's strength of purpose may have served as an example to Jessica, who adored her – and her credo that with sufficient will one can follow any course one chooses.

There was also her belief, now held with the passion of a disciple for a messiah, in her lover's ideology. As she said later, 'It was this which gave me the courage to survive ostracism, the anger of my parents (who did not allow Debo or Decca to come and see me), the disapproval of absolutely everyone – the fonder they were of me, the more they disapproved. But in a strange way I think Kit* and I both knew it was "pour la vie" and that we should always love one another.'

Diana's parents were shocked, angry and wretched. It was a time of dramatic rows, weeping, sleepless nights for Sydney and furious outbursts from David. 'That man will just cast you aside!' he shouted miserably at his favourite daughter. 'If he does, that's too bad,' responded Diana. All her friends implored her to reconsider. 'Please don't do it,' said John Sutro. 'You are leaving something wonderful for something not only unknown, but hopeless.' All of them knew of Mosley's reputation, most of them believed that after three months he would discard her.

In the circles in which the Guinnesses and Mosleys moved, sympathy was solidly behind Bryan and Cimmie. Irene Ravensdale wrote in her diary of 'the hell incarnate beloved Cim is going through over Diana Guinness bitching her life'. Irene, like her sisters, believed that Diana would never have contemplated leaving Bryan had she not intended to marry Cimmie's husband. Nothing, in fact, was further from Diana's mind. She had accepted absolutely and unquestioningly Mosley's statement that he was never going to leave Cimmie; she did not want to break up his marriage but to be free of her own. Nor, disingenuous as it sounds, did she realise Cimmie's sufferings. As she wrote years later, 'Our mutual friends like Bob [Boothby] either didn't know or didn't think it worth while to tell me.' Nothing would have made her give up Mosley but, had she known what agony Cimmie was going through, she might have modified her behaviour in various ways.

From Diana's point of view, Mosley had been unfaithful from the very start of his marriage and even if she gave him up would continue to have liaisons with other women. If Cimmie had come to terms with all these other affairs, she reasoned, why should she not do so with Diana? What Diana did not understand was that Cimmie had realised that her husband's feelings for Diana were far more profound than in any previous liaison and that, irrespective of physical fidelity, Cimmie would be sharing him on the deeper level which he had assured her was hers alone.

At Swinbrook, Nancy could hardly believe that Diana intended to go ahead with her decision to leave Bryan – or indeed would be allowed to:

* Because her brother's name was Tom, Diana called Tom Mosley 'Kit'.

> Mitty [Tom] is horrified and says that your social position will be *nil* if you do this. Darling I do hope you are making a right decision. You are so young to begin getting in wrong with the world, if that's what's going to happen.
>
> However it is all your own affair & whatever happens *I* shall always be on your side as you know & so will anybody who cares for you & perhaps the rest really don't matter.

She followed this two days later on 29 November with a gloomy forecast: 'Oh dear I believe you have a much worse time in store for you than you imagine,' and the comment that Tom thought the £2,000 a year she had requested from Bryan was not nearly enough. 'It will seem tiny to you.'

Neither Diana's father nor her father-in-law was prepared to see the marriage break up without a fight. They agreed that the only thing was to appeal to Mosley himself; surely, as a gentleman born, he must see that his duty was to influence this young girl to *stay* with her husband, not to leave him?

The two old gentlemen, as Mosley and Diana called them (both were in their early fifties), telephoned Mosley and made an appointment to see him. Diana was terrified that her lover would be persuaded, for her sake, to give her up. Mosley had no such intentions. When Randolph Churchill asked, 'What are you going to do?' his response was, 'I suppose wear a balls protector.'

When David and Lord Moyne arrived at the Ebury Street flat they plunged straight into their argument. 'You simply must stop this. You are ruining a young marriage, with young children, and this is awful. She's only twenty-two – how can you do such a thing? You must give your word you will not see her any more.' Mosley refused, saying only, 'Diana must be allowed to do what she wants.' That night at dinner he repeated the entire conversation to her.

In December, Cimmie and Tom Mosley rented a house at Yarlington, in Somerset, for the Christmas holidays (Savehay Farm was let). Here they held a family Christmas, with Irene, Baba and the three Metcalfe children. Fruity Metcalfe was Father Christmas and the children discovered his red cloak and hat hanging up in the downstairs gents ('Look – Father Christmas has left his clothes behind!'). On New Year's Eve they gave a fancy-dress dance, for which a large house party stayed, including

Diana. Cimmie had asked her husband's mistress to stay, perhaps because she believed that, by integrating Diana into their circle as she had done with her husband's other women, she could somehow neutralise her effect, perhaps because she was persuaded to do so by Mosley or perhaps, even, so that she could study them together for herself at first hand.

To venture into that house alone, to face her lover's wife and the icy, hate-filled courtesy of the other two Curzon sisters, required astonishing effrontery. Everyone knew Diana was 'Tom's new girl' – though not everyone knew she was in the process of leaving her husband. Diana not only brazened out the situation but made no secret of her feelings. 'DG and Tom v. irritating,' recorded Georgia Sitwell, also in the house party, adding, 'Tom revealed all to me at lunch.'

The next day, Diana took the lease of a small house of her own, 2 Eaton Square, though because of its rundown condition she was not able to move into it for a month or two. There was just room for Diana, her two children and four servants – cook, nanny, house-parlourmaid and lady's maid. It had been made available to her at a low rent by the Grosvenor Estate on condition she repaired and redecorated it (for which she was given a grant) but its real attraction was its closeness to Mosley's flat two minutes away round the corner.

Chapter XII

When Diana left him, Bryan went through agonies of misery and humiliation. The break-up of the house at Cheyne Walk, where only a few months ago they had given a party that seemed to celebrate their joint happiness, was the public admission that his marriage had failed. He could hardly bear to undertake the necessary domestic arrangements – and, as those around him indignantly sympathised, why should he? 'For her own sake and as far as possible for the harmony of the Cheyne servants, Mrs Guinness to tell them herself,' ran a scribbled note to his secretary, Miss Moore. 'Dear Mrs Guinness,' she wrote on 6 January 1933, 'Would you mind seeing the servants yourself about the dissolution of Cheyne Walk and consequent future arrangements? They are so hurt that it has come through me and not through their mistress. We all want the road to be as smooth as we can make it and it would help very considerably if you would do this.' At the bottom Bryan had pencilled a note to Diana. 'The servants are resenting your going so much and the tiny flat will send them off the deep end again. I shall be thankful when it comes to an end.'

He could still hardly believe that Diana's decision was final, even though she found him a flat, 143 Swan Court, and was decorating it. 'Darling, I really do think you are kind,' he wrote, when she offered to arrange his books for him. 'I do love you for it. I mean, I would if you wanted me to.'

His father was made of sterner stuff, urging his son to stand up for himself and tell Diana that her behaviour was unacceptable. Before leaving for three weeks in Switzerland, Bryan wrote from beneath the parental roof at Grosvenor Place – probably to his father's dictation – to protest at her public appearances with the man whom he believed had stolen her from him.

> Darling Diana,
>
> I would be very grateful if you would *not* give your lunch party at the Savoy Grill as I am lunching there with Reggie myself. I tried but failed to get through to you on the telephone to say this. I cannot consent to your associating with Mosley, either at Cheyne Walk or anywhere else. I do not mean that I shall send a policeman to fling him out but I cannot in any way condone your meeting him. Love from
>
> Bryan

Unity was not allowed to stay with her sister in London, partly because of the Redesdales' disapproval, partly because the ambiguity of Diana's marital status made her an unfit chaperone. Debo, much younger, was too preoccupied with the hunting, riding, skating and animals that took up most of her waking thoughts to do more than simply accept Diana's mysterious absence. Yet the Redesdales had not completely given up hope that Diana might be persuaded to return to Bryan. Sydney still saw her from time to time and on 12 January wrote a final appeal.

> I was so glad I was able to see you the other night and to realise what I did think was the case: the affection you have for Bryan through and beyond everything. His worst fault seems to be a too great fondness for you and perhaps you on your side are too impatient. Do, I beg you, think well before you throw away what is worth while and good for what is nothing and bad. Surely with the amount of real affection you and B have for each other you should be able to get on together if both of you try to. It really is worth trying. As to the other, even if it were possible, it could only end badly, and quickly too. I believe your own good natures which you both have so largely will bring you through. It has been a great trouble to us and to everyone, I believe, who knows you both. When I talked to B about it, he took all the blame. That isn't fair, and you would both have to alter your ways a little, I dare say, but it must be worth it.

Perhaps, thought the Redesdales, the opinion of someone Diana's own age might influence her. Her best friend Cela, who shared the general view that Diana was making an appalling mistake, had developed pleurisy and was to be sent out to Mürren that January to join her aunt and younger brother. Would Cela ask Diana if she would come out to Mürren with her? Once away from daily contact with Mosley, thought her parents, Diana might come to her senses and realise the enormity of what she was doing.

Cela was surprised though delighted when Diana cheerfully accepted her invitation – for her part, Diana felt a break from the rows with her parents and the guilt she was feeling every time she saw Bryan's white, miserable face would be welcome. She had sworn Nancy to secrecy over her whereabouts, so it was only when Diana herself sent him a wire to say she was thinking of him and giving her address that Bryan was able to write to her, as he lay in bed with flu at Swan Court. 'I have not written to you before because I thought you did not want me to write . . . It is very bad for us to think of each other. We must stop it. All the dye will be washed out of my eiderdown if I go on like this. Still I think a little annual tear on the 30 of January can be allowed . . .'

He felt too rotten to go down to Biddesden for the lawn meet of the Tedworth; Pam 'did the honours' instead. Each fresh arrangement was painful, such as writing to Diana to let her know that the superb Biddesden cook, Mrs Mack, wanted to go to Eaton Square with her and was prepared to accept the lower wage Diana would offer ('it is much better she should go to you contented than stay with me disgruntled'). There was also the miserable and humiliating business of explaining to those who had seen Diana and Mosley lunching together, or who had asked Diana and Bryan to dinner parties, that he and Diana were now separated. His novel *Singing Out of Tune*, about a couple whose marriage goes wrong and who separate, was due to come out soon; he begged her to tell everyone that it was *not* their story.

No efforts by Cela and her aunt could alter the steadiness of Diana's purpose. She treated the ten days in Mürren simply as a well-earned respite. 'I have read five books and learnt to waltz on the ice since I have been here and I am feeling like heaven in rude health,' she wrote to Roy Harrod on 30 January.

When Diana returned, she went to see Bryan immediately. On 1 February he had scribbled her a note about the Mitfords' great friend Mrs Ham, known for her delight in catastrophe. 'She was doubly delighted to find illness joined to matrimonial disaster. I need hardly say she is borrowing the car tomorrow to visit a friend who is dying of cancer.' The amused tone of this made Diana think Bryan might, after all, be beginning to accept the situation, and she relaxed into her usual warm, affectionate and good-natured manner. To Bryan, her friendliness gave a glimmer of hope. Late on the night of Sunday 5 February, after she had left him in Swan Court, he wrote a last, passionate appeal. It was dignified, moving and realistic; few women could have read it without tears or second thoughts. On Diana its effect was to intensify her desire for a quick, clean break.

> I was afraid that while you were in Switzerland you would decide to return to me because you couldn't face the material situation. That has

not happened and this is all to the good. But now that you have seen me, do you not find any impression made on a heart that absence may have made grow fonder without its being aware of it. Don't you find that I am the one you will miss most? Are you sure – are you really sure that the circle of our love is complete? Don't you feel any impetus along a second circle round the same centre?

Are you positive that you love Tom more than me? You say you cannot do without him – but can you do without me? I have already pointed out his strength, which lies in the sacrifices you would have to make for him. I have already begged you to leave that out of account in weighing us up. His weakness may lie in his power and his charm, which enable him to do without you. If you must take Biddesden into account against me, you should also take the 11 and more feathers in his cap against him. You were my ewe lamb. He may turn from his flock to take you. He may value you but one day it will be too late.

So think, think, day and night. I have loved you so much and with such an unusual love. I have been so entirely devoted to you. But you must search yourself and inspect the inmost corners of your feelings. I am the one who gave you Jonathan and Desmond. Think, think, whether I did not get some lasting chord in you that will always be there belonging to me. Think of the time when you were coming round from the chloroform and the bottom of your mind was confused with the top – what did you find then? Was it me you wanted – that was natural?

Who did you want as you came to? Was it me or Tom? I don't know, perhaps it was Tom. Perhaps I am arguing against myself. Certainly he has captured your sexual fancy. Certainly he has captured your imagination. But which of us rules the unknown areas of your subconsciousness, where your most real emotions dwell? Which of us would you most want to breathe in your ear if you were dying? (neither, says you).

It would not be easy to build up a life together again, in fact the easiest course from a material point of view, certainly from mine, is to make the breach final. Yet I have this strong feeling that we both belong to one another, that we are bound by our mutually broken virginity. Are you sure that there is no side of you which feels itself my wife as I feel myself your husband? . . . what I hoped and decided would be a good sign would be if you seemed to warm towards me when you saw me and that you seemed a little to do. Perhaps it was only pity and kindness and sentimental memories but see if it was not anything more. Ask yourself the question night and day, I beg you. Love Bryan.

Instead, she stepped further away, moving into the Eatonry (as her Eaton Square flat quickly became known). The grant from the Grosvenor Estate paid for a new bathroom as well as for £200-worth of white paint which transformed its dingy interior. Bryan's generous allowance, made to her on marriage, easily covered the £300 a year rent. He also gave her the Cheyne Walk furniture – they had spent such a brief time there that they had not finished furnishing it – but she refused to take his present of 'The Unveiling of Cookham War Memorial'. 'It is such a marvellous picture,' she told him, 'and it is not right that I should take it.' Sadly, Bryan removed it to hang at Biddesden. She did, however, accept the manuscript of *Vile Bodies* since Bryan said he felt that it really belonged to Jonathan.* She managed to have the Aubusson carpets she loved in two of the small rooms by dint of swapping her own large one for two smaller ones. The tiny dining room, with its red velvet banquettes which could seat six, was usually full. Mosley was constantly there. '*So* marvellous with the boys,' Diana would say. 'He knows what to say to children and jollies them along and makes things interesting for them.'

To Cimmie, a Diana who had left her husband was a menacing figure. In those days few women would have taken a step that would render them so *declassée*. This open flouting of convention appeared a blazing public statement of intent. Diana's abandonment of Bryan must have made it seem to Cimmie as though she were utterly sure of her lover – or that Mosley had already secretly decided he would leave Cimmie and marry her. Whichever alternative was true, Cimmie felt that the husband she loved so much was no longer hers and she suffered terribly.

For Mosley, Diana's solitary state was an irresistible compliment. That such a young and beautiful woman should have sacrificed her marriage, an immense fortune and a position at the peak of society simply to make herself more available to him was the final, conclusive proof of his virility and attractiveness.

Bryan did his best to adjust to the situation. Finding Biddesden intolerably lonely, he made one of the trips abroad he so enjoyed, this time to the Greek islands. 'I don't know how far "sad memory" will "bring the light of other days around me" for I am determined to control dismal associations as far as I can,' he wrote to Nancy on 28 April.

During April Cimmie and Tom Mosley had been to Rome to take part in an International Fascist Exhibition – Mosley and seven of his men represented the BUF, with a black banner emblazoned with a small Union Jack and the

* In 1984 Jonathan sold it at Christie's for £55,000.

fascist symbol. During the visit, Mosley was received by Mussolini, and flatteringly invited by Il Duce to appear with him on the balcony of the Palazzo Venezia.

It was a heady moment. As the excited roars of the crowd rose to their ears, Mussolini's flamboyant gestures and frowning, blue-chinned posturing no longer seemed ridiculous but the response of the central actor to his country's love of spectacle, an affirmation that here, truly, was the leader. In any case, Mussolini's theatrical style and exaggerated masculinity appealed strongly to Mosley, for whom politics and sexuality were inextricably mixed. 'Fascism is the greatest creed that Western civilisation has ever given to the world,' proclaimed Mosley's editorial in the next issue of the *Blackshirt*, the BUF weekly paper.

On returning, Mosley went straight to see Diana. By now Cimmie was desperate, so much so that one day Diana asked Mosley why Cimmie minded about her so much more than the other women he had had affairs with. 'She probably didn't know about most of them,' replied Mosley. Then it struck him that it might help to take Cimmie's mind off Diana and reassure her that nothing would affect their marriage if he made a clean breast of them all. He went home and did this. 'But they are all my best friends!' wailed the unfortunate Cimmie.

She was too upset that night to accompany her husband to the dinner party to which they had both been invited, so Mosley went by himself. There, to his relief, he found their friend Bob Boothby. Would Boothby, he asked, leave the dinner party early and go and see Cimmie who was, he explained, miserable and needed comforting?

'What have you done to her now, Tom?' asked Boothby.

'I've told her all the women I've been to bed with since we've been married,' said Mosley.

'*All*, Tom?' asked Boothby.

'Yes,' replied Mosley. 'Well – all except her sister and her stepmother.'*

Boothby duly left early and did his best to convince Cimmie that it was her and her alone whom her husband really loved.

Perhaps it was only coincidence that Cimmie's final decline began within days of Mosley's devastating frankness, or that she could not face the luncheon party on 4 May given by her close friend Georgia Sitwell,

* One year in St Moritz, Mosley, unable to ski because one leg was so much shorter than the other, had had a fling with the voluptuous Grace Curzon, then in her late forties, while Cimmie was on the slopes.

now revealed by Mosley's disclosure to have been a recent mistress. The pencil note for that day in Georgia's diary reads, 'Cim supposed to lunch but taken ill.'

Cimmie and Mosley spent the weekend together at Denham in an atmosphere of mounting tension with, on Cimmie's side, increasing physical wretchedness as well. On Sunday evening there was a brief cessation of both. Mosley poured them a cocktail and they strolled across the lawn together. One of the housemaids, Elsie Corrigan, was sitting on the wide window-seat of the servants' sitting room writing a letter home on her Sunday evening off; as the Mosleys passed they handed her their empty glasses. She watched them walk, arms about each other, laughing and smiling, into the orchard beyond.

That night they had a terrible quarrel, which left Cimmie shattered. Mosley walked out, drove up to London and went to see Diana, who told him that she had just asked Bryan for a divorce; at Denham, Cimmie wept through a sleepless night. In the morning, as usual unable to bear any rift, Cimmie wrote to her husband, characteristically and generously blaming herself.

> Darling heart, I want to apologise for last night but I was feeling already pretty rotten and that made me I suppose silly. Anyhow I had a star bad night of feeling wretched and this morning was all in with sickness and crashing back and tummy ache. I can't think what it is . . . no temp. so there's nothing to worry about.

She was wrong. That evening, she was rushed to a nursing home with acute appendicitis. The appendix burst and she was operated on the same night.

At first no one was anxious. An appendix operation was commonplace, Cimmie was only thirty-four and seemingly strong – Mosley was so little worried that immediately after the operation had been successfully performed he went to the Eatonry for lunch – among the other guests was Unity, meeting 'the Leader' for the first time.

But despite Cimmie's open-air, smiling, sturdy appearance, her health was poor and had been getting worse. The congenital spinal curvature, inherited from her father, which she had had from childhood, had gradually developed into lower back trouble that became more and more painful. While she was carrying Micky the pain was so severe she had to have morphia to relieve it and after his birth she found she was unable to take any form of active exercise – even sneezing hurt badly. The kidney trouble from which she had been suffering for months refused to be cured and she picked up numerous other infections. Perhaps more relevant still, she had been under intense emotional strain and miserably depressed ever since, almost a year earlier, she had realised the seriousness of her husband's feelings for Diana.

Cimmie's illness was reported in the newspapers and it was from the *Continental Daily Mail* that Irene first learned of it – she had just become engaged at the age of thirty-seven and was in Switzerland with her fiancé, Captain Miles Graham. She wired at once to Mosley, who telegraphed back that all was well and there was no need for her to come home. Instead, she wrote to Cimmie, telling her of her engagement and saying that she wanted her, Cimmie, but not Baba, to come to the quiet wedding she and her fiancé planned in Switzerland – so would she please not tell Baba about it. For by now the relationship between eldest and youngest sister had so deteriorated that the jealousy simmering under its surface had almost reached the point of open expression. Irene, her eyes sharpened by her own love affair, must have realised the depth of her youngest sister's obsession with Mosley. Her intuition was well founded: all that year, Mosley had been juggling not only his wife and Diana but Baba as well, lunching and dining with each several times a week. Sometimes he would see all three on the same day.

Three days later, Mosley wired Irene again, this time to say that Cimmie was critically ill with peritonitis. Irene flew home at once with Miles Graham, arriving on 15 May to be met at the airport by Baba, but was too late to see Cimmie that day.

Cimmie's condition deteriorated terrifyingly fast. The next morning she was too weak even for a blood transfusion; instead, she was given saline injections. Baba, a grim-faced sentinel, sat outside the door of Cimmie's room, refusing to let Irene in – Cimmie, too weak to read her own letters, had asked Baba to read the one Irene had sent from Switzerland, with its fatal sentence, 'I only want you, not Baba, at my wedding.' Irene sat outside with Mosley's mother and Cimmie's maid Andrée, while Baba sobbed and Mosley tried brokenly to speak to his wife. By now all of them had given up hope.

A few hours later, Cimmie was dead. 'Tom came out after a bit, he hugged Baba pathetically,' wrote Irene in her diary, the flame of jealousy flickering as she added, 'If I had not kissed him on the stairs he would have passed me by.'

It was as well she did not know why he was going down the stairs. He was looking for the nearest place from which to telephone in privacy. A few minutes after Cimmie's death, as Irene made arrangements to have the Mosley children collected and their nanny told the sad news, Mosley was ringing Diana to say that his wife had died.

Diana was shattered. 'It seemed a harbinger of unhappiness,' she wrote later. She thought it would be the end as far as she and Mosley were concerned. She knew that he had been genuinely devoted to his wife and she believed that he would be so overcome with remorse and guilt, at the unhappiness he had caused Cimmie and its possible contribution to her death, that he would give her up. 'My heart is full of dark forbodings,' Diana wrote to a new friend, the don Roy Harrod. When Mosley called in at the Eatonry that evening (on his way to Denham) to tell her that they would not be able to see each other for some time, her worst fears seemed realised, even though he assured her, 'It will be all right.'

Cimmie's sisters saw it from a different perspective. 'Oh God what a terrible doom for Tom!' wrote Irene that night, 'and to think that Cim is gone and that Guinness is free and alive and oh! where is there any balance or justice!' They thought then and later that Diana was responsible for Cimmie's death – not, of course, for her illness but for the fact that she refused to fight it, metaphorically turning her face to the wall.

Many of Cimmie's friends also believed that Diana was the direct cause of Cimmie's death; or, as Cimmie's daughter put it, 'Diana destroyed my mother. I don't retract the word destroy. Peritonitis is what killed her but with Diana there, she didn't want to live.'

After seeing Diana, Mosley was driven back to Denham. Irene, sitting at dinner with Miles Graham after consoling as best she could Cimmie's children, Vivien and Nicholas (Micky was still a baby), heard the crashing of the car's passenger door. Before the housemaid Elsie Corrigan, waiting at table (the butler and footman were still at Mosley's London flat), could reach the front door, Mosley came in looking, thought Elsie, 'like a man demented'. Asking her to give him a whisky and soda, he went straight upstairs to the children. Miles Graham, made deeply anxious by Mosley's wild and anguished demeanour, asked Elsie where Mosley kept his guns. One pistol, she told him, was under Mosley's pillow – she saw it every day when she made the bed – and its pair was in the top drawer of his chest of drawers. Graham removed them both.

Many, including Nancy Astor, believed that Mosley's show of grief was theatrical. The maudlin description by the doctor on the case ('how Sir Oswald stood the anxious strain I do not know . . . it may seem a rather strange thing to say but they had a very beautiful time together during their last few days') which appeared the day after Cimmie's death did nothing to counter this.

But there was no doubt that Mosley had loved Cimmie deeply. Even Andrée, Cimmie's maid and confidante, who loathed Mosley for his treatment of her mistress, told Georgia Sitwell that no one could have been more marvellous during Cimmie's final hours. 'He spent every minute with her for a week. He talked to her for hours as she lay dying and Andrée thinks she understood.'

Mosley commissioned a sarcophagus by Lutyens, to stand in a memorial garden in the orchard to which they had walked on that Sunday evening – their last happy moments together. Until this pink marble tomb with its simple inscription 'My Beloved' was ready, Cimmie's coffin lay in the Cliveden chapel a few miles away – it was while canvassing for Nancy Astor that Cimmie and Tom had met each other. On 19 May there was a private funeral service attended only by Mosley, his mother and Cimmie's two sisters; later the same day there was a memorial service at St Margaret's, Westminster, characterised by the fact that the only people in black were members of the BUF in their new shirts – Cimmie's friends had been told not to wear mourning.

Everything that was reported to Diana of these last rites made her more fearful and apprehensive. 'I am very depressed and depressing,' she wrote a few days later to Roy Harrod, '– when I come and depress you on Sunday may I bring Cela Keppel.'

Bryan's agreement to a divorce, in which he would be named as the guilty party, did not cheer her. At dinner parties she looked white-faced and wan; when Mosley came to see her for the first time after Cimmie's death in mid-June, they had a blazing row, provoked by his asking à propos of her forthcoming divorce, 'Well, have you jumped your little hurdle yet?' Cut by this seeming frivolity, she turned on him furiously. 'It isn't "a hurdle" – you are talking of my whole life.' It was a far more serious quarrel than the spats they often had in the early days; Mosley, realising this, apologised at once and they made it up.

Cimmie's death united the two surviving Curzon sisters in a hatred of Diana so virulent that the rivalry between them took second place. Irene realised that Baba was Mosley's preferred companion, while Baba's passion for her brother-in-law, no longer held in check by Cimmie's presence, found more overt expression. Hysterically, she begged Mosley to give up Diana as in some way 'hurtful', as Irene put it in her diary at the end of that June, to the memory of the dead Cimmie. 'Baba fairly let fly about Diana and how she loathes her and she tried to make Tom see that if he went on like this he would be utterly killed and his future smashed as people would not stand for it . . . and he still cannot see what he is doing to beloved Cim through it all.' Sisterly outrage concealed furious jealousy: to console the grieving widower, Baba was to obtain a fortnight's 'leave of absence' from her husband Fruity so that she could motor through France alone with Mosley later that summer. To facilitate this jaunt Irene was to agree to take Cimmie's two older children on a holiday while Micky remained at home with his nanny.

Mosley, well versed in juggling female emotions, managed to convince both Baba and Diana of his need to see each of them. To Baba, as Irene noted, he explained that he had to see Diana occasionally because he could not 'shirk his obligations' and that he had to dine with her 'platonically as he says' from time to time. He persuaded his mother to tell both Curzon sisters that he had no intention of seeing Diana again after the motor trip with Baba – for Baba, this must have seemed a signal that he felt as deeply for her as she did for him. To Diana, whose divorce hearing had been scheduled for June, he said that if it became known that he was seeing a lot of Baba this would make an excellent cover for himself and Diana during the six months between the decree nisi and the decree absolute.* In addition to the possibility of a failed divorce, any scandalous publicity would have been as bad for both of them as if Mosley had been cited as correspondent, so when he went to see Diana after dark, slipping round the corner from his flat to the Eatonry, he would tap on a ground-floor window with his stick to make sure a visit was safe.

Irene, on the emotional sidelines, knew only that Mosley now seemed wrapped up in Baba: 'I pray this obsession with her will utterly oust Diana Guinness.'

On the practical side, Irene had a leading role. She had broken off her engagement and now became the Mosley children's surrogate mother. She sold her house in Deanery Street, had her old dog Winks put down, dismissed her staff and went on a cruise before taking up her new role.

* In those days if the Kings Proctor had evidence that a divorce petitioner had committed adultery and had not disclosed it to the court, he could apply to the court for the Decree Nisi to be rescinded.

She then moved into Savehay, where she supervised the running of the household, with Cimmie's former lady's maid Andrée as housekeeper, and kept the accounts, which had to be approved by the Official Solicitor.

The Mosley children were now entitled to their mother's £10,000 a year from the Leiter Trust in Chicago and, as minors, had been put financially under the care of the Official Solicitor. Mosley contended that without his children's income he could not afford the upkeep of Savehay Farm with its large staff, and it was agreed that the children's income should be used to keep up what was, after all, their family home.

Diana's divorce was due to be heard on 15 June 1933. The day before, three of her sisters, Pam, Unity and Nancy (who had a room at the Eatonry), had come round to tea to give her courage for her appearance in the witness box. What Diana, but none of the others, knew was that Hamish St Clair Erskine, unable to bear the charade of his engagement to Nancy any longer, was desperate to break it off. He had told Diana he thought the only way was to pretend to be engaged to someone else. 'You must be very careful,' said Diana. 'She might do herself an injury.'

After tea the Leader arrived. He greeted the admiring Unity warmly ('Hello, Fascist!') and gave her a badge bearing the fascist emblem. Pam, who disliked him intensely, left, but the others stayed on for supper. As they ate, Hamish telephoned. He asked to speak to Diana, hoping to get her to act as his intermediary, but when the caller was announced Nancy rushed to the telephone. Brutally Hamish broke it to her that he was engaged to someone they had both agreed they thought was awful. Nancy begged him to come round and he agreed. Mosley left hastily, and Diana and Unity took themselves off to a cinema.

Nancy was shattered. 'I knew you weren't IN LOVE with me but . . . I thought that in your soul you loved me & that in the end we should have children & look back on life together when we were old,' she wrote pathetically the same night. The contrast between her life and Diana's must have seemed doubly galling. Here was Diana almost carelessly throwing away something she, Nancy, had never known – a loving husband – for something else equally missing from her own life: a physical passion. For although Nancy, like Pam, disliked Mosley, it was impossible for anyone living at close quarters with Diana not to recognise the sexual and emotional intensity of her relationship with Mosley.

At the divorce hearing the following day, Diana was granted a decree nisi. Bryan did not contest her petition: he had duly gone down to

Brighton and spent the night in an hotel with a professional lady hired for the purpose, bestowing a large tip on the maid who brought breakfast so that she would be sure to remember having seen them together.

The financial arrangements had been settled quickly and amicably. Diana had refused to try to obtain the huge sum that could have been extracted from a very rich man who was technically the guilty party. Nor did she want any part of the marriage settlement from Bryan's parents which had temporarily made her such a rich woman. Her solicitor's comment that 'you are signing away millions' left her unmoved. She did not want to be rich, she said, nor did she want anything that could be described as 'loot'; she merely wanted enough to live on. She gave Bryan back the Guinness tiaras and brooches, keeping only the presents he had bought her since their marriage – a three-strand cultured pearl necklace with a diamond clasp, huge and beautiful Victorian paste earrings and an exquisite bracelet of diamonds and cabochon rubies.

Bryan, anxious to be as generous as he could be to the woman he still loved deeply, asked her to name whatever sum she thought would enable her to live comfortably. She agreed to accept £2,500 a year, in those days an income that could provide two or three servants – Diana had a cook and a nanny – elegant clothes and a car. Bryan, who thought Mosley, even if he married her, would later abandon her, insisted that this income should be hers for life and not cease if she remarried.

After the divorce, Bryan, who wanted to sink out of sight, spent more and more time at Biddesden. He would sometimes have friends to stay for weekends and would invariably ask Pam, still managing his farm, to join them. She would enquire carefully of the parlourmaid, May, who the other guests were, before giving her answer – if the party included Evelyn Waugh, of whom she was terrified, she always refused.

Diana was largely cut off from the grander side of society. Though a few of the well-known hostesses still invited her to stay – one was Mrs Ronnie Greville – to the majority of the older generation she was an undesirable. Quite apart from their hatred of divorce, there was intense disapproval of her behaviour; pity and sympathy for their contemporaries the Redesdales; and shock that anyone should abandon a marriage quite so quickly and for quite so 'shop-soiled' a Lothario. Apart from this, at a time when upper-class young girls were strictly brought up and chaperoned for at least two seasons, the presence of a beautiful young woman (only a few years older than these debutantes) who had left her husband would have seemed a threat to many wives and a temptation to many husbands.

This ostracism passed Diana by completely, since the legion of her own friends remained steadfast. She would go to the cinema with Henry Yorke or Nigel Birch, to concerts with her brother Tom, to Emerald Cunard's evenings. She went as Poppaea to John Sutro's Roman fancy-dress party at the Savoy; she had herself photographed for Cela's brother Derek Keppel, who had worshipped her since he first met her ('My one and only darling Diana,' he wrote from Sialkot, where he was stationed with his regiment, the 13/18 Hussars, 'You are the divinest of all women . . . I do adore you so'). Twice a week she would go with Unity to the Women's League of Health and Beauty, afterwards bringing Unity back to dine at the Eatonry, often with Mosley, evenings which laid the groundwork for much of Unity's later obsession with the fascism of Nazi Germany.

A few months earlier, in the spring, Mrs Richard Guinness – the wife of a distant cousin of Bryan's – had invited her to meet 'a very interesting German'. This was Putzi Hanfstaengl, who worked for the new German Chancellor, dealing with the British press. Diana could speak only a word or two of German, but Putzi, thanks to his American mother, spoke English fluently. He idolised Hitler, whom he had known for twelve years – it was to Hanfstaengl's wife and sister that Hitler had fled after the failure of the 1923 putsch, and few men were closer to the new Chancellor than Putzi. He told Diana that if she came to Germany for the Nuremberg Rally that September, he would introduce her to Hitler.

Chapter XIII

After Cimmie's death Mosley flung himself into establishing his new party in the public consciousness. Meetings began with fanfares, spotlights, the fascist salute and a fascist processional. In the expectant pause that followed, Mosley, dramatic in black shirt, breeches and riding boots, strode alone to the platform to the strains of 'Salutation to the Leader', gazed round with an air of arrogant, commanding menace until the shouts of the crowd died down, and then launched into a rousing oration. Special Blackshirt marching songs were composed, which today sound absurd and Germanic with their exhortatory earnestness and underlying violence. 'The streets are still, the final struggle's ended/Flushed with the fight we proudly hail the dawn/See! Over all the streets the fascist banners waving/Triumphant standard of a race reborn,' ran one favourite.

The first large march was on 1 July 1933. It began outside the Eatonry – a private tribute from the Leader – and wound its way via Whitehall down to the Embankment with the now familiar mixture of para-military discipline and what sounded suspiciously like a call to arms. Though the 1,000-strong band consisted mostly of young and enthusiastic Blackshirts it included 100 girls, also in black shirts but with grey or black skirts. All were inspected by Mosley before setting off. At the Cenotaph the command 'Eyes right!' was given and the officers gave the fascist salute.

There were already several small fascist groups in Britain and from time to time amalgamation had been suggested. But this had always foundered on ideological or financial differences between the organisers. Mosley was determined that any fascist movement in Britain would be headed by him and the march on 1 July was intended as a pre-emptive statement that the BUF was *the* fascist party, and Mosley the only Leader. He first tried negotiating with various of these groups, but had as little success as his predecessors. Soon he had the excuse for more direct action.

On 23 July Diana had a narrow escape at a communist rally in Hyde Park. It was the only time she opened her mouth at a public meeting. The Nazis were already persecuting communists and those with left-wing sympathies. Speaker after speaker recommended cutting trade links with Germany as a protest. Diana first heckled the speakers, then, when some of the crowd started singing 'God Save the King', held up her arm in a fascist salute. It was a silly and provocative gesture and the people at the meeting rounded on her but luckily for her a tough young Mosley supporter pulled her to safety.

A few hours later, a van carrying members of one of the smaller, rival groups, the British Fascists Ltd, drove past BUF headquarters shouting abuse. This 'provocation' was exactly the pretext the BUF needed for a violent attack on their smaller rival. A fifty-strong mob of Blackshirts rushed round to the headquarters of British Fascists Ltd in Harrington Road, smashed the basement and ground-floor windows and poured in. Rapidly and systematically, they wrecked the offices. It was done with brutal efficiency and so fast that, although the police were summoned as the first Blackshirts climbed through the broken windows, by the time they arrived a few minutes later all the invaders had disappeared. Two of the leaders of the British Fascists Ltd had to be taken to hospital, a man with injuries caused by a 'fluid' thrown into his face and a woman beaten about the head. The cost of putting the damage right coupled with the fear of a repeat performance contributed substantially to the liquidation of British Fascists Ltd shortly afterwards.

There was in any event a drift away from these smaller organisations to the BUF. For the young men who formed the bulk of the membership, Mosley's militaristic approach, magnetic personality and sense of theatre were far more glamorous. Implicit in the camaraderie, the discipline, the uniform and the sense of being one of a tough, well-drilled cadre was the promise of aggression and battle. Like soldiers, they were kept physically active, with games, jujitsu and drilling. The marches, the rousing and violent, if ludicrous, songs which accompanied them, all contributed to the idea of a political movement literally on the move – and ready, if need be, to fight for its goals and ideals. One who joined the BUF at that time was William Joyce (better known later as Lord Haw Haw), who served for three years as Mosley's propaganda chief.

English Fascism, it should be repeated, did not then (in the early 1930s) have the connotations it rapidly assumed under the Third Reich. Many intelligent people felt that the conditions imposed upon Germany by the

Treaty of Versailles were far too oppressive. If Hitler, who achieved the Chancellorship of Germany on 30 January 1933, could put Germany back on her feet again, the general feeling among many was good luck to him. In Geneva the Disarmament Conference ground slowly on; at home the pacifist movement was growing. Britain was still in the grip of the Depression, preoccupied with its own woes and inward-looking. Unemployment had barely started to fall. Many travellers who returned from Germany did so with stories of a country given new heart again, of a lowering of the pervasive unemployment, of a mood of optimism.

At that time, Hitler's most uncompromising statements were often buried within turgid reams of German prose. As early as 1931 he had said (in his *Guide and Instructional Letter for Functionaries of the National Socialists*, published on 15 March): 'The natural hostility of the peasant against the Jews, and his hostility against the Freemasons as a servant of the Jews, must be worked up to a frenzy.' Few in Britain would have been aware of such a vicious piece of political dogma – or have taken it seriously. In general, it was the British left wing, and particularly those on the far left – themselves often regarded with suspicion as 'Bolsheviks' – who were most conscious of the dangers. But disturbing rumours were beginning to trickle out. Harold Nicolson, lunching at the Café Royal on 4 May, heard from Peter Rodd, just back from Germany, that conditions there were of 'complete terror'.

One German in a hundred was Jewish. Think, Putzi told Diana, of the other 99 per cent of Germans who make up the *real* Germany. 'Hitler will build a great and prosperous Germany for us Germans. If the Jews don't like it – they can get out.'

It was enough for Diana. Like many people of her class and time, she had a vague mistrust of Jewish influence in high finance – even an ardent Zionist such as John Buchan made the villains in his novels 'sinister international Jewish financiers'. Diana, conscious, as she had been since the General Strike, of the ever-growing gulf between rich and poor, had a generalised antipathy to 'the City' – in which she knew there were a number of Jewish financial houses. 'One felt the City was feathering its nest while three million unemployed were starved,' she wrote later. 'You can have no conception of how totally divided the "two nations" were then.'

Much more powerful than her anti-semitic prejudice, however, was her predisposition in favour of Germany, common to many, perhaps most, upper-class English families until the outbreak of the First World War. Some were linked by blood or marriage to the German aristocracy, many were fluent German speakers. They venerated German poetry, philosophy and music and they sent their daughters to be 'finished' in the musical and cultural ambience of Munich and Dresden. It was a tradition that lingered on, aided by the fact that aristocratic German families impoverished by the First World War were delighted to take in the sons and daughters of those who had themselves completed their education in Germany.

In Diana's family, this tradition was particularly strong. Her paternal grandfather had adored the music of Wagner and been an intimate of the Wagner family. Her brother Tom, the greatest single influence on Diana's cultural judgement, was a fluent German speaker and had brought back rhapsodic accounts of the music, the people, the architecture and the beauty of Germany from his (pre-Hitler) visits there. In common with most of their friends and many politicians, both Diana and Tom believed that Germany had been unjustly penalised by the victorious Allies at Versailles. Now, when Hitler appeared to be achieving economic and social miracles in the country which she already admired through policies similar to those advocated by her beloved Leader, her ability to blot out unpalatable truths or questions came to the fore. The Nazis were a raw, young party unused to government; if a few people were harmed in the cause of the greater good – well, so be it.

As the summer of 1933 drew on, Nazi discrimination developed into straightforward brutality. The first concentration camps were founded; into them went those whom the Nazis considered political undesirables – Jews and communists, Jehovah's Witnesses and Freemasons, gypsies and homosexuals.

Flights and expulsions began as oppression moved into the arts, science and education. Jewish conductors such as Otto Klemperer and Bruno Walter were deprived of their batons, the music of the composer Kurt Weill was proscribed and Jewish actors were allowed only to play 'negative' roles – thieves, murderers, gangsters or swindlers. Teachers had to join the National Socialist Teachers' Union (Jews, of course, were excluded) and Jewish students as well as Jewish professors were driven out of universities; the experimental physicist James Franck, a Nobel prizewinner, was forced to take a 'voluntary' leave of absence because he was a Jew. At the University of Berlin there was the Burning of the Books – all those by Jewish authors were thrown into the flames. Amid the howling of the Nazis, the works of the world's greatest scientist, Albert Einstein (a Swiss subject), were reduced to ashes because of both his race and his left-wing sympathies.

Little of this was seen by most visitors to Germany. They were shown the new housing and the spanking new maternity clinics, where smiling blonde mothers brought healthy, flaxen-haired babies for treatment. They were told about the praiseworthy social welfare schemes, the new hope and happiness of the people.

In August Mosley and Baba went off on their planned motoring tour through France. The launching of this illicit liaison – if they had not physically been lovers before, they became so almost immediately – had the blessing of almost everyone in Baba's family except Baba's husband Fruity. Mosley's mother, the highly conventional Lady Mosley, terrified lest 'the horror' (as Diana was known among them), now divorced, would acquire an unshakable hold on Mosley, urged Baba to set off with him. Irene, who had felt 'sick with horror' when she discovered that Mosley had been dining in London with Diana, gave Fruity 'a good stern talking to in his jealousy and brutality to Baba over her seeing Tom', warning him that such behaviour would cause a deep rift between him and Baba. The unfortunate Fruity, no match for the Curzon sisters in full cry, had to subside. Irene herself took Mosley's two elder children, Vivien and Nicholas, on a cruise to Madeira and the Canary Islands; the youngest, Michael, went with his nanny to the Isle of Wight.

Diana, at a loose end, trying to believe that Mosley had no more than a fraternal interest in Baba but aware of Baba's infatuation with him, could not bear the thought of a month spent solitary and brooding in London. She had already been to Germany three times, once with Bryan and twice to visit Tom in Berlin; Putzi Hanfstaengl was back in Munich and she decided to take up his offer to introduce her to Hitler.

In one of the shifts of relationship so common between the Mitford sisters during their lives, Diana had become increasingly close to Unity. Her sister's arrival in London and her frequent visits to the Eatonry on her way home from her art classes in Vincent Square meant that they saw more of each other than ever before, while Unity's obvious admiration for Mosley reinforced their alliance. Diana now suggested that they take a holiday in Bavaria together. In Munich, she said to Unity, they could go to museums, look at the architecture and listen to the music for which the city was famous.

Unity, who had hoped that Diana would take her to France, was won over when she heard of Putzi Hanfstaengl's promised introduction to Hitler. Unity asked for nothing more.

While Bryan, at Pool Place, was writing, 'I hardly dare look out of the windows or go for a walk. Everywhere is haunted – the ditches, the hedges, the ploughed fields, the Downs in the distance, the sea near at hand,' Diana was in Munich, trying to contact Putzi. He proved elusive, but eventually she managed to track him down. He told the sisters that if they wanted to meet Hitler they should come to Nuremberg for the Parteitag (Party Congress) and he would meet them at the station.

Putzi noticed immediately that the two attractive young women were made-up to the eyebrows in a manner which conflicted directly with the newly proclaimed Nazi ideal of German womanhood – the blonde, blue-eyed fräulein with face bare of all cosmetics, a sharp contrast to the heavily lipsticked mouth fashionable in the 1930s.

On their way to the Deutscher Hof Hotel, where Hitler was staying, there were so many frank comments from passers-by that the embarrassed Putzi had to duck behind a building with the girls. He pulled out a large, clean handkerchief and said, 'My dears, it is no good – to stand any hope of meeting him you will have to wipe some of that stuff off your faces.'

They did so, but it was only a temporary yielding. 'It's no good,' exclaimed Unity as they sat waiting, 'I can't possibly do without lipstick.'

Several times daily during the four days the Parteitag lasted she asked Putzi when he would introduce them to Hitler, but the nearest they got was the lounge of the small hotel where he was staying. Official after official passed by their table until eventually Hitler sent word by Hess that he was too busy to come out and meet them.

The Parteitag, which began on 1 September, was the first major Nazi rally after Hitler came to power. Many of his followers had believed that it would be the last, as the so-called 'immutable points' of Party doctrine had proclaimed for twelve years that, once power was achieved and consolidated, the Party could be dissolved. But the theme of Hitler's speech, 'The State is the Party and the Party is the State', was an ominous pointer to what was to come. There were few foreigners to hear it among the 1,000 privileged guests in the main grandstand; Great Britain, France, Poland, Russia, Spain, Czechoslovakia, Holland had all refused invitations to send an official representative.

On the ancient town of Nuremberg almost 400,000 Party members converged, brought from all over Germany by 400 special trains. There were 80,000 Hitler Youth alone – a smallish proportion of the 3.5 million boys aged 12–18 already enrolled. Files of brown-uniformed Storm Troopers, knapsacks on their backs, marched from the station to the seven enormous camps set up around the town. Five hundred men slept in each huge tent. The streets were hung with bunting and blood-red Nazi banners emblazoned with black swastikas. The tread of marching feet was everywhere, the streets were filled with the sound of military bands. Uniformed men rapped out staccato 'Heil Hitler!' salutes as they passed each other. The parades, the speeches, the marching were ominous, terrifying and relentless, part revolutionary triumph, part political circus, but mostly, and unmistakably, a display of militaristic might. After it, Duff Cooper said, 'The whole of that country is preparing for war on a scale and with an enthusiasm that are astounding and terrible.'

Hitler's speech was on Sunday 3 September. Diana, unable to understand a word, recorded that an 'electric shock' ran through the multitude at his appearance. It was a dizzying atmosphere: the thunderous, rolling cheers, the waves of crashing, reverberating Sieg Heils, took on the incantatory quality of obeisance to a godhead. Unity, wildly excited, determined to return the following year and, somehow, to meet her hero then.

The sisters parted, Unity back to the confines of Swinbrook, Diana travelling on to Rome to stay with her friend Lord Berners.

Diana's parents were under no illusions about the Nazis. They were appalled and furious that she had taken Unity to the Parteitag. 'I suppose you know without being told how absolutely horrified Muv and I were to think of you and Bobo accepting any form of hospitality from people we regard as a murderous gang of pests,' David wrote to her on 7 September. 'That you should associate yourself with such people is a source of utter misery to us.'

Diana, reading these trenchant and uncompromising views in the golden Italian sunshine, was unmoved. She was still on extremely bad terms with her parents and if anything glad that she had annoyed them; and she discounted all their views as old-fashioned and prejudiced. She had found the Nuremberg Rally not only extraordinarily interesting but living evidence that fascism, the chosen political creed of her beloved Leader, could replace despair and chaos with hope and efficiency. What she saw of Italy only confirmed her in her view of the benefits of fascism: here was another country in which it was a proven success.

In any case, she was far too busy absorbing the beauties and splendours of Rome, to which the cultured and witty Gerald Berners was the perfect guide. Every afternoon, after he had finished his morning's work, he would take her sightseeing. They walked or drove from one beautiful building to another, or wandered through museums and galleries. In the evenings, there would be dinner at some splendid villa: Lord Berners was a popular figure in Roman society, enjoying its combination of style and vivid, catty gossip. The visit concluded with a leisurely drive to Paris at the beginning of October.

While Diana was in Rome, a small but significant book, the *Brown Book of the Hitler Terror*, compiled by a body called the World Committee for the Victims of German Fascism and with an introduction by Lord Morley, was published by Victor Gollancz, a respected publisher. It described how many of the top German scientists had been driven out either for left-wing leanings or for 'reasons of ancestry', and it detailed acts of brutality and torture, often with accompanying photographs of the victims. An appendix listed 250 murders perpetrated by the Nazis since 3 March 1933. The Mitford family reacted to it according to their political leanings: Jessica seized upon it, while Unity and Diana ignored or dismissed it.

Diana also dismissed, though affectionately, Bryan's pathetic little note from Swan Court on 6 October, the first of many urging her to return to him. 'If the *situation* ever changes you will let me know, won't you? A wire saying "Come" would enable me to make preparations to take you to China. All this is madness because it doesn't arise and is only a way of sending my love.'

That autumn the BUF bought the lease of the Whitelands Teachers' Training College in Chelsea and turned it into their headquarters, renaming it the Black House. As well as the Party administrative offices, it housed a small printing press which turned out BUF posters and leaflets, a theatre and a large restaurant and canteen. These were always crowded, since the Black House also served as a barracks for young Blackshirts, the number varying between 50 and 200, who lived there under strict military-style discipline, drilling in the large courtyard.

Steadily the Party hierarchy took shape, attracting further followers. One of these was John Beckett, by now coming round to fascist doctrines. His final conversion occurred when Robert Forgan, a man he much admired who was now Deputy Leader of the BUF and its Director of Organisation, assured him that Mosley 'took his mission extremely seriously and was now grown up'.

Mosley's Blackshirts were becoming ever more pugilistic towards what they saw as 'the opposition'. While Harold Nicolson was learning at the Travellers Club that 'the anti-Jewish atrocities in Germany were far worse than even the American papers supposed', Mosley's young Blackshirts were breaking up a meeting of the Imperial Fascist League, the last British fascist group which could be viewed as serious competition.

The fracas was quickly over. With rubber truncheons, knuckledusters and chairs, these fit, well-trained young men made short work of their so-called political allies. Henceforth it was clear that any group which wanted to promote fascism in Britain had to put itself under Mosley's leadership.

Chapter XIV

Nancy, who had suffered desperately when Hamish had jilted her, was in love again within weeks. This time, the object of her affections was infinitely more suitable in one respect. There was no doubt, no doubt at all, that Peter Rodd was heterosexual.

He was a clever, good-looking man of impeccable family: his father, Lord Rennell, was a successful diplomat, his mother was a woman of formidable uprightness. Peter was a great disappointment to them. He was sacked from job after job, he was constantly in debt and his frequent love affairs ended in tears. Possibly his worst drawback from a Mitford point of view was that, although intelligent and well informed, he was a bore; conversation to Peter meant the lengthy imparting of the copious facts with which his mind was so well stocked.

But for Nancy, capable of endless self-deception in her search for love, he appeared like an answer to prayer. During the last miserable months of the Hamish affair she was constantly at the Eatonry: the sight of Diana, carelessly chucking away a glamorous life with one adoring man, exuding the physical glow of shared passion with another, hugging and cuddling her two small boys, must have brought her own single, solitary and childless state sharply home to her. She longed to get married and have children, but though she had had a number of love affairs ('shop-soiled' was how Peter's sister described her) all had been unsatisfactory. Despite her looks, her *savoir-faire*, her funniness and wit, Nancy lacked Diana's fundamental self-assurance, while neither friends nor family were safe from her malicious barbs. On the whole her happiest relationships were with the homosexuals who surrounded her in a kind of *cordon sanitaire*. Now, with Peter Rodd, she was going to be part of the married world to which she had always aspired – and free finally, from Swinbrook and all that it stood for. The moment she was engaged (on 18 July) she left home, going to the room Diana gave her at the Eatonry.

It was perhaps hardly surprising that Nancy briefly became an ardent fascist. If gratitude for Diana's support and kindness had not accounted for it, there was the influence of Peter Rodd. Thrilled by the movement's declared intentions of helping both the poor and the unemployed, he had already joined the BUF. He and Nancy bought black shirts, fervently embraced fascist dogma and attended meetings, clapping enthusiastically.

One of the biggest of these meetings was at Oxford Town Hall on 3 November 1933. Mosley had chosen the university town because he was anxious to capture the country's youth. For some time the BUF had been establishing youth sections, divided into senior and junior groups, and these had now begun to penetrate the public school system. Youth Groups had been established in Winchester, Stowe, Beaumont and Worksop College. It is difficult to avoid drawing a parallel with the Hitler Jugend. But what Mosley may not have known was that the Home Office was carefully monitoring such groups as BUF membership grew.*

As the audience filed in to the Oxford meeting, one of the 100-odd Blackshirt stewards ran his hands over the pockets of anyone who looked as if he might be carrying some form of hidden weapon and, as soon as the meeting had started, any interruptions were summarily dealt with by forcible ejection from the hall. Nancy and Unity were there. 'The hall was full of Oxfordshire Conservatives who sat in hostile and phlegmatic silence,' wrote Nancy to Diana, adding that there had been several fascinating fights, as the Leader had 'brought along a few Neanderthal men with him and they fell tooth and (literally) nail on anyone who shifted his chair or coughed'.

The tough, well-armed Blackshirts invariably got the best of such scrimmages, attacking individual offenders in groups of two or three. At the Oxford meeting, heads were banged on stone floors, faces gashed with knuckledusters and objectors thrown down the stairs. Two of the victims needed hospital treatment and the President of Ruskin College took several sworn affidavits from undergraduates who said they had been roughly handled.

Nancy was married on 5 December at St John's, Smith Square; Jonathan and Desmond, aged two and three and wrapped in cashmere shawls against

* A Home Office memorandum of 15 February 1934 gives the names of the boys in charge of these groups at their respective schools.

the cold, were two of ten little pages dressed in white satin. Nancy and Peter, who had very little money – she always claimed they lived on her Bridge winnings – took a small house in Strand-on-the-Green.

At the same time, Diana became pregnant. Abortion was a crime, but for a pregnancy only a few weeks advanced an understanding gynaecologist would recommend curettage, which required only two or three days in a nursing home. It was this that Diana now underwent. The last thing that either she or Mosley wanted at this stage was an unplanned baby. Still on bad terms with her parents, she told her mother only that she was having a minor operation.

She recovered to what for her were two pieces of good news. On 15 January 1934 her divorce from Bryan became final and, on the same day, Lord Rothermere, proprietor of the *Daily Mail*, publicly offered his support to the BUF with a signed article in his paper. 'Hurrah for the Blackshirts!' ran the headline. 'Because fascism comes from Italy, shortsighted people in this country think they show a sturdy national spirit by deriding it,' he began, before continuing that in every country fascism stood for the party of youth. 'The Blackshirt movement is the organised effort of the younger generation to break the stranglehold which senile politicians have so long maintained on our public affairs.'

He followed this up with a full-page blast on 22 January. 'Give the Blackshirts a Helping Hand!' thundered the *Mail* and its owner. 'Sheer ignorance is an obstacle that every new political development has to face at its start. The hysterical abuse and misrepresentation at present being poured out on the British Blackshirt movement have had their counterpart at each successive stage of our constitutional progress.' In Germany and Italy, he said, he had seen peace and prosperity, confidence and enthusiasm. 'Our young men and women are no less patriotic, courageous and capable of organised effort . . . I appeal to them not just to stand idly by but to take a share in this great task by giving the Blackshirts a helping hand.'

In conclusion, he declared that the Blackshirts would respect the principles of tolerance which were traditional in British politics. 'They have no prejudice either of class or race.' It was an opinion he would soon change.

The backing of an influential national newspaper was exactly what Mosley needed if he was to have a chance of securing votes in a general election. So far, the BUF's electoral record had been dismal; not even a local councillor had been elected. If political power was to be more than a mirage, seats in the House of Commons were essential – and seats in sufficient number to bring about the sweeping changes he envisaged. The ten points of his fascist policy (see Appendix A) spoke not only of patriotism, the need for everyone to serve the country and the expansion of trade with the Empire (it was these points, coupled with Mosley's promise to combat unemployment, which had attracted Lord Rothermere) but also of state intervention and protection on a grand scale ('the total exclusion of foreign goods . . . no alien will enter this country and take a British job and Blackshirts will deal not only with poor aliens but also the great alien financiers of the City of London'). There was a hint only of Mosley's plan to change the format of government by reducing the parliamentary process to an Inner Cabinet of five – with himself, naturally, as the unquestioned Leader.

Soon after Lord Rothermere's heartening support of the Blackshirts, Mosley had had a recurrence of the phlebitis which troubled him all his life. This time it was so severe that he was temporarily incapacitated and his doctors advised rest. He went with Diana to the warmth of Provence, staying in a rented house, to recover.

Although Diana's family knew where she was, Mosley's had no idea that she was with him. He was still conducting a raging affair with Baba and wanted his sisters-in-law to think that he was travelling by himself. Because of the secrecy in which he had visited Diana over the past six months between the decree nisi and the decree absolute, Irene and, more importantly, Baba had been lulled into the agreeable belief that he had given up any serious thoughts of Diana. Both the Curzon sisters were convinced that Mosley was now obsessed with Baba, who had frequently expressed her loathing of Diana and her conviction that no one would take him seriously as a future leader if it were known that Diana had left her husband for him and that their affair was continuing. Mosley reassured her, telling her how her suspicions hurt him and that he dined with Diana only occasionally, and then platonically.

It was a tribute to Mosley's unscrupulousness, charm and powers of dissimulation that he was able to juggle the emotions of the four women closest to him, at least two of whom were deeply in love with him. All wanted to help the man at the centre of their lives in any way they could. Irene, anxious to show loyalty ('I cannot get over Tom's consideration to Baba'), presided at fascist fund-raising meetings; his mother was head of the Women's Section. The atmosphere at Savehay was one of domestic calm, in which even the sight of Mosley walking across the lawns with his arm round Baba was accepted by Irene: *anything* was better than 'the

Guinness'. All this would be jeopardised if any of them knew that, far from giving up Diana, increasingly she was becoming part of his life.

In pursuit of the clandestine, no detail was too small to overlook. Unity, for instance, was forbidden to wear the fascist badge Mosley had given her: the faithful could identify it as a personal present and mark of special favour from the Leader and thus it might lead back to Diana. Cimmie's friends, however, were not quite so wide-eyed as her sisters. 'Went to see Baba and made mischief, deliberately, with her about Diana Guinness,' wrote Georgia Sitwell in her diary that spring.

To Diana herself Mosley explained, with truth, that he had a horror of her becoming actively involved in politics. He felt, he told her, that Cimmie might have died partly because her health had been weakened by constant rushing round to meetings, speaking on draughty platforms and travelling in all weathers. He cited the miscarriage Cimmie had suffered two days after being elected to parliament and he implored Diana to stay outside his political life. She was only too happy to comply; though fascinated by politics, by now believing unquestioningly in everything he said, the idea of making a speech or campaigning filled her with horror.

She had plenty to fill her time. She listened to music with her brother Tom, and went to the cinema with friends. She stayed with Lord Berners at Faringdon, where the pigeons were dyed saffron yellow, bright pink and turquoise. The Russian mosaicist Boris Anrep, one of the Bloomsbury circle, depicted her as Polymnia (the muse of sacred music and oratory) in his group of the Nine Muses on the floor of the National Gallery, and artists clamoured to paint her. That winter Henry Lamb painted her portrait, calling it 'Lady in Blue'. It hung at the Leicester Galleries ('everyone recognised that Mrs Bryan Guinness, who spent the greater part of the afternoon at the exhibition, is the model', reported the *Evening Standard*). Adrian Daintrey, who introduced her to Proust, also painted her, as did Tchelichew, brought over to England to design the sets and costumes for a ballet. The Tchelichew portrait, which depicts Diana and her children with flowing golden hair, blue eyes and blue shadows on their pale gold faces, proved to be her favourite.

It was on the visit to France that Diana acquired the pet name that Mosley used all his life. Driving past a field, he pointed at a chestnut carthorse with blond mane and tail and said, 'Look at that horse with a yellow mane, just like yours.' It was, she told him, a Percheron; from then on, he often called her Percher (pronounced Persher).

* * *

The lovers returned to a scene full of promise. BUF membership was now up to 17,000 and rising. Mosley himself felt restored to full health and ready for the arduous round of speeches, meetings, heckling – and, increasingly, violence.

At larger meetings, Mosley was usually the speaker. Others were W.E.D. Allen, Robert Forgan and William Joyce, whose oratory at a meeting at Paddington Baths so impressed John Beckett that 'I ordered a Blackshirt uniform from Forgan's tailor'. Beckett himself was soon co-opted as a fascist speaker and member of the executive, on a salary of £5 a week. Most of the other officials were of a much lower calibre: men with little or no political experience who spoke at street corners or in local halls for a weekly wage of anything from ten shillings and their keep (at the Black House) to £2.

Many of the meetings were characterised by confrontations with the communists. Shouting and heckling were commonplace, but as BUF membership grew these gave way to more forceful tactics. Sometimes Red Front members would obtain seats in the front row, bursting into the Internationale as the BUF speaker strode to the platform and chanting it continuously in order to prevent him being heard. When Blackshirt stewards came to eject them, fights would break out.

On 7 June 1934 the biggest and most important meeting to date took place in the huge Exhibition Hall at Olympia. Diana had given a dinner party for it, asking Nancy, Vivien Jackson, Lord Berners and the young man who was living with him, Robert Heber-Percy. To her distress, a sudden attack of gastritis prevented her going with them to the meeting.

She missed a dramatic few hours. At first everything seemed as usual, with blackshirted stewards standing at the end of every row of seats, turning this way and that as they watched for any sign of 'trouble' in the 12,000-strong audience (this included 2,000 Blackshirts, about half of whom were stewards). There were also 750 police in reserve: Special Branch believed that Mosley's communist opponents were planning to throw the main power switch at a crucial moment, blacking out microphone and lights and probably causing chaos, riots and a stampede.

As the minutes passed, expectation built up. Suddenly, fanfares sounded from hidden trumpets and arc lights flooded the central aisle. The fascist anthem struck up and Mosley, flanked by four blond young Blackshirts, strode into the brilliant pathway of white light. He was followed by a platoon of Blackshirts carrying banners. As the crowd cheered and shouted, arms upraised in the fascist salute, the procession marched slowly to the daïs.

Here Mosley, one thumb tucked negligently into his wide black belt, chin raised as he gazed commandingly round the hall, waited until the noise had died down before his own arm shot up in salute.

Just as his mouth opened to speak came the interruption. Three young men and a girl standing with them began to chant: 'Hitler and Mosley, what are they for?/Thuggery, buggery, hunger and war.' It was the anti-fascist response to the Blackshirt chant 'Two, four, six, eight/Who do we appreciate?/M.O.S.L.E.Y. Mosley.'

The quartet had scarcely shouted the last word before they were submerged in a sea of black-clad stewards. The girl screamed, and fighting began. Many of the audience had come prepared for violent hostility to the BUF. Philip Toynbee and Esmond Romilly had both bought knuckledusters that afternoon from an ironmonger in Drury Lane. Now they threw themselves on the stewards' backs. They were quickly overwhelmed and separated; Toynbee, bruised and bloodied, was thrown out into the street, where he was picked up by communist sympathisers and taken to a makeshift first-aid post. About sixty other people were taken to hospital, where medical evidence pointed to the use of knuckledusters and razors.

The Olympia meeting marked the peak of the BUF's success in Britain. After it Mosley was even invited by the BBC to speak on the wireless. It was the last time for thirty-four years that such an invitation would be extended.

For Nancy the violence she had witnessed caused a complete volte-face. Overnight her fascist convictions fell away. By the time an article she had written a few weeks earlier, extolling fascism as 'the hope of the future', appeared in the July issue of *Vanguard*, Nancy was busy explaining the perniciousness of its ideology. True to form, she immediately began to use everything she had learned, filtered through the prism of ridicule, as material for her next novel.

Nancy's defection heralded a much more important one. Lord Rothermere had become alarmed by the worsening public image of the BUF, exacerbated by the bad publicity arising from the Olympia meeting. Now he withdrew the support that had been so crucial. The correspondence which he printed in the *Daily Mail* of 19 July was friendly, but clearly marked an irrevocable break.

Mosley's letter, dated 12 July, appears to be an attempt to put the best possible gloss on the divergences which had appeared in their private conversations. 'We Blackshirts are fascists. We hold the new creed of the modern world . . . it is our task to convert the British people to the new faith. You, on the other hand, are a Conservative and would like to see a revived Conservative Party.' He went on to state that no Jew was admitted to membership of the BUF because 'they have bitterly attacked us and they have organised as an international movement, setting their racial interests above the national interests'.

Rothermere's reply, on 14 July, was a polite but vigorous rebuttal of the Mosley version of fascism.

> As you know, I have never thought that a movement calling itself 'Fascist' could be successful in this country, and I have also made it quite clear in my conversations with you that I never could support any movement with an anti-semitic bias, any movement which had dictatorship as one of its objectives, or any movement which will substitute a 'Corporate State' for the Parliamentary institutions of this country.

His assistance, he declared, had been given in the hope that Mosley would be prepared to ally himself with the Conservative forces to defeat socialism in the next election. He concluded with a final broadside:

> I have never thought that the political situation here bears any resemblance to the political situation in Italy or Germany. In each of these countries Parliamentary institutions were largely of exotic growth, whereas in England they have, since the time of Queen Elizabeth, exercised the real decisive influence.

Rothermere's public rejection was a devastating blow. If not the *coup de grâce*, it was certainly the turning point in the fortunes of the BUF. Membership, which with the assistance of the *Daily Mail* had reached a high point of 50,000, immediately began to decline. Mosley's Jewish opponents were confirmed in their antagonism and the trades unions in their belief that if Mosley gained power he would suppress them – especially since his more extreme speakers, such as William Joyce, were already stating that freedom of speech could not be tolerated in a fascist state. Mosley himself told his followers that Rothermere had withdrawn his support because some Jewish advertisers had intimated that they would no longer advertise with him if he supported British Union, and fascist graffiti declaring 'Jews want War' were daubed on walls. With this polarisation came an increase in violent confrontations and the loss of many of the more moderate members. Equally inevitably, those who did join were often those for whom opportunities for physical aggression were the main attraction.

August brought a welcome switch to private life. It was time for Mosley to resume his parental role. Every year, he and Cimmie had taken a large house by the sea somewhere in Europe, usually on the Mediterranean, where they spent a month with the children. This August, Baba and Irene, with the assistance of Mosley's mother, had settled on Toulon. They found a white house with a large terrace, above a rocky shore.

The Toulon house saw the first cracks in the domestic contentment so carefully established at Savehay. When the Curzon sisters learned that Diana was to spend the first two weeks there before Baba arrived for the remainder of the holiday, there were furious and emotional scenes, tears and reproaches, while Mosley protested that she was 'just a friend'.

For the Mosley children, it was their first meeting with Diana on anything other than a fleeting basis. That first holiday none of them was aware that she stood in any special relationship to their father – weekends at home, and all their family holidays, had always been crowded with their parents' grown-up friends. When she arrived, accompanied by her maid, they thought of her as no more than 'Dad's friend Mrs Guinness'. She left the day before Baba's arrival, flying to Ravello to stay with her art-connoisseur friend Edward James in the Villa Cimbrone, where they swam, visited the Greek temples at Paestum and dined in the garden under the stars.

She returned to London in time for the arrival of her children from Ireland; they had been staying with Bryan at Knockmaroon, where he was entertaining the usual August houseparty. 'I wonder what Pansy and Dig [Yorke] reported to you,' he wrote. 'I long to see you and tell you all the stories.' Pansy Lamb wrote to say how well the children were and that 'we only went to the theatre once, and that was made tolerable by a delicious dinner Mr Peake, one of the brewers, gave us at the Kildare Street Club. It was flowing with champagne and delicacies and we all went to the Abbey pretty tight.'

Diana was a loving and conscientious mother who spent more time than was then usual with her two boys. When they were small they came into her bed every morning, where she taught them both to read. She had tea with them every day and often lunch as well and she would go up to the nursery and play with them during odd hours in the day. On their nanny's day out she bathed and put them both to bed instead of – as was the usual custom – leaving this to the nurserymaid. If she had a fault as a mother, it was favouritism; it was impossible for her to conceal that Jonathan, her first-born, was her most adored. Fortunately, Nanny's passion was for Desmond.

Bryan and Diana had been constantly in touch since their divorce. He longed to see her; she was determined to retain his friendship. He confided in her about the book he was writing, she found him a cook. He asked her if she liked the sound of the birthday present he had chosen for her, 'a watercolour by F. Nicholson, circa 1830 . . . an eighteenth century kind of a view of a ruin, with bridges and people on the bank, lovely trees hanging richly over them all'; if not, he would choose her something else. 'I must only add one thing: that you are still the only person I know who makes me cry when you go away in the train.'

It was the same when, wretched and lonely, he sought diversion in travelling. 'I am very lonely and sad,' he wrote in May 1934 from the *Empress of Australia*.

> There is no one at all to be in love with, and the journey seems to stretch endlessly ahead. The ship is comfortable in the extreme and the food excellent but there is no one to be fond of or even to admire. I miss you so much. Sometimes they play our tune from Carmen at dinner, and I water the soup with my tears.

Earlier, back at home, he wrote to her from Swan Court on 6 September to say how sad he was to miss her by such a narrow margin. 'I expect you are now at Nuremberg.' She was.

Chapter XV

Diana had left Ravello at the end of August. Instead of going home, she went direct to Munich, to join Unity there. The sisters had found the first Nazi Party Congress the previous year so exciting that they were determined to go again. The 1934 Parteitag was to take place in the first week of September.

Since her visit to Munich in the autumn of 1933, Unity had bombarded her parents with requests to allow her to go back. The Redesdales were naturally unwilling: it was an era when most girls lived at home until they were married, and Unity was only nineteen. But she said, with truth, that she longed to learn German, a plan of which both Sydney and David approved. From Unity's passionately single-minded viewpoint, this was preparation for the hoped-for moment when she would meet her hero (Hitler did not speak a word of any other language) rather than cultural aspiration. Accordingly, in May 1934, she was installed by Sydney in the house of a respectable elderly baroness who made ends meet by taking in young foreign girls of good family.

Unity soon had a circle of friends. As well as the other English girls staying with the baroness, she met young Germans through Putzi Hanfstaengl's sister Erna. She also met a young artist, Derek Hill, who had been in Munich studying theatre design since the previous autumn. Pam, who had known Derek Hill when he was a schoolboy, had told him to look up Unity, and the two of them often went sightseeing or walking in the mountains together.

It was through Derek that Unity had her first sight of Hitler at close quarters – at six p.m. on 11 June in the Carlton Teeraum.

Derek knew that on Mondays Hitler often patronised these tearooms on his way from Berchtesgaden to Berlin. When Derek's mother and an elderly aunt from Scotland were visiting him in Munich, they asked if there were any chances of seeing 'this new man everyone's talking about'. He told them the best chance was at the Carlton Teeraum and took them there the following Monday.

The small party sat sipping tea and waiting. Suddenly, when they had almost given up hope, Hitler, followed by Goebbels, Hess and various henchmen, swung into the long room. Derek, who like all her friends knew of Unity's obsession, went straight to the telephone. 'The Führer's here – if you want to look at him you'd better come quick.'

Unity was pathetically grateful. 'This is the kindest thing that's ever been done for me in my life. I'll never forget it.' She rushed out, jumped into a taxi and arrived in a state of breathless excitement. She was trembling so much as she stared at Hitler that she was unable to drink her chocolate and had to hand her cup to Derek. Later, he was to recall a more bizarre manifestation of Hitler's extraordinary charisma: his mother and aunt, strongminded, apolitical Scotswomen, were so affected by this sight of the Führer that they gave the Nazi salute as they left the Teeraum.

This glimpse of Hitler among the palms and teacups of the Carlton so increased Unity's obsessional devotion that henceforth she would ascribe to him a kind of saintliness. Even his brutal massacre of his close friend and associate, Ernst Röhm, and eighty-odd other leaders of the SA (on what became known as the Night of the Long Knives less than three weeks later on 30 June) elicited from Unity only pity for the Führer. 'It must have been so dreadful for Hitler when he arrested Röhm himself and tore off his decorations,' she wrote. On Röhm's photographs in her scrapbook she drew black crosses and scrawled the word 'Schwein!'

By the time Diana arrived in Munich, Unity, back there again after the summer holidays at Swinbrook, spoke German with fluency. Though this year there was no Putzi to help them get into events during the Parteitag, they had the luck to run into one of the original members of the Party, whereupon Unity was able to tell him their problem. He solved it by sending them with a note to the accommodation office of the Parteitag, which fixed them up with the necessary tickets and a room in a small hotel.

If the first Party Congress had impressed the sisters, the second was to do so more profoundly still. It was the one featured in Leni Riefenstahl's film *Triumph of the Will*, made to celebrate the glories of a risen Germany embarking on its thousand-year Reich.

From its opening shots of the Leader's aeroplane, descending through the clouds like the silver chariot of some Teutonic hero-god to the strains of Wagner, there was but one theme: the invincibility of Germany. The eagles, the spearlike flagstaffs with red and black banners hanging from them, the menacing, helmeted figures silhouetted against firelit darkness, the juxtaposition of stillness and movement, the marching columns, their goosestepping suggestive of a robotic implacability trampling all beneath it – all declared an unconquerable Master Race. When Hess, in one of the speeches heralding the Leader, proclaimed, 'A country that does not maintain the purity of its race perishes,' a hundred thousand throats roared their belief in Aryan supremacy.

The Parteitag opened at eleven a.m. on Wednesday 5 September, in the Luitpold Hall. Its huge auditorium was decorated in the colours of blood and fire, against a background of sacrificial white. The pillars supporting the high roof were draped in red, with twining yellow and orange artificial flowers. White canvas covered the walls and festooned the roof. To stirring marches, banner after scarlet banner was borne in, each inscribed with the name of a different German city, until all were ranged under a huge red flag, a black swastika at its centre, hanging on the far wall. As each flag appeared, the 30,000 audience rose with the Nazi salute and a shout that drowned the music.

Hitler, of course, was well aware of the power of uniforms and hierarchy in a country where even town hall officials were known by their titles. He understood the potency of symbols, of dramatic momentum – he never spoke until one or more of his associates had already roused the crowd to a frenzy – and, knowing the German temperament, he did not fear the ridicule often evoked elsewhere by fascist theatricality.

Every appearance was carefully staged. As the Führer entered from the back of the auditorium, powerful spotlights cut through the darkness to focus brilliant light on his solitary figure. To a fanfare of trumpets and accompanied at a respectful distance by Goering, Goebbels and Hess, he marched to his place on the platform. As he raised his right arm in salute, arc lights sprang on, bathing the white canvas in dazzling light, glittering from polished buttons, from eyes shining with excitement, from the flash of teeth in mouths open and shouting. 'Heil Hitler!' thundered out as from a single, gigantic throat. The next moment, the orchestra's conductor, silver hair gleaming against his chocolate dress suit, raised his baton for the overture to *Die Meistersinger*, Hitler's favourite piece of music.

The parades outside, in the Zeppelin Meadow three miles from the centre of the city, were equally dramatic. All day long columns of men marched, wheeled, saluted, listened, roared their loyalty and finally marched back to Nuremberg in torchlit procession. All were in uniform –

Storm Troopers, Nazi officials, detachments from the army and navy and the so-called civilians. First came the Soldiers of Labour, an ostensibly peaceful force who learned drill along with digging, their spades taking the place of rifles. As Hitler saluted, the command 'Shoulder your spades!' rang out and 52,000 blades flashed as they rose from foot to shoulder of the earth-brown Labour uniform.

On the last day, Hitler arrived at twilight, taking his place on the grandstand in a blaze of white light. On the far slope of the huge field, powerful searchlights lit up a force of 181,000 men. Every ninth man held a scarlet banner. As the silent army advanced to fill the field, the flagbearers marched through them and forward until they arrived beneath the podium on which Hitler stood. Here, vivid as a pool of blood in the spotlights, the flags were dipped in salute to the fallen as the poignant, mournful air 'Ich hatt' einen Kameraden' stole out through the growing darkness. Then came Hitler's voice, magnified by loudspeakers, speaking of Germany's unconquerable strength and declaring that she would willingly hold her hand out to all those foreign nations who would accept it. It was euphoric, it was overwhelming, it was like being caught up in a whirlwind.

For Diana, enthralled by the spectacle, touched by the pulsing electricity of excitement but frustratingly unable to understand the German language, the scenes she had just witnessed crystallised a growing urge to find some meaningful occupation to fill her life while Mosley stumped the country on speaking engagements. With his encouragement, she decided to learn German. This also meant that she would see plenty of Unity who was now the sister to whom she was closest. Munich was cheap, there was opera, there was skiing, there were museums. Diana took a flat, complete with Biedermayer furniture and a cook found for her by Unity, who moved in straight away and enrolled in a German course for foreigners at Munich university.

While Diana was settling into her Munich life, the BUF was holding a Blackshirt Rally in Hyde Park. Although it was the largest such open-air parade they had yet mounted, Mosley was uneasily conscious that, now that Rothermere had withdrawn his support, membership was declining. It was time to inject fresh impetus – in the form of another leaf torn out of the German book. There was no better moment to do this than at the next large meeting: at eight p.m. on 28 October at the Royal Albert Hall.

A large audience had gathered to hear the Leader speak. The BUF had sent letters to potential supporters, inviting them to apply for tickets (these cost from 1s to 7s 6d) so that they might witness 'yet another stage in the advance of fascism'. This would turn out to be Mosley's declaration that anti-semitism was henceforth to be one of the main planks of the movement. 'A dynamic creed such as fascism cannot flourish unless it has a scapegoat to hit out at, such as the Jews,' he told his sister-in-law, Irene.

On the day of the meeting, the fascists assembled at their King's Road headquarters and at 5.45 set off for the Albert Hall, down wide streets in which there was no obstacle to their purposeful marching. They went along Sloane Avenue and Pelham Street, past South Kensington underground and up the Exhibition Road, arriving at the south door of the Albert Hall. They entered straight away, to prepare the stage and take up position.

This time, there was a much greater police presence than at the Olympia meeting in June. Large numbers of anti-fascists had gathered in Trafalgar Square in the afternoon and they had regrouped in Hyde Park for further speeches and demonstrations before arriving at the Albert Hall. The police expected trouble and were there in strength to deal with it: there were 1,236 constables, 88 mounted police and almost 200 officers, under the command of a Superintendent.

Pam, sitting near the front, found it difficult to conceal her distaste for what she called 'ridiculous play-acting behaviour' as she watched Mosley march up the main aisle, arm outstretched in the fascist salute, to the chanting of the BUF's Teutonic-sounding anthem, 'Mosley, Leader of Thousands!'

'This great meeting tonight is another milestone on the irresistible march of the Blackshirt movement to power,' he began, to loud applause.

> In the past few months [it] has passed through various attacks, misrepresentations and physical violence such as no other movement in the history of this country has been called on to survive . . . the Press of this country combined with the old hens of Parliament have succeeded in creating in the minds of the public the impression that Blackshirts are responsible for disorder on the streets and disorder at meetings simply from the fact that fights have taken place at our meetings.

He went on to repudiate charges of violence, to advocate protectionism and, finally, in a rousing peroration, to attack the Jews:

> The power of organised Jewry is mobilised against fascism. They have thrown down their challenge . . . the way in which Jews attack fascism is when there are six Jews to one fascist. The point we can prove is the

> victimisation of employees of Jewry, men and women who have been dismissed for no better reason than that they were Blackshirts.

Next came the familiar charge that the Jews controlled international finance before his unequivocal conclusion:

> They admit they owe allegiance not to our Empire but to friends, relatives and kith and kin in other nations and they know that fascism will not tolerate anyone who owes allegiance to a foreign country. They have been striving for eighteen months to arouse feeling in this country to raise the banners of war with a nation with whom we made peace in 1918. We fought Germany once in a British quarrel. We shall not fight Germany again in a Jewish quarrel.

It was horribly reminiscent of the conclusion of a recent speech by Rudolf Hess: 'The country that does not maintain the purity of its race perishes' – and the Blackshirts were quick to take up its theme. It was not long before they were marching through the East End to the chant of 'The Yids! The Yids! We've got to get rid of the Yids!'

Some reacted with outrage, some with ridicule. One of the latter was Nancy, whose natural weapon was mockery and who had to the full the novelist's instinct to transmute personal experience into copy, as well as a habit of basing her characters on clearly identifiable family or friends. Her brief passage with fascism had given her plenty of material, which she had been busy pouring into a novel, *Wigs on the Green*. Its heroine, Eugenia, a pretty, fair-haired eighteen-year-old who spouted fascist ideology on the village green though still obedient to Nanny, and who rode a horse called Vivian Jackson (Pam's husband Derek had a twin brother called Vivian), would have been recognisable only within the family circle as Unity – but its unseen fascist supremo Colonel Jack with his band of loyal Jackshirts could have been identified by anyone who read a newspaper.

With the fun of writing past and the book nearing publication, Nancy was becoming worried at the effect it would have on Diana. On 7 November she wrote placatingly:

> I am calling it the Union Jack movement, the members wear Union Jackshirts & their Leader is called Colonel Jack. But I shall give it to you to edit before publication because although it is very pro-fascism there are one or two jokes & you could tell better than I whether they would be Leaderteases.

She was soon to learn what Diana thought.

After watching, with Unity, a ceremony on 9 November commemorating the deaths of sixteen men in one narrow Munich street during the 1923 putsch, from which stemmed Hitler's rise to power, Diana gave up her flat in Munich and went home to London. Over the course of a long luncheon with Nancy at the Ritz she detailed her objections to the Leaderteases, making her displeasure coldly clear.

Unity, still in Munich, moved into a students' lodging house. Her fatal meeting with Hitler was only three months ahead.

Chapter XVI

By a freakish chance, Diana and Unity already knew one of the most intimate secrets of Hitler's life: that he had a mistress. They had even met the woman who, in her last hours, would become Frau Hitler – Eva Braun, shy, blonde, pretty and twenty-three years the Führer's junior.

Eva Braun's existence was then virtually unknown. The German people, like the rest of the world, had never heard of her. She was a shadowy figure kept by Hitler so severely in the background that even Frau Wagner,* closer to Hitler than almost anyone and with whom he stayed for the Bayreuth Festival, knew nothing of her. Only a handful of the most senior Nazis had met her and between them was a conspiracy of silence. No one mentioned her; when Party dignitaries arrived she had to remain closeted in her room (next door to Hitler's bedroom) and she was so fearful of Hitler's displeasure if her presence became known that she did not dare leave the house. 'I might meet the Goerings in the hall,' she would say.

Eva was a sweet, simple girl who literally worshipped Hitler – always referred to in her diary as He. She loved clothes, cosmetics, jewellery and glamour. She was well brought up, with a convent education, and she was athletic: she swam, skated, danced and played tennis. She worked in the shop of Hitler's 'court' photographer and constant companion, Heinrich Hoffmann, and it was here, one evening in 1929, that Hitler had first seen her, perched on top of a ladder reaching for some dusty files. In the early days he had confined himself to dropping in from time to time with flowers or chocolates for 'my lovely siren from Hoffmann's'. As the months passed, chocolates and roses were replaced by invitations to supper and visits to the opera – but always with others. It was not until 1932, when Eva was twenty and Hitler forty-three, that they became lovers.

* The English daughter-in-law of the composer and an early admirer of Hitler.

At first, their affair was sporadic. A few months after he came to power he gave her a set of bracelet, earrings and ring in tourmalines for her twenty-first birthday, but despite Eva's hopes this expensive present did not signify commitment. Long periods would pass without him sending for her and by the autumn of 1934 she was deeply depressed.

Finally, on 1 November, she could bear her ambiguous position no longer. With her father's pistol, she attempted suicide, shooting herself in the neck and severing an artery. Before she lost consciousness, she telephoned her doctor to say what she had done and he arrived in time to save her.

She could not have contrived anything designed more to appeal to and impress her lover. Her despairing gesture vividly echoed the suicide of the young girl with whom he had earlier been obsessed, his niece Geli – daughter of his half-sister Angela – for whom he had had an incestuous passion. He now determined to take responsibility for Eva and from that moment the tentative affair crystallised into a permanent liaison. The following summer he set her up in an apartment in Munich, close to his own, sending for her whenever he came to the city.

Appearances were still maintained – Eva shared the Munich apartment with her sister and when she went to Berchtesgaden she stayed discreetly at an hotel rather than at Hitler's private retreat, the Haus Wachenfeld (renamed the Berghof in 1935).

The Haus Wachenfeld was the first home Hitler could have called his own. For years he had spent holidays in the small mountain town of Berchtesgaden; wearing *lederhosen*, he would wander through the hills amid the peace and the spectacular views of this part of the Obersalzburg. At first he stayed in the Pension Moritz, registering under the name of Herr Wolf;* then, in 1928, he was able to rent the Haus Wachenfeld for 100 marks a month. Angela kept house for him. She disapproved violently of Eva, whom she called 'die blude Kuh' (the stupid cow).

Marriage to Eva was out of the question. 'The worst feature of marriage is the creating of rights,' Hitler would declare. 'A highly intelligent man should take a primitive and stupid woman. Imagine if on top of everything else I had a woman who interfered with my work!' As for children, he believed that to have any of his own would present insuperable problems.

* In the early days of 1923 and 1924, Hitler was known as 'Wolf' by close friends and associates; hence Wolfsburg, where the Volkswagen, the car he devised for the people, was manufactured.

Diana and Unity had realised Eva's relationship with Hitler immediately they met her. It had simply been a case of putting two and two together. Unity, in her unceasing exploration of every avenue that would lead her to the Führer, had got to know Heinrich Hoffmann. One day, visiting the photographer, she had looked up and seen Eva, whose looks so struck her that in a letter to Diana she described this 'lovely girl' she had seen working at the rear of the shop. Shortly afterwards, she saw Eva, beautifully dressed, being driven past in a large and gleaming white Mercedes; immediately, she jumped to the conclusion that this pretty but humble photographer's assistant must be the mistress of someone important. At the 1934 Parteitag, she realised whose.

Because the tickets Unity and Diana had been given were vouched for by one of the earliest, most trusted members of the Party, they were in the most exclusive section of the stands, reserved for the highest Nazi officials. In the seat next to them was Eva Braun. A few friendly questions and the sisters knew the truth.

The meeting with Hitler that had become the chief goal of Unity's life was finally achieved on 9 February 1935, at a small restaurant called the Osteria Bavaria. Unity had been planning for it for months. She had familiarised herself as much as possible with the Führer's routine, she had questioned everyone she met who might be able to give her a 'lead' as to his habits, in her increasingly fluent German she had spoken to the guards, custodians and doormen who saw him daily.

His routine was surprisingly simple. In Munich, he lived at 16 Prinzregentplatz, a square in one of the most expensive neighbourhoods in the city. His flat, in a corner house, was on the second floor. Its hall was lined with bookshelves, several good pictures – a Cranach, two Brueghels and a portrait of Bismarck, his hero – hung in the long narrow drawing room with its bird's eye maple furniture. At the far end was a marble-topped table, behind which he sat when receiving guests. He was looked after by the former soldier-servant of a general and the man's wife, once a lady's maid. (He had moved into his official residence in the Chancellery in Berlin at the end of 1934, giving his first dinner party there on 19 December – Lord Rothermere and his son Esmond Harmsworth were among the guests in the 100-foot dining room with its red marble pillars, blue and gold mosaic ceiling and Gobelin tapestry from the museum in Munich.)

In Munich, Hitler usually ate out in the middle of the day. Unity quickly discovered that he often lunched at the Osteria Bavaria, an artists' café full of the drawings and watercolours that Hitler, himself a watercolourist, loved. He would usually arrive at about two-thirty and often later. He was an insomniac, reliant on sleeping draughts, and his whole day was geared round the late rising caused by his inability to sleep before three or four in the morning; he was also naturally unpunctual. His entourage was usually the same: his constant companion the photographer Hoffmann, Martin Bormann, Reich press chief Otto Dietrich, the Bavarian *gauleiter* Wagner and an aide. They would drive up in two black Mercedes and make straight for their regular table, in a corner of the room shielded by a low partition. In good weather they ate outside in the small courtyard.

When she knew that Hitler was in Munich, Unity would lunch every day at the Osteria Bavaria, lingering over her cup of coffee until, with luck, he strode in. She had begun this routine at the end of 1934, often persuading Diana to come with her. At Unity's insistence, they would arrive before the Führer's party came in and stay until after it had left – often not until five o'clock. As Unity had carefully chosen a table that Hitler had to pass on the way to his own, this meant that she had two close-up glimpses of him.

Unity was impossible to overlook. A tall, striking, well-dressed blonde, her scarlet mouth and silky powdered complexion contrasted vividly with the scrubbed faces of the women around her as she sat alone at her corner table, her huge blue eyes fixed on the Führer as, after gazing lengthily at the menu, he ordered his customary ravioli – in a meat-eating, coffee-drinking culture, he was a vegetarian, subsisting largely on pasta, eggs, salads and fruit, and drinking herb tea. It was not long before he asked one of the waitresses who she was, but, to Unity's annoyance, the Christmas holidays intervened.

She returned to Munich in January 1935, accompanied this time by her father, whom she took immediately to the Osteria Bavaria. Hitler was there ('Farve has been completely won over by him and admits himself to have been in the wrong until now'), but her own acquaintance with the Führer seemed to have progressed no further than a nod of greeting on both sides.

Then, on a dank February Saturday, everything changed. Yet again Unity had lunched at the Osteria Bavaria and was sitting over a cup of coffee hoping that Hitler would appear. At three o'clock he came in with his usual band of cronies, sat down – and ten minutes later sent the manager to her table. To Unity, his words, 'The Führer would like to speak to you,' were a summons from Olympus.

Hitler, with his customary politeness to women, rose as she arrived at his table, saluted and shook hands with her. Introducing her to his companions, he asked her to sit next to him and then, in her own words, 'I sat and talked to him for half an hour . . . I can't tell you all the things we talked about.' One of them was *Cavalcade*, which he told her he thought was the best film he had ever seen; another was that international Jews must never again be allowed to make two Nordic races fight against one another.

After 'the greatest man of all time' had left, Unity wrote ecstatically to her father, 'I am so happy I wouldn't mind dying.' She also wrote at once to Diana, begging her to come to Munich immediately so that she, too, could meet Hitler.

Diana was almost as excited as Unity. She spoke to Mosley who urged her to go. She left at once for Paris – she and Mosley often went to Paris and during one weekend there they had seen, in a Champs-Elysées showroom, a little Voisin car so pretty and elegant that Diana immediately craved it, and Mosley had ordered one for her. It was now ready and she determined to drive to Munich despite vile weather conditions, though she took the precaution of taking a Voisin mechanic with her. It was fortunate that she did: on the way they ran into such snowstorms and drifts that at times they were in real danger.

Within a day or two of Diana's arrival she was introduced to Hitler. Though years later she was to say, 'Meeting Hitler ruined my life – and that of my husband too,' at the time his personal magnetism overwhelmed her, and she found her inability to speak more than a few words of German immensely frustrating. There and then, she determined to redouble her efforts to learn the language.

Physically, there was nothing about Hitler to account for his extraordinary, almost mesmeric, personal magnetism. Except for his obsessive neatness and peculiarly white, well-shaped hands, he was unremarkable to the point of ordinariness – a man of about five feet nine inches, with a clear skin, fine dark hair and gold-filled teeth. When not in uniform his clothes, said Randolph Churchill, 'had all the unpretentious respectability of the German or Austrian middle class' – grey or dark blue suits, not very well cut, worn with soft-collared white shirts and, instead of an overcoat, a mackintosh. 'Oh, he's so sweet, in his dear little old mackintosh,' Diana would coo. His most striking feature was his eyes, of a greyish blue so dark that contemporary observers often mistook them for brown, dull and opaque when in repose, piercing and vivid when he was speaking to a crowd or an individual.

His great gifts were a superb memory, ferocious concentration and an intuition so highly developed it could be called psychic. Once, flying through fog from Stettin to Hamburg, in the teeth of the pilot's opposition and with no evidence save his own conviction, he ordered the pilot to turn round and fly back the way they had come because he believed the compass was wrong. They landed at Kiel with ten minutes' fuel to spare – had it not been for Hitler's 'intuition' they would have run out of fuel far out over the Baltic. Unsurprisingly, he believed in astrologers, consulted the mystic Gurdjieff and had complete faith in the power of instinct. He often told the story of how, during the First World War, an 'inner voice', so distinct it sounded like a military order, told him to move to a different place while he was eating dinner with his fellow soldiers; he got up, walked twenty yards down the trench and sat down. Next moment a shell burst over his friends, killing them all.

This talent for seeing beyond the surface and into the inner emotional core gave him a magnetic charm that affected, among others, Anthony Eden and Lloyd George, who described him as 'a very great man'. To the mass of the German people he was a spellbinder, a saviour, a political magician who had led them out of bankruptcy and despair, who had given them better homes and food, who had built them roads and maternity clinics and given them a transcendent sense of the power that would arise out of unity – in short, the Leader, whom they trusted absolutely. His oracular pronouncements, his deification of the Fatherland, his ability to persuade his compatriots that they, too, were supermen, were all calculated to appeal to a nation resentful of the loss of its former greatness, angry and humiliated by defeat and still smarting under the harsh terms of the Treaty of Versailles.

At the same time, his position of lofty isolation allowed him to distance himself from the brutalities of the regime – especially as many of its foulest crimes were perpetrated in such secrecy that many Germans knew nothing of them, while those who did could convince themselves that it was not the Führer who was responsible but his subordinates. As late as February 1939 the British Ambassador in Berlin, Sir Nevile Henderson, was exonerating Hitler from responsibility for the pogroms (in a letter to Lord Halifax, the Foreign Secretary).

Hitler himself realised that his mystic Wagnerian ideal of the patriotic hero appealed to the German desire both to dominate and be dominated. 'Every individual, whether rich or poor, has in his inner being a feeling of unfulfilment,' he said.

> Slumbering somewhere is the readiness to risk some final sacrifice, some adventure, in order to give a new shape to their lives . . . the humbler people are, the greater the craving to identify themselves with a cause bigger than themselves, and if I can persuade them that the fate of the German nation is at stake, then they will become part of an irresistible movement, embracing all classes.

His effect on the German people has often been compared to a kind of mass hypnosis. Jung has a different explanation. 'The psychopathology of the masses is rooted in the psychology of individuals,' he wrote less than two years after the Führer's suicide.

> As early as 1918, I noticed peculiar disturbances in the unconscious of my German patients which could not be ascribed to their personal psychology. Such non-personal phenomena always manifest themselves in dreams as mythological motifs also to be found in legends and fairytales throughout the world. I call these mythological motifs the 'archetypes'.

In his German patients, Jung found again and again archetypes expressing primitiveness, violence and cruelty, 'to which they were susceptible because the German mentality has a marked tendency to mass psychology. Moreover, defeat and social disaster had increased the herd instinct in Germans.'

When these archetypes occur *en masse*, said Jung, they draw individuals together as if by magnetic force:

> Thus a mob is formed. Its leader will be found in the individual who has the least resistance to such archetypes, the least sense of responsibility and because of his inferiority the greatest will to power . . . However confused the government or political situation around them, most people have a natural orientation towards order. But in those in whom the collective unconscious has long been permeated with feelings of weakness, deprivation and powerlessness, the archetypes of order are missing – and the unconscious is open to invasion by the archetypes of darkness. Hitler was the exponent of this 'new order' and that is the real reason why practically every German fell for him.

On women the effect of this rather plain, ordinary little man was uncanny. From them he received adulation of the sort reserved for film stars and royalty. Some would turn up at Berchtesgaden almost naked under their coats to offer him their virginity; others would try to throw themselves under his car in the hope of being first injured and then comforted by the man they regarded as part prophet and part Leader. On his birthday, hand-knitted socks and jerseys would arrive from all over Germany, while love letters, proposals and marriage licences awaiting his signature poured in daily by the sackful, invariably couched in language of the utmost respectfulness. 'Herr Führer Adolf Hitler, a Saxon woman would like to bear your child . . .' wrote Friedel from Hartmannsdorf. 'Please, dear Führer, let me come to you,' begged Luise from Hof. 'I'm having a front door key and a key to my room made for you, Adolf. In the next letter, you'll get the first one and in the letter after that you'll get the room key,' wrote Margarete from Königsberg. None was answered – indeed over-persistent admirers received a warning from the police – though birthday wishes received a printed note of acknowledgement.

His policies towards these adoring creatures smacked of contempt, a wholesale restructuring of women's role that amounted to regarding them as a different, inferior species, whose sole duty was to look after their men and bear future soldiers and mothers of the Reich. The emancipated, independent woman was seen as an agent of degeneracy; in Nazi Germany women were thrust firmly back into the home by a combination of chivvying and inducement. Courses in domestic science and biology were compulsory in female education, all forms of birth control were frowned on, slimming and dieting discouraged as possibly contributing to a lowering of fertility, and a generous system of marriage loans introduced.

There were, of course, conditions attached to this liberality. Those applying for these loans had to undergo stringent examination to ensure that they were racially pure. Not surprisingly, there was a thriving black market in documents proving Aryan origin. The ideal German woman was a statuesque Gretchen with a rosy, cosmetic-free complexion and several small blond children clustering round her skirts – these to be of a style and shape advocated by the new German Fashion Bureau, set up in 1933 under the honorary presidency of Hitler's favourite Aryan blonde, Magda Goebbels (who never wore such clothes herself).

In his private life, Hitler was fascinated by women. He enjoyed female company and had a great eye for a good-looking woman. His physical preferences were for the stereotypical Aryan blonde – tall, blue-eyed and full-figured – as typified by his mistress Eva Braun and by Magda Goebbels (though he did not realise that Frau Goebbels bleached her darkening hair to achieve his preferred shade of 'Aryan' gold-blonde). His obvious pleasure in women's company did not, however, stop the abounding rumours that he was impotent or of ambiguous sexuality. Popular rumour credited him with only one testicle.* 'His over-compensation for the inferiority complex of an impotent masturbator was the driving force of his lust for power,' summed up Putzi Hanfstaengl, who had known him for years.

His behaviour towards the women who worked for him or whom he met socially was different from his attitude to men. He was invariably gallant, he would kiss a woman's hand, first smiling into her eyes before he bent low over her hand to touch it with his lips. He would never sit down if a woman was standing, he would usher women out of the room first and he would always speak to them in kindly, almost caressing fashion. He would shout and rage at his generals, but he was always patient and gentle towards his female secretaries, paying compliments and asking about their families. Not surprisingly, they were devoted to him. The good-looking actresses and cabaret artistes whom he invited to dinners were seated on his right and treated with a flirtatious gallantry that roused the jealousy of Eva Braun.

Most of these women, however, made only one or two appearances. Much more serious from Eva's point of view was his growing friendship with the worshipping Unity. Eva scented a serious rival in this handsome, self-assured blonde from a social milieu which she, Eva, found terrifying but which clearly fascinated the Führer. Eva's patience was monumental, and she was well accustomed to hanging about – or as she put it, 'kicking her heels' – but being kept waiting on account of affairs of state was one thing, preoccupation with a possible new mistress another. 'Herr Hoffmann lovingly and tactlessly informs me that he has found a replacement for me,' she wrote in her diary on 10 May:

> She is known as the Walkure and looks the part. Including her legs . . .
>
> I shall wait until 3 June, in other words a quarter of a year since our last meeting, and then demand an explanation. Let nobody say I'm not patient.
>
> The weather is magnificent and I, the mistress of the greatest man in Germany and the whole world, I sit here waiting while the sun mocks me through the windowpanes.

* The Soviet commission charged with establishing the facts about Hitler's death found that: 'The left testicle could not be found either in the scrotum or on the spermatic cord inside the inguinal canal nor in the small pelvis.' On the other hand, this defect was not in the medical records kept by Hitler's own doctors.

Chapter XVII

Put baldly, a friendship between the most powerful man in Germany and an unknown young foreign woman picked up in a restaurant sounds extraordinary. But the idea that Unity might have been 'planted' never seems to have entered Hitler's head, despite his paranoid suspicion of political plots. Hitler never feared assassination attempts – during the Parteitag, when he noticed that the pavements had been cleared for this reason as he drove through the narrow Munich streets, he at once rescinded the order so that people could see him more closely.

Unity and Diana were exactly the type of beauty that Hitler admired, so much so that both sisters got away with wearing the make-up of which he so disapproved. They were young, upper-class and socially at ease. Hitler was always impressed by women who had the sophistication, assurance and polish which he knew he himself lacked – he did not, for instance, know how to dance. Among the Party wives, his favourites were the two most socially adept: Annaliese Ribbentrop and Magda Goebbels. Diana, independent, self-confident, elegant in her Paris clothes, familiar with many of the great cities of Europe, a music lover with a family history of reverence for Hitler's own favourite composer, and with the social gloss conferred by six years of constant entertaining, epitomised such women.

Above all Diana, who had never been afraid of anyone in her life, was not, unlike most of those around Hitler, in awe of him. Unity, too, though she regarded him as the incarnation of her deepest beliefs, had all the Mitford confidence. Hitler enjoyed their unafraid manner; they appealed to his penchant for the English aristocracy; they were pretty, stylish, frank and amusing. Chatting away with all the Mitford inconsequence and insouciance, they made him roar with laughter. It was not long before he was inviting them to tea, to dinner at the Chancellery, to festivals, to see films or just to come and talk; and he gave them each a swastika badge with a facsimile of his signature on the back. For Diana and Unity, twenty-four and twenty respectively, it was all very exciting, and 'sweet Uncle Wolf' did not seem at all the ogre depicted in the British press. 'The truth is that in private life he was exceptionally charming, clever and original, and that he inspired affection,' Diana wrote many years later.

The truth also was that she found Hitler fascinating. To understand his appeal, it is necessary to look at him from the standpoint of someone living at the time. With hindsight, he is a man of incalculable wickedness, with the blood of millions on his hands, the atrocities he ordered depicted in all their hideous detail for the world to know; then, these black deeds were in the future.

People who met him could understand the magnetism and charisma that inspired blind faith and devotion in the German people. Those near him spoke of how he aroused their protective instincts. This monster had a modest and simple demeanour, such that, even when his more upright followers could not shut their eyes to some terrible misdeed, they somehow persuaded themselves that it had been done not at their Führer's behest but by his advisers. General Wilhelm Groener, whose record as War Minister was one of moderation, thought of Hitler as someone who was essentially decent in contrast to many of those around him. Lloyd George referred to him as 'the George Washington of Germany' and a 'bewitcher'. Robert Birley later described him as the most extraordinary phenomenon in all European history.

Perhaps the most telling example of the unique spell he cast was his effect on Field-Marshal Erwin Rommel. Rommel, apolitical and admitted by everyone, British and German alike, to be of undoubted integrity and a soldier to his fingertips, was a faithful believer in Hitler's genius. 'For a hardbitten, successful and principled man to succumb, as Rommel periodically did, to the enchantment of Adolf Hitler, implies strong magic,' writes Rommel's biographer David Fraser, who recounts numerous occasions when Rommel's tactical and strategic judgements were overturned by Hitler's charm.

It was a magic that overcame rational thought, logic or suspicion. Even after Hitler had flatly rejected Rommel's military advice that it was no longer possible to hold Alamein, even after he had stormed at Rommel and accused him of the worst of soldierly sins, cowardice, Hitler was still able to win Rommel back. Fraser writes:

> On 8 May, the day after the British captured Tunis, Rommel was ordered to report to Hitler in Berlin. Hitler wanted to reassure Rommel, to renew the bonds of trust and devotion which had held Rommel. He succeeded. Two months later Rommel, still with Hitler, was commenting to von Mannstein, whom he had just met for the first time, 'I am here for a sunray cure and soaking up sun and faith.' Both of them understood that the sunray lamp was Hitler.

Like Rommel, Albert Speer, the Goebbelses and countless others around Hitler, Diana found the German leader's charm overwhelming. When she realised that Hitler had become deeply fond of her she was not only touched but immensely flattered. When she also realised that he was very attracted to her – though he never made any overt move – their relationship was further strengthened by this subterranean sexual link.

It was exhilarating to enjoy the friendship of the man on whom the eyes of the entire Western world were fixed. 'What is he *really* like?' Diana's friends would ask her. Winston Churchill, who had just written (in *Great Contemporaries*) that Hitler had 'succeeded in restoring Germany to the most powerful position in Europe', invited her to lunch to ask her about Germany's leader and she was able to tell him that Hitler found negotiating with democracies disconcerting. 'One day you are speaking to one man, the next day to his successor.' She found Hitler's aura of power profoundly exciting or, as she put it, 'He was the person in control – the person that everyone was interested in.' Hitler was, as he told her again and again, a great admirer of the English. Because she could now speak German, she could communicate with him spontaneously and therefore directly, picking up nuances, jokes or asides lost through the medium of the interpreter.

Diana stayed in Munich only two days after the first meeting; the purpose of the visit had been achieved. She drove Unity to Paris, where Bryan and the boys were staying in the rue de Poitiers flat, for Jonathan's fifth birthday (on 6 March), remaining there with the children after Bryan had gone back to Biddesden and Unity had returned to Munich. Diana was still in Paris when, on 18 March, the darker side of her lover's fascist credo made another appearance. In Leicester, Mosley made a second, more overt anti-semitic speech.

'For the first time I openly and publicly challenge the Jewish interests of this country commanding the press' – this was a none-too-veiled dig at Rothermere's defection – 'commanding the cinema, dominating the City of London, killing London with their sweat-shops,' he began, continuing in a vein so outspoken that Julius Streicher, the Nazi for whom anti-semitism was vocation and religion, sent him a telegram of congratulation.

Mosley's reply, belated because of his absence travelling, was published in the 10 May issue of *Die Sturmer*, the newspaper Streicher had founded the year before to further his campaign against the Jews. It was all that the most virulent anti-semite could have wished. 'I value your advice greatly in the midst of our hard struggle. The power of Jewish corruption must be destroyed in all countries before peace and justice can be successfully achieved in Europe. Our struggle to this end is hard, but our victory is certain.'

In April Mosley had met Hitler for the first time. It was a private meeting and Hitler gave a luncheon for him; the guests included the Kaiser's daughter, the Duchess of Brunswick, Frau Winifred Wagner and Unity. For Unity it was a milestone: her first formal invitation from the Führer – Hitler, unaware of her connection with Mosley, had asked her as a fellow guest who shared both Mosley's nationality and his fascist convictions. Afterwards, Mosley described his host as 'ein ganzer Kerl' (quite a man); later still he wrote, 'He seemed to me to be a calm, cool customer, certainly ruthless, but in no way neurotic. I remember remarking afterwards if it be true he bites the carpet, he knows to a millimetre how far his tooth is going.'

Now that they were on terms of friendship, Unity's obsessive infatuation with Hitler was increasing, and with it the simplistic, headstrong urge towards extremism that was so much part of her character. 'Hitler is so kind and so divine I suddenly thought I would not only like to kill all who say things against him but also torture them,' she wrote in June.

She had also dispatched a letter to *Die Sturmer*. It was a gesture of crass stupidity that was to brand her publicly as a Nazi sympathiser and racist in the eyes of her compatriots.

Unity adored Streicher, though in true Mitford fashion she thought him 'a terrific joke'. But there was nothing amusing about the sentiments she expressed in his newspaper. 'The English have no notion of the Jewish danger,' she wrote. 'Our worst Jews work only behind the scenes . . . I want everyone to know I am a Jew hater.'

It was the letter of someone whose personal ignorance of such matters was equalled only by her violent, unreasoning prejudices. To Streicher it was a propaganda coup; and though he did not publish her letter until July he got in touch with Unity at once.

Streicher had been a loyal worker from the inception of the Nazi Party

– his membership number was 2 (Hitler's was 7). His mild manner and appearance concealed an obsessive fanaticism devoted to the total removal or elimination of Jews from the Fatherland. It was impossible to speak to him for five minutes without becoming aware of these views. When, on 29 March 1933, Hitler made him Chief of the Central Committee of the Defence against Jewish Atrocity and Boycott Agitation, he had a virtually free hand; from then on, a policy of terror became inevitable. Though Streicher himself neither needed nor wanted justification for his actions, he was aware that anything that could help present them to the outside world in a more favourable light would please the Führer. The fervent enthusiasm of a young, photogenic, aristocratic and obviously Aryan foreigner might help to sway the opinion-forming classes in her own country as well as impressing the German people.

He invited Unity to a midsummer festival near Nuremberg; 'we marched through the crowd, band playing, between cordons of S.A. men with torches, to the speakers' stand,' she wrote. Streicher addressed the crowd; during his speech, he turned towards Unity and read out from her as yet unpublished letter:

> They [the Jews] never come into the open, and therefore we cannot show them to the British public in their true dreadfulness. We hope, however, that you will see that we will soon win against the world enemy, in spite of all his cunning. We think with joy of the day when we shall be able to say with might and authority: England for the English! Out with the Jews!

Diana, reading Unity's outburst in the London papers, regarded it with the eye of one accustomed from the nursery onwards to Unity's headstrong convictions and her desire to shock. Diana was neither alarmed, nor outraged, nor even mildly disgusted; it was, she thought, 'a piece of silliness – a wild thing to say'.

Her sympathies were largely with her sister. Though she would never have dreamed of admitting that she was anti-semitic – she had far too many Jewish friends for that – Diana was so deeply committed to Mosley and his ideological attitudes that his views on 'the Jewish problem' were also hers.

She was not so indulgent towards another sister. Nancy's novel, *Wigs on the Green*, was due out on 25 June. After their luncheon at the Ritz Nancy had removed some of the material to which Diana objected but, as publication date approached, she became increasingly apprehensive about Diana's reaction to her parody of fascism. A week before the novel's appearance she wrote to Diana explaining that she had found it impossible to eliminate everything Diana and Mosley had disliked, but that she had removed almost three entire chapters relating to her hero Colonel Jack.

'There are now, I think, about four references to him & he never appears in the book as a character at all. In spite of this I am very much worried at the idea of publishing a book which you may object to.' She went on to argue that the sort of people who read her books were exactly the sort Mosley did not want in his movement and – rather disingenuously – that her book was far more in favour of fascism than against it. 'Far the nicest character in the book is a fascist – the others all become much nicer as soon as they have joined up. But I also know your point of view that fascism is something too serious to be dealt with in a funny book at all.'

She was right. Diana, believing that her beloved Leader had been held up to ridicule, was coldly angry. Sentences such as, 'If your husband is an Aryan you should be able to persuade him that it is right to live together and breed; if he is a filthy non-Aryan it may be your duty to leave him and marry Jackshirt Aspect,' or, 'I'm sure Hitler must be a wonderful man. Hasn't he forbidden German women to work in offices and told them they never need worry about anything again except arranging the flowers? How they must love him,' were, Diana felt, a frivolous rubbishing of everything she and Mosley believed in. For the next few years, relations with Nancy were strained to non-existent.

Two weeks later, as Diana was dressing for a dinner party one hot evening in late July, her telephone rang. It was Mosley. They had planned to drive down to Savehay together on her return from the party, leaving at about midnight; now he said that he could not leave London until the following morning.

It was to be their last weekend together for two or three weeks, as Mosley was going away with his children, with Andrée to care for them, for their annual month's summer holiday together. Through Irene, Mosley had rented a villa at Posillipo which belonged to Sir Rennell Rodd, the British Ambassador in Rome. It stood on the cliffs – there were 1,000 steps down to the beach – looking across the Bay of Naples to Vesuvius, with Capri lying like a jewel in the blue waters below. With the family for the first fortnight would go Baba, the intensity of whose love affair with Mosley showed no signs of dying down.

Only after everything was arranged had Mosley revealed that, the day after Baba had left, Diana was to join them for the following two weeks ('that miserable Diana Guinness follows Baba out there', noted Irene). To Diana, he had explained that it was essential to keep the children's aunt happy and Diana, who knew that she was the one Mosley truly loved, cheerfully fell in with this plan.

The evening of the telephone call, Diana put down the receiver, having agreed with Mosley that she would drive down to Savehay by herself that night. It was too hot to sleep in London, she explained, and the air at Savehay would be fresher and less oppressive; would he ask them to leave a door unlocked for her? She finished changing into her coolest, and coolest-looking, outfit, a long skirt and coat in cream silk, and went off to her dinner party, where she sat next to Lord Beaverbrook. Enthralled by his conversation and thirsty because of the heat, she drank champagne liberally. Back at the Eatonry, she quickly collected a few clothes, picked up her spaniel and set off.

She had got only to the five-road intersection of Lowndes Street and Cadogan Place when she was hit by a large Rolls-Royce. It splintered the side of the little Voisin and flung Diana forward into the windscreen, which broke with the force of the impact.

She was pulled half-conscious out of the car by two young policemen passing on their beat. She presented a horrific sight. Her dress was soaked with blood, which stood out so vividly against the pale silk that two women who were passing thought she must be dead. 'Poor thing!' she heard them murmur in her confused state. 'So young, too.' The dog, unharmed – though she did not realise this – was whimpering in the back.

She was taken by ambulance to St George's Hospital at Hyde Park Corner, enquiring constantly if her dog was all right. 'Yes, he's been taken home,' she was told soothingly (in fact, he spent the night at the police station which, with the shock of the crash, left him nervous for the rest of his life). At St George's the cuts on her face were quickly stitched up by a young doctor in the Casualty Department. Because the fine thread had been locked up for the night, this was done with thicker thread, an agonising process that felt as though rope was being pulled through raw flesh. Terrified that Mosley would read of her accident in the morning papers and fear the worst, she demanded to ring him and, after a battle, was allowed to do so at two a.m.

Next morning her former father-in-law, Lord Moyne, sent for the country's leading plastic surgeon, the brilliant New Zealander Harold

Gillies (later to become a national figure for his work restoring the burned faces of Battle of Britain fighter pilots and others). Horrified at the thick thread used to stitch up her wounds, which would have left permanent scars, Gillies removed it. He sewed her face temporarily into position with two huge stitches and, telling her that she must not talk or laugh for a week, admitted her straight away into the London Clinic. Here he deftly stitched each wound with the finest thread, working wherever possible from the inside, with such skill that in a few months there was no trace of even the smallest scar.

As she lay in the clinic, face painful and swollen, letters poured in. 'I am reading the news with a confident faith in your speedy recovery,' wrote her dinner partner Lord Beaverbrook. Decca, surrounded by her files of the *Daily Worker* in the Swinbrook sitting room she shared with Unity, wrote to 'Darling Cord' to sympathise. 'I do hope you're better now and not in too much agony – it sounds too frightful.' It was one of the last affectionate letters Diana would receive from this younger sister. Cecil Beaton, out of touch all summer, wrote, 'My last vision of you was a radiant one, in a large picture hat, but you were whisked on as the lights went green before I had been able to catch your aquamarine eye.'

Most compelling of all was the letter from Mosley in Posillipo, where his thirty-foot, three-cabin yacht, the *Vivien* (named after his daughter), lay in one of the bay's inlets. 'Hurry up and get better, as this place is lovely – no great horrors have been revealed except the ancient truth that "Rodds never wash". We . . . run up the steps now, saying, "Won't they be fun when Diana arrives!"'

The image of sun, sea, sand and love conjured up by this brief message made Diana, already bored and wretched, determined to leave the clinic despite strict orders to stay there until her face had finally healed. But how to get out? She was too weak to manage it alone and every request was brushed aside by both staff and her visitors.

Then came inspiration – her father. When she begged him for help, he agreed. It was just the sort of plot he loved, outwitting authority in the shape of doctors and nurses and at the same time pleasing his favourite daughter.

Her escape, they decided, should take place at the time of day it would be least obvious: when the day and night nurses were changing shifts and everyone else was asleep. Accordingly, David arrived at five a.m., to find Diana dressed and waiting. To lessen the chance of detection, they did not take the lift but walked down the stairs, Diana clinging to the banister on one side and her father's hand on the other. Once outside, he drove her through the empty streets to Croydon, where she sent Mosley a telegram to announce her imminent arrival. From Croydon she flew to Marseilles, thence to Italy in a seaplane, finishing her journey by train.

She arrived at the villa in the middle of a grand dinner party at which the Crown Prince and Princess of Italy were the guests of honour. Baba had known the Crown Princess, daughter of King Albert of the Belgians, since childhood: during the First World War, her father Lord Curzon had given a home to the Belgian royal family at Hackwood. When the footman entered the dining room to announce Diana's presence, Baba, furious, was restrained from a recriminatory outburst only by the presence of her guests. Seething with jealous fury, she had to keep the party going while Mosley went to see Diana, who had gone straight to bed. 'I didn't realise you were coming,' he said, though without anger. 'I'm so sorry, but I did send a telegram,' said Diana placatingly. A few minutes later the telegram arrived.

When the last Altissima had gone, Baba turned on Mosley, moral outrage serving as a pretext for livid anger. Mosley managed to pacify her by telling her that he was furious with Diana. He had, he told Baba, wired Diana not to come until the Thursday, but she said she had not received his telegram. He continued that he had commanded Diana to stay in her room and not dare to appear as Baba would not meet her and he himself did not wish to speak to her. 'Poor Diana has had a bad car accident,' he concluded, 'and she has only come to recuperate.'

Next day he took Baba and his children for a three-day trip in the *Vivien* to Amalfi. The beauty and glamour of the cruise, Mosley's charm and undivided attention, their lovemaking during the nights they spent in a clifftop hotel while the children slept in the boat, convinced Baba of his devotion. When she finally took a taxi from Amalfi to the airport to catch her flight, the status quo that so suited Mosley both temperamentally and physically had been satisfactorily restored.

Back at Posillipo, Diana, left in the charge of the servants, was installed in the bedroom just vacated by Baba.

At first her presence took Mosley's children unawares. One morning Nicholas, devoted to his aunt and accustomed to visit her in her bedroom after breakfast, was walking down the passage to her room when he was stopped by Andrée. 'Where are you going?' she asked. 'To say good morning to Auntie Baba.' 'It's not Auntie Baba in there, it's Mrs Guinness,' replied Andrée bitterly.

At the time, neither Vivien nor Nicholas thought the presence of Mrs Guinness on holiday was particularly unusual. Both of them knew that their father was very fond of her, yet at the same time they were aware that Auntie Baba also loved him but did not like Mrs Guinness. They did not know, and they did not want to know, what it all meant; neither discussed it with the other. Better to treat it all as just another aspect of 'grown-up' behaviour and go on enjoying the holiday.

What they did understand, and appreciate, was that their father was extraordinarily happy during those weeks, full of jokes, fun, quotations and ideas for games.

As Diana recovered, she began to come down to the beach, sitting under a sunshade as the faint scars on her face healed. The children approached her cautiously. Both were very conscious of her beauty; she, for her part, was determined that they would like her. She taught them poker and Bridge, she sang plaintive German songs while accompanying herself on a 'squeezebox' accordion so old it had buttons instead of piano keys.

Unlike their mother, they noticed, she never argued with their father. Sometimes he would imitate her voice and her swooping Mitford intonations and she would smile. If she disagreed with him, she would simply close her enormous blue eyes, and both she and Mosley would laugh. When she opened her eyes again, the conversation would have changed and whatever was disagreeable would have disappeared.

On Mosley's return home, his affair with Baba continued as before. Its effect on the Metcalfes' marriage was devastating: Fruity sank into a slough of jealous misery, while Baba vented her disenchantment with her husband in displays of temper and rudeness. 'Fruity almost in tears', 'Baba in a vile mood', 'Fruity in a pitiful condition, half-crazed', are typical entries that autumn in Irene's diary.

Chapter XVIII

Mosley returned from the Posillipo holiday to a BUF with a steadily declining membership and finances dwindling even faster. With a subscription of only one shilling a month – dropping to 4d for the unemployed – little money was coming in and, after the initial enthusiasm, funds from rich supporters had ceased to appear. At the same time, the costs of the Black House, the salaries of officials and the general running expenses remained the same. The previous year, these had been estimated at between £60,000 and £80,000.

Some of this money came from Mosley. He was a rich man in his own right and thanks to his children's inheritance from Cimmie the upkeep of his home cost him nothing. Even so, the BUF needed more than he could provide.

Where the rest of the money had come from was more of a mystery. Mosley himself was deliberately vague about the sources of BUF funds, often declaring that he had 'no knowledge of the financial side' of the movement. There were persistent rumours – vigorously denied by Mosley – that the BUF was largely funded by Mussolini; and Special Branch, during one of its investigations, discovered that sizeable payments from abroad were made into BUF accounts. Most of these came from a Swiss bank in a monthly package made up of £5,000-worth of notes of different currencies and denominations, and were paid to Bill Allen, an old friend of Mosley's and a BUF member whose agency, with its European links, could have produced a valid reason for such payments. Special Branch therefore came to believe that the BUF was largely funded from Italy, concluding from this and the steadily decreasing membership that fascism in Britain was no threat. Sixty years later, documents discovered in the Italian archives proved that Il Duce was indeed subsidising British fascism from 1933 to 1935.

On 25 August 1933, a letter headed 'Secret Mission' was sent in the diplomatic bag to the Italian Ambassador in London, Count Dino Grandi. It informed him that 'by courier leaving Rome on Saturday 26th inst the second instalment of £5,000 is being transmitted for Sir Oswald Mosley'. This money, in five packages, was made up of mixed currencies – pound notes, dollar bills, Reichsmarks, Swiss and French francs – and, continued the letter, 'the packets containing the aforesaid currency notes are without any identification and are secured by seals of no significance'. The clandestine nature of these donations was made clear in Grandi's instructions from Rome: the monies must be delivered into Mosley's hands 'in whatever form Your Excellency considers best and in such a way that the consignment is made secretly'. They were slipped to Mosley secretly by a Dr Enderle, a specialist in such tasks. The second 'drop', of the same value, made up in the same way, was performed by Enderle's nephew, who had left Rome for the purpose on 29 August, in case anyone might have been watching the first. A third payment was made in October.

After Mosley had visited Mussolini on 9 January 1934, the payments increased dramatically. On 24 January a courier left Rome carrying £20,000. 'Mosley asked me to express to you his gratitude for sending the conspicuous sum which I have arranged this day to pay over,' wrote Count Grandi to Mussolini. 'Not long after his return from Rome he came to see me . . . he told me that the talk with you had "enriched" and illuminated him.' The donations stopped abruptly in 1935.

One reason may have been the hostile influence of Grandi, who had realised that the BUF would never be a force to reckon with in British politics (not a single candidate gained a seat in the general election of 1935). He advised Mussolini that 'with a tenth of what you give Mosley, I feel I could produce a result ten times better'.

Mosley's immediate reaction to the cessation of the Italian funds was to give up Black House, with its complex of training facilities, and move the BUF headquarters to more modest premises in Great Smith Street. Joyce and Beckett, considered key men, were given more prominent positions. At the same time, there was a drive to halt falling membership.

So strong had become the feeling against Mosley's political stance that Princess Alice, approached to open the day nursery built in Kennington in memory of Cimmie, said she would accept only on condition that Mosley was not present. Irene rejected these terms saying that Cimmie had loved Mosley all her life, and the Princess relented.

Mosley had always declared that when he came to power he could cure unemployment, so the BUF concentrated its recruitment drive on the north of England and the Midlands, where unemployment was highest. He was also becoming increasingly outspoken in his anti-semitism. When in Germany for his meeting with Hitler in April, he had been interviewed for the *Frankische Tageszeitung*, which declared (on 24 June), 'Mosley very soon recognised that the Jewish danger may well work its way from country to country, but fundamentally it poses a danger to all the peoples of the world.'

In Germany, anti-semitism was becoming steadily more brutal. When the Nuremberg Laws were enacted in September 1935 the gates on the long avenue that would lead inexorably to the Final Solution swung wide open.

The Nuremberg Laws gave constitutional authority to the repression already being practised. It became a crime to marry anyone with Jewish blood, Jewish names were erased from war memorials and most basic rights were removed from Jews. The fact that Trotsky and Karl Marx were Jewish was constantly brought up, to reinforce Hitler's often-stated claim that Germany had a 'right' to the lands of the east and that the Slav peoples were born to be slaves. Elsewhere fear of Germany grew when German troops, in breach of the Treaty of Versailles, reoccupied the demilitarised west bank of the Rhine in March 1936, to be greeted by cheering crowds. Diana, who had just spent a week with Unity in Munich, now travelled with her to Cologne to greet Hitler and applaud this 'victory'. Hitler asked them both to join him for the Bayreuth Festival and the Olympic Games later in the summer.

Few missed the chance to draw a parallel between the attitude of the BUF and that of German National Socialists towards both communists and Jews – especially when the BUF changed its name early in 1936 to the British Union of Fascists and National Socialists, known for short as BU (as it will henceforth be referred to). In fact, Nazi anti-semitism and the Mosley version were not quite the same: where the Nazis preached racial purity, Mosley's emphasis was economic. The country's economic ills, said Mosley, were due to the financial machinations of international Jewry. But to the furious crowds, whether left wing or Jewish, which increasingly surrounded a Mosley march or attended a Mosley meeting, this was a distinction without a difference.

Their opposition came to a head in the two famous 'battles' of 1936: the meeting at the Carfax Assembly Rooms in Oxford on 14 May and the street riots known as the Battle of Cable Street five months later.

The Carfax meeting was an opportunity for the academic intelligentsia to make their feelings known: Mosley, by carrying his views into the heartland of liberal thought, had even seemed to invite protest. The presence of tough trade unionists from the Cowley motor works, along with Mosley's own enjoyment of a rough house, ensured that the protest would take a physical form.

Diana, who knew where Mosley was speaking, stopped at the Assembly Rooms on her way from Swinbrook to London. Aware that there might be violence, she sat to one side. The front rows were occupied by dons, their wives and some undergraduates. Frank and Elizabeth Pakenham sat in the fourth row; also present were Richard Crossman, Philip Toynbee and his uncle Basil Murray (the son of Professor Gilbert Murray).

The meeting began quietly. Mosley strode up to the platform alone, dramatic in his black sweater, trousers, leather belt and (according to Elizabeth Pakenham) 'expression of haughty challenge'. His young blackshirted stewards, rubber truncheons at their sides, were ranged along the walls. As Mosley began to declaim the fascist message, the Oxford contingent in the front rows shuffled their feet, opened newspapers and ostentatiously rattled the pages. Mosley, quick-witted as usual, promptly remarked, 'I'm so glad to see the young gentlemen catching up on their lessons because I hear they're rather behind this year.'

There were a few shouts. Mosley, who was clearly out to needle the socialists, raised his voice and shouted to the workers from the Cowley factories sitting at the back of the hall, 'I know you Ruskin fellows, with your sham guardee accents!' There was a roar of rage and someone called out, 'Red Front!', a popular communist slogan at the time. 'Be quiet!' shouted Mosley, as the audience applauded and the stewards banded together and moved towards the platform. 'Red Front!' yelled several voices and Mosley shouted, 'If anyone says that again I'll have him thrown out forthwith!'

There was a brief pause. Then Basil Murray stood up. He was the mildest of men, but to fascist eyes his appearance was against him. Bespectacled and bearded, in rather shabby clothes, he looked the picture of a communist agitator. In a high-pitched academic voice he cried, 'Red Front!'

'Out with him!' said Mosley and the stewards moved forward, grabbed Murray and threw him out.

As pandemonium broke loose, Diana pressed herself against the wall. Several dons, among them Crossman, leapt on to the platform and the Cowley busmen charged forward, picking up whatever chairs were not screwed to the floor to use their iron legs as weapons. As the Blackshirts and the busmen battled with chairs and rubber truncheons, clouds of dust arose. The violence had erupted so suddenly that many people thought a fire had broken out and mistook the dust for smoke, so that cries of 'Women on to the platform!' added to the confusion.

After the melée had died down, Mosley spoke for an hour and took questions for a further hour.

Despite the anticipation of trouble, the police were nowhere visible: all the arrangements for the meeting had been left to Mosley. The next day, Basil Murray was charged before magistrates with a breach of the peace and fined £2. Frank Pakenham wound up with bruises over his kidneys where one of the stewards had stamped on him, a cut on the head and a headache that after five days was diagnosed as concussion.

As the summer wore on, the marches and tussles grew more violent and the BU ever more militaristic on the Nazi pattern. The appearance of the Leader at meetings was heralded by a stirring fanfare of bugles and drums, then came Mosley escorted by his guard of honour of picked men, then the women's group (none of whom, following the Nazi pattern, was allowed cosmetics), then the largest body of men marching three deep.

Attitudes, often largely motivated by fear – of communism, of fascism – were polarised still further with the outbreak of the Spanish Civil War on 14 July. Friendships ruptured; and fringe politics became more extreme. After one large BU meeting at Hulme Town Hall, bricks and stones were thrown at the new fascist headquarters nearby, breaking every window, and telephone wires were cut so that the local police could not call for reinforcements.

Some families were politically divided, among them the Mitfords. Nancy and her husband Peter Rodd, who had moved steadily leftwards after their earlier brush with fascism, supported the Republican Spanish government (later Nancy went to Perpignan to help the refugees fleeing from Franco). Tom Mitford, whose education had been completed in Germany and who loved its music, culture and philosophy, now began to view German fascism as the only shield against communism. This change of attitude heralded the beginnings of a *rapprochement* with Mosley, of whom hitherto he had violently disapproved: as he saw it, Mosley had persuaded Diana to leave her husband and family life for the ambiguous situation of a known but unacknowledged mistress.

Most affected of all was Jessica. With a clutch of older sisters, Jessica had sharpened her wits since childhood by listening to the opinions of older people and by reading everything she could lay her hands on. Sydney and David never tried to censor their children's reading and Jessica devoured everything from weeklies such as the *New Statesman* to the influential younger left-wing writers such as Stephen Spender, W.H. Auden, Christopher Isherwood and John Strachey. The childish conflict with Unity, both of them taking opposite sides as a matter of course, had now become real. Too young to have been a butt of Nancy's hurtful teasing, Jessica adored her eldest sister for her wit and intelligence; when Nancy took the side of the Spanish communists and later announced her decision to leave for Perpignan it was exactly what Jessica, the convinced communist, longed to do.

Instead, she was brought up to London for her season, and the nearest she could get to leaving the nest and rebelling was sneaking off on her Sunday walk to listen to the communist speakers at Hyde Park Corner. When left-wing stalwarts sang, 'Class conscious we are and class conscious we'll be/And we'll TREAD ON THE NECK of the Bourgeoisie!' on their way to the Hyde Park Labour Festival, the eighteen-year-old Jessica marched with them.

Chants, slogans and songs were an integral part of all the fascist demonstrations. Their primary use was to keep the marchers in step and thus present a more disciplined and therefore awesome aspect. They also served to keep up spirits and weld disparate individuals into a group; but most of all they were employed to provoke. Songs such as 'Crush the tyranny of traitors/Vested power and Marxian lie/Moscow-rented agitators/Strife and chaos we defy,' chanted by tough young Blackshirts could be guaranteed to bring forth a hail of bricks, as could the East End marches, with their even more offensive marching chant, 'The Yids! The Yids! we've got to get rid of the Yids!'

Mosley's increasing anti-semitism ('this alien force that rises to rob us of our heritage') was in part a deliberate recruiting ploy. He realised that a political creed which preached constant action needed an easily targeted enemy if it were to keep up its momentum. 'One must have a scapegoat,' he told his sister-in-law Irene. To Jewish friends he would say disingenuously that his speech next day would contain phrases they 'would not like'.

His tactic was highly successful. The fall in BU membership had been arrested and had even begun to rise again (by November it would be over 15,000). The new recruits, largely young men who liked nothing better than marching, chanting and the chance to use muscle, were exactly the sort he liked, presenting an appearance of well-drilled toughness while enhancing the image of potent, virile sexuality that was so essential to his political as well as his private persona.

Words, of course, were not enough. To give his followers the action they craved, Mosley led them to where they were sure of finding it – the East End of London with its sizeable Jewish population. His pretext was that Jewish slum landlords were exploiting Jewish and other tenants. To his audiences he appeared simply to be preaching anti-semitism, with the result that virtually every meeting ended in a riot.

Matters came to a head in what became known as the Battle of Cable Street, on 4 October 1936. That Sunday, a fascist anniversary was to be celebrated by a number of meetings in different parts of the East End. Simultaneously, there would be a grand parade in Royal Mint Street, where BU members were to assemble for inspection by Mosley at three-thirty, followed by a march down Cable Street through the East End, picking up supporters from the various meetings in Stepney, Bow, Bethnal Green and Shoreditch.

Down one side of Royal Mint Street was a huge hoarding; the other was lined with houses and shops, many owned by Jews. The police, who knew there would be trouble, had put men at the various places where the fascist meetings were to be held, with the largest number in Royal Mint Street. In all, nearly 2,000 policemen were on duty.

When the first fascists began to arrive at the 'parade ground', in twos and threes, from about one p.m. onwards, they found a bunch of angry young men and women shouting and jeering. Soon Royal Mint Street was seething. The police kept order as the growing crowd became more hostile, throwing small missiles and the marbles that caused police horses such difficulty as their hooves slipped on them. Mounted police charged with raised batons again and again. By three o'clock about 1,900 fascists, most in the Blackshirt uniform, had arrived. Weapons had grown uglier, from lumps of scrap iron to chair legs wrapped in barbed wire, which were used vigorously as the occasional demonstrator managed to break through the police cordon.

Meanwhile in Cable Street preparations to stop the march were going on. A builder's yard in nearby Wellclose Square was broken into and a lorry loaded with bricks driven into Cable Street, where it was overturned as a barricade. In a moment ladders, cardboard boxes, old pieces of furniture and rubbish of all descriptions were thrown on to and round it, rendering the street almost impassable. Paving stones were torn up and piled in a heap, windows smashed and the fragments scattered over the ground to prevent police horses charging.

When Mosley arrived at three-thirty to inspect his men, the booing and catcalls changed into an eruption of fury. By now there was a crowd of thousands in and around Royal Mint Street and Cable Street. The streets crackled with broken glass; the demonstrators, howling with anger, hurled bricks, milk bottles and anything else they could lay their hands on at the police, who stood in their way. Adding to the din were fireworks, let off to sound like pistol shots, and the clanging bells of ambulances, racing to pick up the injured.

Almost at once, Mosley was sent for by the Commissioner of Police, Sir Philip Game, who was standing at the rear of the parade. 'I am convinced that if you carry out your march and hold your meetings, serious disorder is certain,' said Sir Philip. 'I have therefore decided that neither the march nor the meetings can be allowed.'

Mosley, who always complied with police requests, agreed immediately. Sir Philip said that he wanted the fascists moved within five minutes. At three-forty the fascists set off, accompanied by 400 police, via Cannon Street and Queen Victoria Street. At four-thirty they halted on the Embankment and were dismissed. About 1,000 walked off to the Black House (their headquarters in the King's Road), where Mosley addressed them.

There was similar but milder trouble at the other fascist meetings. Altogether, there were 85 arrests – 6 fascists and 79 anti-fascists – 30 civilians were taken to hospital and 40 police were injured, of whom 2 were taken to hospital. Mosley issued a statement castigating 'the corruption and decadence' of a government which could not manage to secure free and uninterrupted speech for his movement wherever it chose to hold its meetings. 'I take the view that the fascist doctrine is as un-English and unwanted as the communist doctrine,' commented Sir John Simon, the Home Secretary. 'But the duty of the authorities is to do all they can with complete impartiality to retain freedom of meeting and speech.' The direct result of that Sunday afternoon of provocation and violence was the Public Order Act.

Chapter XIX

After Cimmie's death, Mosley had not been able to bring himself to think of marriage for a year or two. As for Diana, she did not care whether they were married or not as long as they were together; the only point of marriage, she thought, would be if they had children, for she did not believe that children should be illegitimate. But as it became increasingly clear that they were going to spend the rest of their lives together, the first step was to find a home.

The Eatonry was far too small and its *raison d'être* had, in any case, been its proximity to Mosley's Ebury Street flat. Now that he was spending so much more time in the Midlands and the north of England, it made sense to have a base there – and a home in the country, she thought, would be much nicer for Jonathan and Desmond than the Eatonry. She began to househunt for somewhere within reach of Manchester – many of Mosley's speaking engagements were in the Manchester Trade Hall – and Liverpool.

It did not take her long. She asked a friend who worked in a nearby estate agent if he had anything to let in that area. 'Yes, but it's a white elephant,' he replied. When he showed her the photograph of Wootton Lodge, in Staffordshire, she said at once, 'I *must* look at it.'

She and Mosley drove up one spring morning. Twenty-five miles from Derby they entered a winding drive one and a half miles long. Tall beeches, newly green, stood on each side. As they turned a corner, they emerged from the trees to see an exquisite early seventeenth-century house facing them. It embodied all the classical features that were Diana's passion – stone, lakes, pediments, trees, lawns, foliage, all grouped together in a perfectly modulated harmony. The house, with its elegant façade, its two rows of deep sash windows and eighteenth-century panelling in almost every room, stood on a rocky knoll. In front a terrace descended by shallow stone steps to a lawn; at the back more lawns sloped down to two lakes. As they drove up, the owner, Captain Unwin, VC, the hero of the *River Clyde* in the Dardanelles landing, was waiting on the steps.

'How do you do, Mrs Guinness,' he said courteously, then turned to Mosley, 'and Sir Oswald too, I see.' Wootton had become too much for him, said Captain Unwin, it was so large and he was getting old. Diana had fallen in love with the house at sight, and almost at once she and Mosley agreed to rent it for £400 a year.

Wootton stood in what had once been a lead-mining area and the hills around were riddled with shafts from disused mines. The countryside was wild, beautiful and heavily wooded, and the land was poor. The house, large and icy cold, was completely unmodernised, with only two bathrooms. Diana installed the central heating they both considered essential to survive the Derbyshire winters, but they deferred the idea of more bathrooms, planning to put them in if, as they hoped, they could eventually buy.

Wootton was almost entirely Diana's house. Although Mosley paid the rent and the wages of the two gardeners, all other costs were paid by her, largely from Bryan's allowance of £2,500 a year. She found the usual excellent cook, this time a man called Grimwood, who was so devoted that he got up early so that Diana could have freshly baked rolls every morning; he was paid the then enormous wage of £90 a year, from which he purchased for £15 a small car for visits to the cinema in Ashbury six miles away. The butler, James, had originally been footman in Cimmie's Smith Square household and had been persuaded by this promotion to come to Wootton with his wife; and there were two or three young housemaids at £20 a year each. Diana also had to continue paying the £300 a year rent for the Eatonry, since it was on a repairing lease and she could not get rid of it (not until the lease finally expired at the end of the war did her tenancy come to an end).

Mosley gave up the expensive Ebury Street flat for a much cheaper one at 129 Grosvenor Road, on the Embankment, with two enormous rooms, one with pillars down the side, and several small ones. Originally a garage, it had had a brief incarnation as a fashionable night club, the River Club; but when the leaders of society returned to their more usual haunts in Mayfair its enormous dance floor and velvet sofas were abandoned. Diana painted the largest room pale blue and its pillars white. It cost a lot to heat but the rent was low; when they stayed there they 'managed', as Diana put it, with a butler and a parlourmaid.

Furnishing Wootton was a problem, as Diana had little disposable cash and the pretty furniture from the Eatonry hardly filled one of Wootton's large rooms. Once again, Bryan came indirectly to the rescue. She sold the exquisite diamond and ruby bracelet he had given her during their brief marriage, writing to tell him what she had done. It fetched £400 – enough to furnish Wootton completely.

Ironically, much of what she bought was the furniture she had grown up with. Two or three weeks earlier, her father had sold Swinbrook – typically, as his children pointed out, at the bottom of the market – retaining only the Old Mill Cottage next to the Swan public house in the village. Most of the large family for which Swinbrook had been built had flown the nest: Nancy was married to Peter Rodd, Diana leading her own life, Tom was living in London as a young barrister, Unity was away most of the time and Pam had married the brilliant and extremely rich scientist Derek Jackson. When David sold the furniture from Swinbrook at auction in Swindon, most of it was bought by Diana – for a song. She acquired a superb Hepplewhite mahogany canopy bedstead, its posts carved into foliage, for ten guineas, antique mahogany dressing tables for £4, an inlaid Sheraton bow-fronted sideboard for £14 and more mundane articles like beds for a few shillings apiece.

Though not essentially a countrywoman, Diana was blissfully happy at Wootton. It was as unlike her London life as it was possible to be, but it was her first home with Mosley, who also loved it deeply. On summer evenings he would shoot rabbits with a .22 rifle or fish for trout in the lakes; after dinner they would play classical music on Diana's gramophone. In the winter her sons would toboggan down the steep slopes of nearby Weaver Hill. In the spring the woods were filled with bluebells.

From Mosley's point of view the move had even more to commend it. With Diana now so far from Savehay, his sisters-in-law need know even less about her. Both Irene and Baba had known since the previous year that Mosley was still seeing Diana and there were occasional furious rows. 'This is intolerable, impossible!' Irene would cry dramatically from her position as unofficial mother to his children. 'You promised not to see Mrs Guinness again!' Sometimes Mosley would placate her, telling her soothingly that his occasional glimpses of Diana meant nothing; at other times he would angrily rebuff her, shouting that his life was his own. But both Irene and Baba hoped that a Diana buried in the depths of the country would, with luck, lose much of her power over Mosley.

Much as she loved Wootton, Diana's peripatetic life continued. She went constantly to Munich to see Unity as she had determined when she first met

Hitler, and since she had taught herself German, chiefly through reading books, newspapers and poetry, she could now chat away as fluently as Unity. It was not long before Hitler confirmed his invitations to both sisters to the Olympic Games to be held that August in Berlin, to the Wagner Festival at Bayreuth afterwards and then to the Parteitag in September. Within only a year, they had become part of the inner circle of companions on whom he depended for relaxation and entertainment.

Unity now virtually lived in Munich, her life tracking Hitler's. Over a five-year period she saw him 140 times to talk to (she did not count nods and greetings), with each occasion marked in red in her diary; and the words 'He' and 'Him' spelt with a capital H when they referred to Hitler. She gave him one of the collages she made from time to time from scraps of material, pieces of wool, silver paper and pictures cut from magazines. It depicted Hannibal crossing the Alps.

Yet in many ways he was more intrigued by Diana, whom he saw with Unity when she came to Munich. He would ask the sisters to luncheon in his flat, the food being sent in from a nearby three-star restaurant.

But Diana's friendship with Hitler really grew in Berlin, where Unity seldom went. For the next three years, whenever she arrived in Berlin, she would telephone the Chancery, Hitler's official residence since the end of 1934, and leave a message that she was staying at the Kaiserhof Hotel. If Hitler – who far preferred Munich and the Obersalzburg – was there, there would be a telephone call inviting her to visit him. She would walk across the Wilhelmplatz and would be shown in to see the Führer in his private sanctum. Sometimes these encounters were for lunch, sometimes for dinner, sometimes the two of them were alone, sometimes a few cronies were there. 'At the Führer's. Both the English women there [Diana and Unity]. The four of us ate lunch alone. It was very nice,' runs Goebbels' diary entry for 6 June 1936. Often these meetings were late at night (the insomniac Hitler seldom went to bed before two or three in the morning) and were later regarded with deep suspicion by MI5.

If it was an evening visit, Diana might watch with Hitler one of the films he adored, at other times they would simply chat for hours *tête-à-tête* in front of a log fire. But although it seemed a picture of domestic cosiness, the darker vein of German politics was always near the surface. 'Afterwards again at the Führer's with the Mitfords. He spoke about the Jewish and Bolshevik danger. He will suppress them in Germany. The world can do what it wants,' wrote Goebbels on 22 July.

Their intimacy with Hitler meant that Diana and Unity knew most of the senior Nazis, and often their wives also. Magda Goebbels was Diana's favourite; an immediate *rapport* had formed between them and they often spent the day together. For the Olympic Games, Magda invited the Mitford sisters to stay in the Goebbelses' lakeside villa at Schwanenwerder, about ten miles outside Berlin. Goebbels, who viewed them both, Diana in particular, with suspicion, was less than delighted. 'I had a row with Magda about the visit,' he wrote on 2 August. 'She cried and I was so sorry . . .' The sisters were put up in a small bachelor annexe called the Kavalier Haus instead of the main villa.

Built around 1910, this was plain and comfortable and done up by Magda Goebbels in a style and taste as far from the heavy furniture and plush-swamped décor of the Berlin of the 1890s as she could manage.

The Goebbels lived extremely simply. They would not, for instance, automatically sit down to a meal at lunchtime or dinner, but ate only if they were hungry; this was, Diana imagined, a legacy of the hard times of post-war half-starvation and the collapse of the currency. As the days passed she became even fonder of Magda, whom she came to know well, but she disliked the Doctor. Despite his 'marvellous blue eyes' and 'lovely speaking voice', she sensed an underlying hostility.

She and Unity were taken every day by chauffeur-driven car to the games, which began through a fine drizzle and in an atmosphere which made it plain they were to be a celebration of German supremacy. Thousands of pigeons were released, the first German marathon winner presented a laurel wreath to Hitler and the ceremony concluded with Handel's 'Hallelujah Chorus'. A triumph for the German cause, thought Goebbels, but Diana was quickly bored by the various contests of speed, strength and skill.

The Wagner Festspiel, at which Hitler introduced the sisters to Frau Wagner and her four children, was much more to Diana's taste. There she saw the 'Ring' and for the first time *Parsifal*. When she told Hitler that she liked it the least of Wagner's operas he replied, 'That is because you are young. You will find as you get older that you love *Parsifal* more and more.' After performances they would all have supper at the long table reserved for Hitler in the restaurant nearby.

In August Diana joined Mosley and his older children, this time at Sorrento. It was her second holiday with Vivien and Nicholas, and they were gradually beginning to notice her as someone who was more than one of the floating, ephemeral crowd of their father's friends. It dawned on them that she had the same relationship with their father as their Aunt

Baba. They enjoyed Diana's company: she would play card games with them when their father was having his siesta and she would talk to them as if they too were grown up. She would tease Nicholas, who was passionate about cricket and always anxious to get hold of English newspapers, that the cricket score was the only thing he wanted to read.

At the end of the holiday, she went, as before, to Munich for the Parteitag, to meet Unity and, this time, their brother Tom as well. By now virtually all pretence had been abandoned that this annual peak of the Nazi calendar was anything other than a purely militaristic event. Through the narrow streets of Munich marched 100,000 well-trained, fit-looking German soldiers (over three times as many as had reoccupied the Rhineland six months earlier). It took five hours for them to march past Hitler, standing immobile in the central square, Goering at attention below him.

The ceremonies were similar to the three earlier ones the sisters had witnessed – the same flags hung round the vast stadium, the same seeming wall of blue light round the outside rising to the black sky above, where only the brightest stars could be seen, the same red banners carried by brownshirted *gauleiters*, the same dramatic, solo entrance by the Führer. But this time the speeches in the Luitpold Halle were even more outspoken, with communists and Jews, 'enemies of the Reich', unremittingly denounced throughout hour-long harangues from one high-ranking Nazi official after another. Only when Hitler spoke, though, were the crowds roused to frenzy.

The three Mitfords, as favoured guests, had been given seats in the third row. They were also given the signal honour of a smile of greeting from the platform every time the Führer spoke. Unity, sitting in the third row with Diana and Tom, cheered so much that by the third day she had lost her voice.

Once again, Eva Braun was sitting beside Unity but this time it was unlikely that Eva was jealous of the handsome blonde foreigner. At last she was certain of her position: Hitler had installed her at the Berghof when his sister Angela left to get married. He had even remodelled the house's interior to give Eva a bedroom, boudoir and bathroom next to his own quarters. As resident mistress, she was now securely at the heart of his emotional life.

Mosley had been gradually coming to the conclusion that he and Diana should marry but he was determined that the marriage should be kept secret. He told Diana that remarriage would detract from his political image and he wanted, genuinely, to protect her from possible attack by his most virulent political enemies. What he did not tell her was that secrecy would also enable him to continue his affair with Baba.

Both of them knew that if they wanted their marriage to remain a secret, it could not take place at a London register office, where the press kept an eye on notices of intended marriages. They thought of Paris, then discovered that for any marriage by the Consul, notice also had to be displayed publicly for three weeks and any British journalist who wandered into the Consulate could see it.

Diana, on one of her trips to Munich, asked the Consulate there what the position was in Germany. She learned that there was a reciprocal arrangement whereby German nationals could be married in England by an ordinary registrar and vice versa, again with a three-week public display of notice. But in Munich there was one crucial extra factor: she had the ear of the autocrat who ruled the country. She asked Hitler if he could stop the usual notice being put up and the Registrar was duly instructed to drop it in a drawer.

Magda Goebbels, by now devoted to Diana, organised the wedding, from paperwork to buying the rings and planning a wedding breakfast at Schwanenwerder – much against her husband's secret wishes. 'At Führer's in evening,' he wrote on 17 September. 'Magda there, also Ms Ginest [his spelling of Diana's name varied according to his temper]. She marries Mosley. Beginning of October. The wedding is to be with us, at Schwanenwerder. That's not all right with me. But the Führer wants it to be so. Magda has flung herself into organising it.'

Diana and Mosley were married on 6 October, in the Goebbels' ministerial house at 20 Hermann Goeringstrasse. The Chancellery and Number 20, in parallel streets, were exactly opposite each other, separated by the Chancellery garden, an expanse of lawn so large and with such magnificent trees that it was more like a small park, and the Goebbels' smaller garden.

It was a prospect that drew the eye. Diana, changing in an upstairs bedroom into her wedding clothes – a long, gleaming pale gold tunic worn over an ankle-length black skirt of fine silk velvet – glanced out of the window at the trees, just beginning to turn colour. As she watched a few yellowing leaves float down, the uniformed figure of Hitler appeared through the trees. She watched as he walked across the grass from the Chancellery, his adjutant keeping step behind him. The adjutant's arms were filled with roses, carnations and chrysanthemums; one elbow pressed a large square box to his side. It contained Hitler's wedding present, a signed photograph of himself in a silver frame with the monogram AH surmounted by the German eagle, in a leather case lined with red velvet.

The ceremony took place in the long, low ground-floor drawing room with its simple furniture and curtains. Mosley and Diana exchanged rings and signed their names on the marriage certificate. Mosley's best man was Bobby Gordon-Canning, his friend Bill Allen was one witness and Unity the other. Hitler and his adjutant, Goebbels, his sister and his adjutant and Magda, were the only others there. Despite Goebbels's less than whole-hearted feelings about Diana, he gave her a sumptuous present: a complete set of Goethe's works in twenty volumes bound in pink calfskin.

At the wedding breakfast, a luncheon for twelve at the Goebbelses' villa at Wannsee, with a special vegetarian dish for Hitler, Diana regaled her host with a blow-by-blow account of a scandal still known only to a small circle: the new King of England was obsessively in love with the American Wallis Simpson, who was about to divorce her second husband.

At Mosley's request, no photographs of the wedding, or the luncheon afterwards, were taken – the Mosleys felt that any pictures would inevitably have found their way into the newspapers.

After their marriage, Hitler suggested Diana give him the marriage certificate for safe keeping. She did so (later, its absence was to prove an embarrassment). That evening, the newly wed couple went to the Sports Palace, where Hitler and Goebbels were inaugurating a vast welfare organisation, before dining at the Chancellery (the last time Mosley was to see Hitler).

Two days later, Mosley went back to England, still ostensibly a single man. Apart from the two friends who had been at his wedding, he had told no one, not even his mother. 'If you tell only two or three, they tell two or three – and then the world knows,' he said to Diana.

She, however, insisted on telling her parents, though swearing them to secrecy. It was a chance to break the cycle of anger and disapproval that had built up since she left her first husband. 'They've got all these old-fashioned ideas about marriage,' she said to Mosley, 'and it will be a comfort to them.'

She also told her brother Tom; he had been deeply hostile to Mosley for, as he thought, persuading Diana to leave Bryan and she wanted the two men she loved most to like each other. But Pam, Nancy and her two youngest sisters were kept in the dark.

* * *

Marriage did not alter Mosley's way of life. The speaking engagements continued, often with an hour or two of talk with the faithful afterwards. Worried by the fear of fog or ice on those lonely roads, Diana made him promise that, however late, he was to come into her bedroom and tell her the moment he was back.

Diana spent much of her time at Wootton alone. Her younger unmarried sisters were not allowed to visit her as she was apparently still 'living in sin'. A few old friends came to stay: Jim Lees-Milne was there when the King made his abdication broadcast on 7 December and they listened with tears pouring down their faces.

Diana's new husband spent Christmas not with his wife but at Savehay with Irene, his mother, his children, and Baba, who was still his mistress. Frustratedly longing to be with Diana, he was foul to them all, starting an appalling row at dinner because no green vegetables were served.

This miserable state of affairs was exacerbated by the Curzon sisters' growing disenchantment with Nazi Germany and their brother-in-law's increasingly suspect political orientation. There was also a rumbling disagreement over finance. Earlier in the year, the Official Solicitor had ordered that the children's payments towards the upkeep of Savehay Farm should be reduced; the purpose of his original order, he said, had not been to free their father of all financial liability towards his family so that he could devote his entire income to fascism.

Diana, on one of her constant visits to Germany, missed Mosley dreadfully. 'it is a long time since I wrote to tell you how much I love you because my letters are bad and stupidly written,' she wrote from Berlin. 'But today as my heart is full of love I shall write what is always in my thoughts; and that is that I love you more than all the world and more than life; that to be with you is paradise and each parting from you a little tragedy which does not become less with the years . . .'

(*Above*) Diana's father, David Freeman - Mitford (later Lord Redesdale). He enlisted as a private in the Boer War. Photographed in 1900 in Kimberley, with his dog, Marker.
(*Below left*) Diana's mother, Sydney Bowles. She married David Mitford in 1904.
(*Below right*) Diana aged three.

(*Above*) 1912: David and Sydney with their older children: Nancy, Tom, Diana, Pamela.
(*Below*) 1918: Growing family: Diana, Pamela, Unity, Tom, Jessica on Sydney's lap, Nancy.
Photographed in 1918 at Batsford House, the Redesdale family home.

Asthall Manor where Diana spent her childhood from nine to sixteen. The family moved in 1926 to Swinbrook, a house designed and built by David himself.

(*Above*) Deborah, the seventh and last of the Mitford children, photographed on Diana's lap, 1922.
(*Right)* Diana.

Annual family photograph: Sydney and David with Nancy, Tom, Diana (middle row, with plaits), Pam, Unity, Decca, Debo.

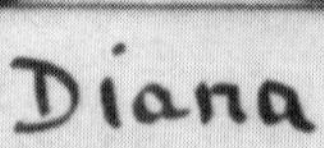

Aged sixteen, in Paris.
Dry-paint etching of Diana by Helleu.

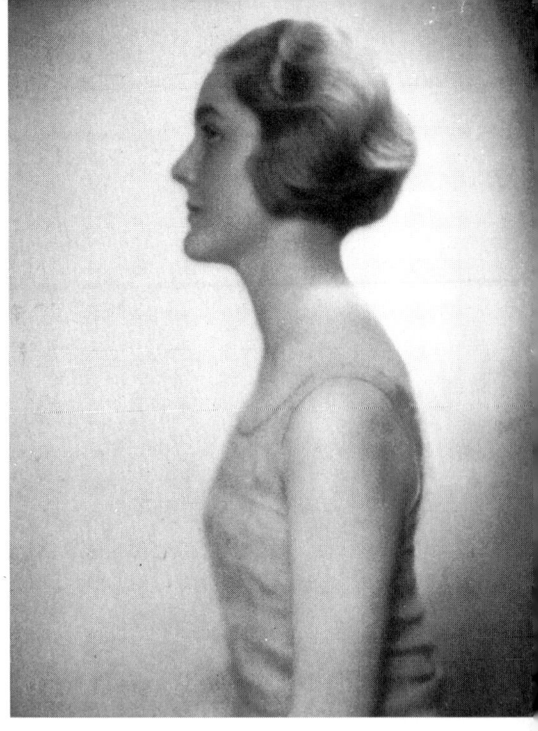

Diana after she became engaged to
Bryan Guinness.

Diana in her wedding dress. She was only eighteen when she married Bryan Guinness on 30 January 1929.

FACING PAGE

(*Above left*) Diana and Bryan two days before their wedding. (*Above right*) On honeymoon in Sicily. (*Below*) Jonathan's christening, 1930. Nanny Higgs holds the baby between Bryan and Diana.

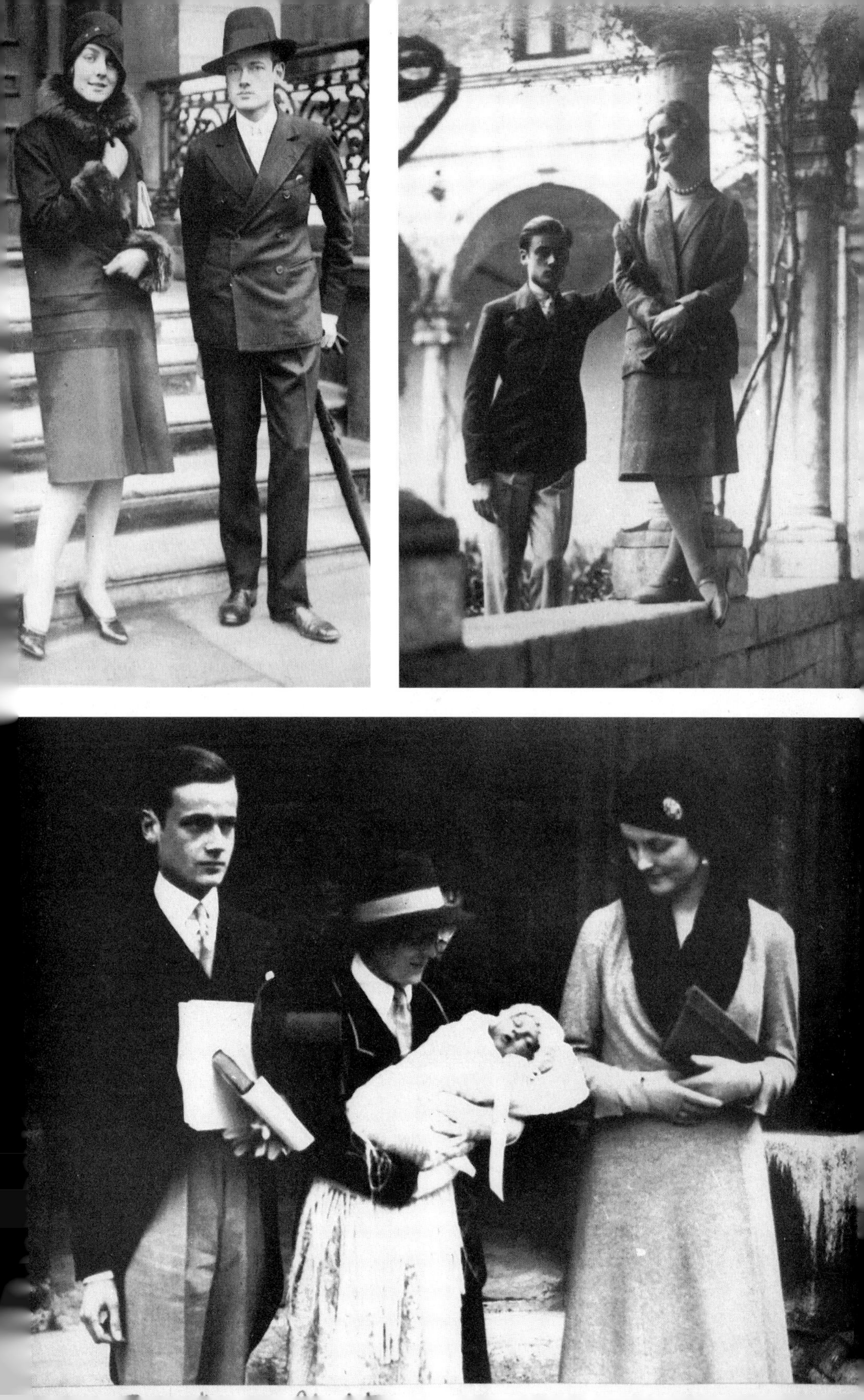

Bryan and Diana at the window of their first home, 10 Buckingham Street.
Diana's brother, Tom, on the doorstep.

Chapter XX

The BU was desperately short of money. For the first time, Diana began working actively for the fascist cause; first, by attempting to raise funds from the Nazis and, later, by negotiating a contract with the German government. In both cases, her leverage was her friendship with Hitler.

Even then, three years away from war, her actions bordered on the unpatriotic. Diana, though, saw her conduct in a different light. When she first fell in love with Mosley, absorbing his beliefs until her whole being was saturated with them, he had told her that she could help him. Now, at last, here was a chance to do so – and in a way that no one else could.

At first, her fund-raising took the form of simple pleas for money to sustain the British fascist ideal. If the journals of Dr Goebbels are to be believed, these requests occurred frequently.

> 24 April 1936. Lady Mitford's [Unity's] sister sent on behalf of Mosley. Wants Germany to give £50,000–£100,000 credit to the Morgan Bank. That would be wonderful, a great blow against falling into financial collapse . . . and with that Mosley would be saved. The Führer will investigate this.
>
> 25 April 1936 . . . at Schwanenwarder we have a discussion with Mrs Giness. She repeated again the request from Mosley and his henchmen.
>
> 20 September 1936. Again to the Führer. Mrs Genest there. Three short films of the Mosley Party. Everything is still very much at the beginning but it will probably work out. We, also, began that way.
>
> 5 December 1936. The Führer has now released the money arranged for Mrs Guinness. As a result there will be peace. But they need so much.
>
> 7 February 1937. Mrs Ginnest wants more money. They use up a fortune and accomplish nothing. I am having nothing more to do with this thing. Refer her to Hanke and Wiedemann.

These constant visits to Germany could not escape the attention of the press. Unity's links with Nazis were already well known; now numerous photographs of Diana, with or without her sister but with Hitler, began to appear. Later these would link Mosley with the highest echelons of Nazi Germany in the public perception.

The reason for the BU's sudden and calamitous lack of funds was not difficult to discover. The disappearance of the Italian subsidies had been followed by the unsavoury publicity engendered by the Cable Street affair. Few rich men, or their businesses, wanted to be seen to contribute to such an extreme and disruptive organisation, and many who had given donations on a clandestine basis were also frightened off.

The Public Order Act, passed on 18 December 1936 in the wake of the disturbance, had dealt the BU a further severe blow. By giving the police the right to call off marches and demonstrations and by forbidding the wearing of uniforms, it neutralised at a stroke two of the key elements of the BU's appeal: its militaristic approach and the chance of constant, provocative action. By the end of 1936 the BU was nearly bankrupt and Mosley's own finances were in a parlous state.

Mosley's first action was to cut expenses by sacking most of the BU's paid staff, including William Joyce and John Beckett. He then mortgaged his estate for £80,000, and put a total of £100,000 into BU funds.

Next he turned to Savehay. He could no longer, he said, afford to pay anything towards its upkeep; he was going to let it – and his children could come to Wootton.

The Curzon sisters were appalled. They did not want their family home to be broken up, and they were horrified by the thought of Cimmie's children being in the same house as the woman they held responsible for her death. It also seemed a blatant flouting of the proprieties since – thanks to Mosley's secrecy over his marriage – they believed Diana was his mistress. 'Baba is writing to him he has no right to put growing-up Vivien in such a position. Oh dear! God help us, help us,' wrote Irene in her diary that June.

Two days later, through his long-suffering mother, Mosley put forward a plan he had no doubt held in reserve. Irene wrote: 'Ma said her proposal from Tom was distasteful and she felt very badly about it. It was this: if I wanted to keep Denham for the children I should take over Tom's £1500 annually and get the children to repay me on coming of age as he could no longer afford to spend any money on them.'

The blackmail worked. To the outrage of her solicitor, Irene complied, agreeing to pay her brother-in-law's share out of her own pocket.

But these were only stopgap measures. What was needed for the BU was a source of regular income. The answer, which could have made Mosley a millionaire, was commercial radio.

Mosley's espousal of this new concept was based on another man's idea – this time, that of his old friend Bill Allen. Allen, a Belfast MP from 1929 to 1931, pointed out that the Conservative MP Captain Leonard Plugge, who had secured a licence from France to found the very successful English-language commercial station Radio Normandy, had become extremely rich. By 1935, British advertisers were spending about £400,000 a year on commercials, a figure which would go up to £1,700,000 in 1938, mainly with Plugge's Radio Normandy and the other major commercial station, Radio Luxemburg.

Mosley saw the possibilities immediately. Commercial radio stations had of necessity to be based outside the United Kingdom, since the BBC held the broadcasting licence for the whole of Great Britain and Northern Ireland and strenuously opposed any attempt to chip away at its monopoly. With a network of stations, ideally to the east, west and south, the whole of the UK could be covered. Bill Allen's radio contacts were excellent: he already employed as a consultant a man who was arguably Britain's top radio expert. This was Peter Eckersley, for many years the BBC's chief engineer, who combined unrivalled technical knowledge with a grudge against the corporation and strong fascist sympathies.

Peter Pendleton Eckersley had been divorced by his first wife in 1930 and was promptly sacked by the BBC, then in the iron moral grip of Sir John Reith. Eckersley, understandably, was enraged. His resentment was fuelled by his second wife, Dorothy, whom he married in October of the same year.

In 1935 Dorothy, a member of the Independent Labour Party until then, suggested to her husband that they take their summer holiday in Germany, where, as she said later, 'We saw something of the social benefits brought about by Nazi policy.' For Dorothy, deeply impressed by the economic and social benefits brought about by the Nazis, it was a Pauline conversion. She became known as a strongly pro-German fascist and a fanatical admirer of Hitler, scattering photographs of him round the Eckersleys' Swan Court flat. She became a friend of Unity's and an intimate of William Joyce, whom she chose as a tutor for her son by her first marriage.

The Eckersleys continued to spend their summer holidays in Germany and in 1937 were taken by Unity, who had been lunching that day with Hitler, to observe the Führer in his favourite Munich tearoom. 'And there,' in Dorothy's words, 'we gazed upon him.' That December she became a daily announcer on German radio, although she soon realised that she was taking part in propaganda programmes directed against her own country as well as those beamed to it extolling Germany.*

Peter Eckersley, almost as enthusiastic about Hitler as his wife, earned £1,000 a year from the Allens. Mosley now paid him a further £2,500 a year (a payment that exposes the dishonesty of Mosley's claim that he could not afford his £1,500 per annum share of the upkeep of Savehay for his children).

Mosley believed that secrecy as to his own part in the proposed radio network was essential. Italy's invasion of Abyssinia had done much to complete the discrediting of fascism begun by Nazi thuggery and he was well aware that not only would many potential advertisers be alienated if they thought they were subsidising the BU, but potential listeners might also switch off. Apart from his talents in radio, Eckersley had one supreme virtue: he was immensely discreet.

As head of the Ulster advertising firm David Allen and Son, Bill Allen had a legitimate reason for an interest in commercial radio, but his two younger brothers were bitterly hostile to any kind of business involvement with Mosley. Bill, as the eldest and most forceful, overcame their opposition and pushed through the partnership with Mosley.

The associates founded a company called Air Time Ltd, which had no apparent connection with Mosley. It was run by a Scottish accountant, James Herd, whom Eckersley had recruited from his post as Deputy Principal of the LCC Commercial Institute in Catford. Like everyone connected with Air Time, Herd had to sign an undertaking that he would never disclose Mosley's involvement. As time passed, this would be further shrouded behind a smokescreen of subsidiary companies – on none of which would Mosley's name appear as a director.

The plan was to set up radio stations that transmitted entertainment, music, news and advertisements in the Republic of Ireland, Belgium, Heligoland and Sark. Since there was little light entertainment on the

* In July 1939 Dorothy Eckersley left England for Germany, remaining there on the outbreak of war and continuing her job until October 1941. On 10 December 1945, at the Central Criminal Court, she was found guilty of assisting the enemy by broadcasting and sentenced to 12 months imprisonment.

BBC, they felt sure there would be a ready market. Sir Oliver Hoare, brother of the Home Secretary, agreed to handle the Belgian negotiations. Colin Beaumont, son of Mrs Hathaway, the Dame of Sark, was a Mosley follower, and believed he could persuade his mother to agree to such a station.

The island of Sark had been granted legal autonomy by Elizabeth I which, Mosley believed, meant that the BBC would be unable to block a station there. Before going ahead, however, it was necessary to find out if this was the case. The task fell to Frederick Lawton*, a young barrister in the chambers of C.T. Le Quesne, KC, a distinguished silk with roots in the Channel Islands.

Mosley had first met Lawton, the clever son of a prison governor, when he had debated at the Cambridge Union three years earlier and had made a great impression on the young man. Shortly after Lawton had been called to the Bar (in 1935) he heard that Mosley was suing for libel and conducting his own case. Lawton went along to listen. Mosley, who won the case, spotted him in court and went up to talk to him afterwards. Learning that Lawton was now a barrister, Mosley began to give him small but regular jobs of work through his solicitors. Lawton, who admired Mosley immensely, was grateful to him for his kindness.

Le Quesne advised that the legal situation vis-à-vis the Channel Islands would have to be investigated in depth. It was not a job for a silk, so he suggested Lawton, knowing he was a protégé of Mosley's. After several visits to Sark and a fortnight at Caen University researching the medieval land law relating to Normandy to find out on what terms the Dame held the land, Lawton advised that the enterprise might, just, be possible. A company called Museum Investment Trust was set up, fronted by Mosley's solicitor Gerald Keith, and in the spring of 1937 an agreement was concluded that, if setting up a station proved possible, there would be a thirty-year contract under the terms of which Colin Beaumont would receive 25 per cent of all profits.

Simultaneously, feelers were being put out about the establishment of a station in Germany, which could be far more important and lucrative than Sark. It could provide an invaluable bridgehead into continental Europe; German technical expertise was famous and, above all, Mosley had a direct line to the German authorities through Diana. He decided to pursue the German possibilities. As usual, the moment he had reached his

* Later the Rt. Hon. Sir Frederick Lawton, a Lord Justice of Appeal.

decision, he acted, telephoning Frederick Lawton. It was a Sunday, very early in the morning and Lawton was newly married, but none of these considerations affected Mosley.

'Could you possibly come up, now, to this house near Ashbourne that I've rented, to discuss this radio business?' he said. 'John will drive you up. He'll be with you directly.'

Hastily Lawton shaved and dressed and within half an hour the doorbell of his flat was rung. Outside was a large Bentley, its driver Mosley's youngest brother John.

The two men talked generalities until John Mosley said to his passenger, 'I have something to tell you. You are going to meet my brother's wife – you didn't know he was married, did you?'

'I thought his wife had died,' replied Lawton.

'She did,' said John Mosley. 'He has married again. But it is secret and no one else must know.'

Lawton was staggered. Only recently he had been summoned by Mosley to the hotel on Lake Vyrnwy to discuss the legal implications of the Sark situation. Lake Vyrnwy was famous for its fishing and Mosley had joined Irene and his children at this Montgomeryshire beauty spot, he to fish, they to walk, picnic and swim, and Lawton had been struck by how close Mosley and his sister-in-law seemed.

They arrived at Wootton in time for lunch. Lawton already knew Diana's brother Tom through the Bar – they were both members of the Inner Temple and usually sat at the same table for lunch – but it was his first social contact of any kind with what he thought of as 'the aristocracy' and he was dazzled. Mosley, at home, was expansive and charming; Diana's appearance overwhelmed him. 'She was so *strikingly* beautiful,' he said later. 'And so nice.' He quickly felt at ease, and his two or three visits there – including one in the autumn, when he went ferreting with Mosley – sealed his admiration for both Mosley and Diana, and when Mosley asked him if he would, if necessary, go to Germany with Diana to assist in negotiations, he was happy to agree.

As had become their custom, Diana and Unity went to the Nuremberg Rally at the beginning of September. This time, their brother Tom went with them. By now smart and comfortable hotels had sprung up to welcome the foreigners whom Hitler was anxious to impress – and who brought with them valuable foreign currency. The cheerful, bustling friendliness, the optimism and enthusiasm of the people, and the ritualistic, compelling drama of the rally itself, affected Tom deeply. Shortly afterwards he joined the BU.

After this annual high point of the fascist year – although Diana did not then know it, this was to be her last Parteitag – Diana turned her attention to the question of a radio station in Heligoland. Her first approach was through the official channels of the Ministry of Propaganda. It was unavailing. Then, on one of her trips to visit Unity, she brought the matter up with Hitler.

At first he was dismissive. On 9 October 1937 she received a letter from his adjutant, Captain Wiedemann. 'I have today reported to the Führer the whole matter concerning the advertising transmissions,' he wrote. 'The chief objection was raised by the appropriate military authorities. The Führer regrets that in these circumstances he is not able to agree to your proposal.'

But Diana did not take this as final. She knew how suspicious the top Nazi officials were, but she knew also that she was a favourite of Hitler's. She had faith in her power to persuade and charm and, for Mosley's sake, she was utterly determined. She continued with her visits, making contact with Hitler whenever she arrived in Berlin and bringing up the subject of the radio station whenever there was an opening. Usually this would be when Hitler was in a mellow mood – there was no company he liked better than that of beautiful, sophisticated and safely married women. As they chatted late into the night in his private rooms, one by one his officials would slip away gratefully, leaving 'Mrs Guinness' with a clear field.

Diana's persistence was the more important because Mosley was suffering an enforced convalescence. At a meeting in Liverpool in October, where the usual high feelings had erupted, he had been hit by a brick. He spent the two months it took him to recover from concussion and cuts to the head at Wootton. Since no one knew that he and Diana were now married, this apparent open flouting of the conventions added to the scandal.

His convalescence also increased the frustration Mosley was suffering because of the delays in getting his radio project off the ground; it was taking far longer than anyone had anticipated. He confided many of his worries about money to Lawton though, as Lawton later discovered, he was not telling the truth: on numerous occasions Mosley reiterated that despite the rumours he had 'never had a penny' from Mussolini. When the truth about the BU's Italian financing eventually emerged years later Lawton was shocked to learn that the man he had admired so much had deliberately lied to him.

Despite this secretiveness, Mosley was now relying to a considerable extent on the young lawyer for business advice. Lawton saved him from at least one potential disaster. So desperate was Mosley for a scheme which would produce a fortune quickly that he became enthusiastic about the claims of a young chemist to have invented a formula for a new 'rejuvenating' pill.

It was just the kind of idea to appeal to a man for whom virility was as important in politics as in personal life. The chemist was a persuasive talker and Mosley foresaw a golden harvest. But Lawton doggedly insisted that before any money was sunk into production of this pill it should be sent to the public analyst for a report. When Mosley finally agreed that this was a necessary first step, the chemist and his pill melted away.

Chapter XXI

While Diana was dashing backwards and forwards to Germany, another of the Mitford sisters was in the headlines. Jessica, youngest but one of the family, had been presented at court in the spring of 1935. She had been bored by her first season and was even more bored by her second. At the beginning of 1937 she ran away with her second cousin Esmond Romilly (she is thinly disguised as Jassy in Nancy's novel *The Pursuit of Love*).

Pretty, clever, sharp and irreverent, Jessica was the little sister with whom Diana had had the strongest bond during their childhood, the protégée whom she had helped and encouraged; in the factors that led up to Jessica's elopement there were echoes of Diana's own life.

Like Diana, Jessica longed to get away from Swinbrook and to break into a different world. Like Diana, she had been deeply affected by the miseries of the 1920s and the Depression of the 1930s and, as with Diana, these had formed her political views. As with Diana, these views were extreme – though to the left rather than the right. Like Diana, she had no difficulty in discounting the horror stories reported as committed in the name of her chosen creed as unreliable propaganda by ideological enemies. Above all, for both sisters, man and cause together crystallised into one irresistible, indivisible whole.

Esmond Romilly, almost a year younger than Jessica, was physically unprepossessing – short, thick-set and round-shouldered, with a square white face and dark hair that lay dankly across his head. But his personality was so powerful that his looks were quickly forgotten. He was clever, fearless, unscrupulous and enterprising, an originator and natural leader. When still at school he had founded *Out of Bounds*, a magazine with the subtitle *The Public Schools' Journal Against Fascism, Militarism and Reaction*. It was designed to encourage public schoolboys to rebel against the system and its material came from the disaffected among them.

Esmond himself was always getting into trouble and quickly became too much for his parents – at one point he was sent to a remand home for six weeks. He and his brother Giles wrote a book (also called *Out of Bounds*), telling the story of the magazine and their adventures in founding it, which enhanced his glamorous reputation among his peers, while his energy, fluency and vitality allowed him effortlessly to dominate friends such as Philip Toynbee. When the Spanish Civil War broke out Esmond took part in the Republican government's successful defence of Madrid in the winter of 1936, fighting gallantly in an action in which almost all his comrades were killed, before being sent home on sick leave in January 1937. He was one of Jessica's heroes before she had even met him; when she did, she fell instantly in love. It was not long before she said she longed to work for the communist cause in the Civil War and Esmond, whose instinct was always to rebel against the conventional, agreed to take her back with him to Spain.

First, though, she had to get away from Swinbrook and her parents' watchful eyes without arousing suspicion. She told her mother that she had been invited to stay in Dieppe with friends known to Sydney, and she got permission to go. She even went to Dieppe, but the 'friends' were, of course, Esmond. It was here he told her that he had fallen in love with her and the escapade became an elopement. Together they set off for Bilbao, having made arrangements for cards and letters to be posted regularly from Dieppe to continue the deception as long as possible.

When the truth was discovered at Swinbrook it was as if a volcano had exploded. David's rage was worse than anyone could remember and Sydney was miserable. She realised, however, that any parental intervention was likely to be counter-productive; far better that Nancy, whose political credentials would be the most acceptable, should try and persuade her sister to return.

Nancy and her husband Peter Rodd met the runaways at St Jean-de-Luz and invited them to a splendid lunch on a destroyer – once on board they could be brought home. But although the Rodds too were working for the Republican cause, Esmond was wary. Despite their 'watering mouths', he and Jessica refused the lunch invitation, and evaded the trap.

To escape further pursuit they moved to Bayonne, staying in a cheap room above a café; here they were found by Sydney, who had set off after Nancy's failure to lure them back. She hoped that if she could see Jessica on her own, she could persuade her of the awfulness of what she was doing. When she arrived at the café, she sent the owner up to tell Jessica that 'a lady' wanted to see her. Jessica came down and immediately told her mother she would not return home. Sydney grasped the nettle and asked to be taken upstairs to meet Esmond.

She was not struck by him and quickly launched into a furious denunciation of his behaviour. 'I told them how abominably they had behaved and informed him what a coward he must be not to have the courage to face Farve and ask to be engaged to Decca – and I said his conduct was what you would expect from a communist.'

Somewhat to Sydney's surprise, Esmond took this meekly, agreeing that they had behaved very badly, 'but it was the only way'. Sydney soon came to the conclusion that encouraging them to get married was the best of the three solutions available – the alternatives were leaving Jessica there unmarried or dragging her home by force. She gave her daughter the box of clothes she had thoughtfully brought with her, dined with the errant pair and then caught her train back.

'They came to see me off in pouring rain,' she wrote to Diana on 27 March. 'So I gave Decca my umbrella as a parting gift. It was terrible to see them go off in the rain. But there again, she seems happy.'

Though at first David would not hear of marriage, he finally agreed when he found that Tom and most of the older generation of the family advised it. Jessica was married in Bayonne at noon on 18 May, in an outfit hastily put together by Sydney, who had brought with her a silk dress from Harrods, and had taken her daughter out that morning to buy coat, hat, gloves and shoes before a hasty visit to the hairdresser. 'Little D looked very sweet,' she reported to Diana.

Diana's wedding present to the sister who had so adored her during their childhood was an amethyst and pearl necklace. This pretty jewel, which would once have been welcomed with ecstatic cries, was not well received. Esmond, so extreme a communist that he called himself an anarchist, disapproved deeply of Diana and her political views and Jessica, madly in love and drawn into his world, soon began to regard her once-favourite sister as a villain. But the schoolroom bond with Unity was too strong to sever and Jessica still saw Unity on her visits to England, meetings that on Jessica's side were conducted clandestinely for fear of Esmond's disapproval.

In December 1937 Jessica gave birth to a daughter, Julia. This event signalled the last communication between Diana and Jessica:* when Diana sent her sister a dress for the baby Jessica wrote back, 'Thank you so much for the lovely dress but please don't send any more presents because Esmond doesn't like it.'

* Apart from a brief meeting over Nancy's deathbed.

The Romillys' ideological correctness had a tragic outcome. When they wanted to go to Belgium for Easter, Sydney suggested that they leave Julia at the Old Mill Cottage with Nanny, 'who would love to look after her'. This drew a horrified refusal from Jessica: 'Esmond wouldn't hear of our baby going to a house with a nanny in it.' Jessica, like the working-class mothers whom Esmond so admired in the Bermondsey slums around them, took Julia to a clinic. Here the baby caught measles, infecting Jessica. Both were very ill; Jessica recovered but Julia, who was only four months old, died.

> Don't know whether you are back at Wootton or not but I write in case you have not heard to tell you the sad and heartrending news that Decca's baby died on Saturday [wrote Sydney to Diana on 30 May]. They went to the seaside the same day, and are now to move, leaving England, she says, for a long time, going first to France. I think it is best they should go right away. I have had such a terribly sad letter from her. It was such a particularly sweet little baby. She doesn't want to see anyone* before she goes.'

Diana, who now knew that she was expecting her own first child by Mosley, was horrified by the tragedy, and miserable when she realised that the last sentence of Sydney's letter was a delicate way of explaining that Jessica wanted no further contact with her.**

Another casualty of Diana's dedication to Mosley was her friendship with Professor Lindemann. 'When I was married [to Bryan] he gave me a beautiful gold watch in different coloured golds,' she wrote later to James Lees-Milne. 'After I married Kit he never liked me again.'

It had become clear that the idea of a radio station on Sark was a non-starter. Although Lawton's researches had shown that permission might, theoretically, be granted by the Dame of Sark, it was obvious that the British government would pull out every stop to prevent what they saw as a 'pirate' station.

* Jessica did, however, see Unity, who had returned from Austria – when the *Anschluss* took place on 12 March Unity had rushed to Vienna to witness Hitler's arrival.

** Jessica and Esmond went to America on 18 February 1939. They said goodbye to Unity and Sydney; Tom, Nancy and Philip Toynbee saw them off at the station. When war broke out, Esmond joined up, training in Canada. He was posted to Britain as a navigator with the rank of Pilot Officer. In November 1941 he failed to return from a raid on Hamburg.

The Belgian option also appeared to be falling through. Mosley and Allen had believed there was an influential Belgian Cabinet minister who could be bribed to persuade his government to give the necessary permission and Lawton visited Brussels a number of times to lay the legal groundwork. But whether because the bribe was not big enough, the Minister's honesty was unassailable or the government was adamant, this deal collapsed too. It was time to put every effort into the Heligoland project.

One evening in May Mosley told Lawton that Air Time was now going to concentrate on the possibility of a radio station in Germany, saying that the suggestion had come from 'a fashionable Berlin jeweller'. Lawton was suspicious and he remarked that, to him, it sounded like the opening move in a confidence trick. He advised Mosley to look at the approaches of the 'jeweller' very carefully indeed, and dismissed the matter from his mind.

A fortnight later his telephone rang late one evening. 'Could you possibly go to Berlin tomorrow?' asked Mosley. Lawton, whose legal career had not yet taken off, said he could. 'I'd like you to meet my wife there,' said Mosley. 'She thinks she will be able to start negotiations with the German government and I'd like you to help her. She'll book you a room at the Kaiserhof, where she's staying. I'd be grateful if you could give her all the help you can.'

Early next morning Lawton was at Croydon airport, where there was a seat on the Berlin plane. It was only the second time he had flown, and the journey at 3,000 feet over Germany – daily occupying more of the world's headlines – fascinated him.

His visit lasted four days and he and Diana lunched together daily. She had little to report to him of her meetings, so entertained him with stories of her family and her childhood. She talked of David and his eccentricities and his tempers, she told the fascinated young man about the Hons cupboard and the private Mitford childhood language of Boudledidge, talking to him in it fluently for minutes at a time. She told him about exhibitions she had seen, of visiting Frau Goering to admire her baby daughter, Edda, born on 2 June and the cause of great excitement – Goering had become a father for the first time at the age of forty-five. She told him the gossip that was scandalising the Nazi inner circle: it had been discovered that the beautiful 24-year-old wife of the newly married Field-Marshal Blomberg, who was sixty, the War Minister and Goering's arch-rival, had been a prostitute and Hitler was demanding that the Field-Marshal give her up. Of politics Diana did not speak, but she made plain her admiration for all that Hitler had done for Germany.

Though her affection for Hitler had grown to the extent that she had a portrait of him hanging at Wootton, Diana found the process of negotiating with his officials intensely boring. Thirty years later she was to write (*à propos* of the government):

> As to the haphazard way in which a Cabinet works, I am sure one can't exaggerate the extent of it . . . But businessmen are even worse. I was once involved for two years in a business venture in which hundreds of thousands of pounds were involved. I can never tell [sic] the frightful boredom one suffered from the drawn-out negotiations, the wordiness, the irrelevancy of much of what was said. One would have to be a real saint to put up with businessmen.

Diana and Lawton returned to Berlin in July 1938. A day or two after their arrival, Diana was able to report during their usual luncheon together that success now seemed likely and they discussed what she should ask for if the question of detailed proposals came up. The following day, she arrived at their rendezvous with the news that she was to visit the Minister of Posts and Telegraphs that afternoon, just before the Ministry closed for the day; would Lawton accompany her? They set off together and were shown into the office of the Minister, a man who had the inappropriate name of Ohnesorge (which means without care). He was short and unattractive, with a ratlike face; Lawton had scarcely recovered from his astonishment that such a powerful country as Germany should have such a peculiarly undistinguished man in this position before the meeting was over. The German government, said the Minister, was prepared to grant Mrs Guinness a concession for a commercial radio station, the details to be negotiated with their representatives. 'Thank you very much, good evening, good night.'

That evening Diana and Lawton went off together for a celebratory dinner at the Hausvaterland, a large building in the Potsdammerplatz with a restaurant on every floor – some, like the one they had chosen, with a variety show. For Lawton, the gaiety and exhilaration of the evening abruptly disappeared when Diana, in the midst of her flow of chat, told him something that was to disturb him for years. She had dined the previous night, she said, with Hitler, Goering and other senior Nazis and the topic of conversation had been what they would do when they took over Czechoslovakia.

That night Lawton lay awake wondering what he should do about this dangerous information. Should he report it to the War Office? Or should he preserve the confidentiality of his client? In the end, he came to the conclusion that the rules of his profession demanded that he remain silent. The next day they flew home.

Once back with the good news that the station was 'on', a new company was formed, Wire Broadcasting Ltd, with offices first at 11 Garrick Street and then in the Adelphi building in the Strand. The company was incorporated on 24 December with capital of £15,000. The Memorandum of Association shows that one of its objects was 'to carry into effect an agreement already prepared and expressed to be made between Peter Pendleton Eckersley and Dr Francis Stafford Clark', who were directors. Another director was Bill Allen; among the shareholders was the firm of David Allen and Sons. Mosley's name was, of course, nowhere mentioned.

Mosley's need for secrecy caused semi-farcical complications. The Allen brothers had agreed to pay him £5,000 for their one-third share in the proposed enterprise and neither they nor Mosley wanted this money to be traceable back to them. Once again, it was to Lawton they turned, Bill Allen handing the money over in used notes to be slipped to Mosley the next day. At this point Lawton's young wife put her foot down. She would not, she said, have £5,000 in cash overnight in their flat. Packing the money in a suitcase, she took it round to her bank and paid it in. The next day it was withdrawn and handed over to a trusted member of Mosley's staff. Soon afterwards, James Herd's Nazi opposite number, von Kaufmann, came over to discuss the proposed station and horrified the discreet Herd by greeting him with a 'Heil Hitler!'

Diana knew nothing of this byplay. Once initial agreement had been secured, her part in the negotiations was over. With only a few months to wait until the birth of her baby, she retired to her secluded life at Wootton. A few friends, including James Lees-Milne and Lord Berners, came to stay and Mosley returned between speaking engagements and visits to London, where he pursued his affair with Baba whose infatuation with him showed no sign of waning. 'All through the thirties it was as if I had two wives,' he once commented.

The chief excitement of the late summer for Diana occurred when Grimwood, the chef, and James, the butler, gave notice simultaneously. There had, been a fight during the preparations for a luncheon party (fortunately the lunch was already cooked) and the hot-tempered Grimwood had hurled all the Dresden plates at James's head. When Diana rang the bell to find out why lunch was so late, the hallboy arrived in the drawing room with the words, 'James is lying on the pantry floor in a pool of blood.' Next morning, on her breakfast tray, were two notes, saying that the warring pair were very sorry but they had to leave her service. Diana, who like most of her friends found it impossible to envisage life without servants, was distraught at the prospect of losing two such good ones and straight away went to see James and his wife in their cottage in the stables, and then Grimwood. By dint of tact, flattery and appeals to their loyalty she persuaded them both to withdraw their notice. 'Managing to get them both to stay was the most brilliant thing I ever did in my life,' she said afterwards.

Her pregnancy was now impossible to conceal. Though she did not care a fig for 'public opinion', she did not want to hurt those near to her. Her family knew the position and she had written to Bryan and told of her remarriage.

Bryan had also remarried. Three weeks before Mosley and Diana had been married in Germany, he had begun a long and happy married life with Elizabeth Nelson. Now with his customary generosity, he wrote, 'Betty and I are so very pleased and excited about your news . . . we won't breathe a word about it.'

Yet for all public purposes Diana was still 'living in sin' and the time was rapidly approaching when, if she did not want a scandal over her 'illegitimate' baby, she would have to announce her marriage to Mosley. As it was, Debo, now grown up, was still forbidden to stay at Wootton – though that August, driving nearby with her future husband Lord Andrew Cavendish and a friend, she paid a surreptitious ten-minute visit, to find Mosley fishing in the lake, Diana, Jonathan and Desmond beside him.

On the other hand, public disclosure would mean that Mosley's involvement in the German negotiations would be guessed instantly. A letter from Bryan on 14 August summed up all the awkwardnesses.

> You didn't say in your last letter whether you have told them [Jonathan and Desmond] about your coming event. Have you? or whether you have mentioned to them your marriage: nor when and how you are going to announce it. I don't know that there is any need to tell me if you don't want to but you may have simply forgotten to answer these questions. Only I would like to know what to tell the boys supposing they hear something and ask questions. Nobody outside your family seems to know of your marriage though there was a wave of rumours, as you know, last year. Wouldn't it be easier to announce the marriage

before the birth? Or are you going to put 'To Lady Mosley, a son' and leave it to people to realise that you have been married?

Finalising the radio contract seemed to be dragging on endlessly. One reason was that all meetings had to take place on neutral ground. Mosley was well aware that if anyone came over from Germany, British intelligence would immediately become suspicious. Taking Lawton with him, he would meet the German side in Paris. For Lawton, struggling to make a living at the Bar, the luxury of the Crillon and the Georges V hotel, as well as the excitement of Mosley's company, made these trips an exotic experience he remembered all his life.

The German negotiators were an odd pair. The deal was in the hands of an astute and knowledgeable businessman called Dr Johannes Bernhardt. Accompanying him, presumably so that he could report direct to the Führer, was Hitler's aide, Captain Wiedemann, a tall, thin, former professional soldier who looked like the popular idea of a Prussian officer and whose main talent was his outstanding horsemanship. Fritz Wiedemann had originally met Hitler during the First World War, when he was Adjutant of the Bavarian infantry regiment in which Hitler had been a corporal. It was Wiedemann who had recommended Hitler for his decoration (he received the Iron Cross, second class). In 1935 Hitler had appointed him to his personal staff.

Mosley courted these men in a way familiar to anyone who has ever experienced corporate hospitality. A connoisseur of both food and wine, he would invite them to a sumptuous luncheon, with everything of the best, in a private room to preserve secrecy.

A further softening-up operation was a tour of the Paris night clubs. At the request of the two Germans, the first stop was the Sphynxe, Paris's most fashionable brothel (shortly to be shut down). Lawton's eyes opened wide at its Egyptian decor and the Cleopatra-handmaiden attire of the girls, but the party stayed only a quarter of an hour before leaving for the equally well-known but much more reputable night club, the Schéhérazade. The first to leave, at four-thirty, was Lawton, who ill-advisedly returned in his dinner jacket to the Crillon by metro, encountering hostile jibes and jeers from the early-morning workmen. Mosley did not return until much later, but showed little sign next day of his sleepless night.

That September Diana, seven months pregnant, did not attend the Parteitag or the Bayreuth Festival which preceded it. Unity, who went to Bayreuth as usual, developed pneumonia afterwards and was put in the Bayreuth Clinic. So close to Hitler had she become that he sent his personal physician, Doctor Morell, to look after her – a mixed blessing: Morell was appallingly ignorant and inefficient but for Unity this was yet another example of the Führer's supreme kindness.

Even Unity, blindly worshipping as she was, understood that Germany and England were now on a collision course. Like Diana, she had heard Hitler talking of the invasion of Czechoslovakia – since the *Anschluss* surrounded on three sides by Germany – despite his public denials of any such intention. As she lay in the clinic, she came to a decision: should there ever be war between her own country and the one she now loved so deeply, she would kill herself. She told Diana of this, as she told her of everything.

When Sydney came to Bayreuth to see her, Unity persuaded her mother to accompany her to the Nuremberg Rally. From having been bitterly opposed to Hitler and everything he stood for, Sydney was completely won over. Hitler, always overly impressed by British aristocrats and, thanks to Ribbentrop, convinced that they wielded much political influence, had turned the full force of his charm on her, to such effect that she would never afterwards hear a word against him.

When Sydney returned home, she went to Scotland. With some of the proceeds of the sale of Swinbrook, the Redesdales had bought the small island of Inch Kenneth, in the Inner Hebrides. The only buildings on it were the ruins of an abbey and a beautiful, rather grand eighteenth-century house. Sydney furnished it with her usual taste, buying furniture cheaply at auctions. No chest of drawers cost more than £3 10s, with elegant eighteenth-century tallboys for £7. She put her grand piano in the drawing room, where she hung pencil sketches by William Acton of her six daughters, framed in red brocade.

The island itself was flat and green, beneath the high cliffs of the Grievan rocks. The weather was often appalling. Nancy described it in horror-struck terms.

> There is a perpetual howling wind, shrieking gulls, driving rain, water water wherever you look, streaming mountains with waterfalls every few yards, squelch, squelch if you put a foot out of doors (which of course I don't). The only colour in the landscape is provided by dead bracken and seaweed which exactly matches it. No heather, black rocks shiny like coal with wet, and a little colourless grass clinging to them.

Sydney loved it.

She was there when Unity went down with a bad bout of flu after the Parteitag. This time it was David who went to collect his daughter and bring her home. He too had become a firm Hitler supporter. 'Farve really does adore him in the same way we do,' wrote Unity to Diana on 12 September. 'He treasures every word and every expression.' When, on Friday 30 September, the Munich Agreement was signed, it was proof positive to the Redesdales that Herr Hitler's intentions had all along been peaceable. Irene took the two eldest Mosley children to Heston aerodrome, where they stood in the rain to watch Chamberlain's triumphant return, waving the piece of paper that bore the German leader's empty promises. Diana and Mosley listened to the news in their cavernous Grosvenor Road flat.

Eventually the broadcasting contract, drawn up by Lawton, was finalised. It was dated 18 July 1938, the day on which agreement had finally been reached. It was a complicated affair. As, under Nazi laws, only a German company was permitted to receive a broadcasting licence from the German government – though once granted it could later be transferred to a British company – the Germans agreed to finance the construction of the station and to cover all operating costs over a period of fourteen months. Programmes, recorded music and announcers would be provided by another company, Radio Variety, to be set up specifically for this purpose (on 12 April 1939) and to sell advertising. Radio Variety had a capital of £5,000; Eckersley and Herd were co-directors, each holding 2,500 shares, and the Allens were to sell the advertising. The British would pay their share of the construction costs, plus interest of 5 per cent on the German capital involved; whatever profit remained would be divided 55/45 in favour of the Germans.

There were various clauses designed to save both sides from either hostility or ridicule. Section B of clause four stated that 'programmes shall contain no matter which can reasonably be construed as political propaganda or cause offence in Greater Germany or Great Britain'. Under this, even news bulletins were forbidden. There was, however, no doubt that by playing German music and presenting 'educational talks' on German life, the impression of a cultured and reasonable nation would be subtly conveyed – especially as the Germans insisted on a clause forbidding any jokes about the Nazi regime by British comedians. They backed this up with a clause stating that a Nazi official should be present in the studio during such programmes to pass or if necessary censor whatever went out on the air.

The agreement was finally signed on 9 November, in Paris, for the German side by Bernhardt and Kurt von Schroeder, a prominent investment banker who was President of the Cologne Chamber of Commerce and a member of the International Chamber of Commerce and who had embraced Nazism just before Hitler's rise to power. The British signatories were Diana, Mosley and Bill Allen. They stayed at the Crillon – Diana, who was expecting her baby at any moment, had her monthly nurse with her.

'Now there is no reason to keep our marriage a secret any longer,' said Diana to her husband. 'And the moment to announce it is when our baby is born.' Two days later, on 26 November at 129 Grosvenor Road, Diana gave birth to their first son, Alexander.

Chapter XXII

For Mosley's children and his sisters-in-law, learning of his marriage through the newspapers was devastating.

Baba Metcalfe was in the train on the way to Paris when, idly opening her paper, she saw that her lover of the past five years had for two of them been deceiving her on an epic scale. She was of course well aware of Diana's existence but believed, as Mosley had told her, that his relationship with Diana had become one of obligation and convenience rather than of passion. Baba had always been the sister closest to Cimmie, sharing her thoughts and emotions, her passion for Mosley, identifying herself with Cimmie's anguish over Diana. When she followed Cimmie into Mosley's bed it had seemed in some sense right and a victory over the hated 'Guinness'. The discovery of this double betrayal was a body blow. She never forgave either of them and for the rest of her life would refuse to speak of Diana.

It was almost as bad for Mosley's two older children. Like their aunt, they learned of their father's marriage through newspaper headlines.* Vivien, aged seventeen and being 'finished' in Paris before her first season, saw her father's name in *Paris Soir*. She read the story with incredulity, saying to her friends, 'I don't believe this paper – they've got it wrong.' But when she received a telephone call from her father, saying he was coming to Paris to see her next day, she realised it must be true – and that Diana must actually have been his wife during the Mediterranean holidays they had all spent together for the last two summers. She felt foolish, angry and bitterly hurt.

By the time Mosley arrived the next day, she had gained sufficient control

* The national press picked up Mosley's announcement in the 1 December issue of *Action* that he had been married to Mrs Diana Guinness for just over two years.

of her emotions not to reproach him; acceptance of his behaviour, she had learned, was vital if they were all to have any kind of family life. He took her out to dinner and they had a stilted conversation, Vivien choking down her pain and resentment. Her former friendly feelings for Diana turned to hatred, and it was months before she came to terms with the situation and once more enjoyed Diana's company.

Nicholas was equally taken by surprise. He was on his way to the seven-thirty a.m. 'early school' during the last fortnight of the Eton winter half when he too caught sight of a newspaper headline. When he got back and went in to breakfast his friends asked him if it was true that his father and Mrs Guinness were married. 'No, no,' said Nicholas. 'My father would have told me. You know how the press make things up.' But after breakfast a telephone call from Irene confirmed that it was so and that his father was writing to tell him. Confusedly Nicholas sought to save his face with his friends by telling them that of course he had known all along but had promised not to say anything until his father had tipped him the wink. To his father he wrote that he was upset at not being told and could not think why he had been kept in the dark.

'Sweet Nick had noticed Mrs Guinness's huge size at Wootton in August but had gallantly not breathed a word tho' agonised at the situation. Oh why did Tom not tell us all at Lake Vyrnwy?' runs Irene's diary, adding, 'Poor Ma looks a liar to all the B.U. who will never believe she did not know.' Irene herself behaved with great dignity. 'I told Tom I did not want to go over the tale of his marriage. I only wanted to say the loyalty of his children after those cruel blows was amazing and he must never betray it. I was thankful Baba was out of the country.'

The ménage at Denham, already resentful of Diana as an interloper, were confirmed in their opinion. Nanny Hyslop, while conceding that Mosley's behaviour had been 'very naughty', regarded Diana as the incarnation of evil and believed that she was totally to blame for seducing Mosley into an adulterous passion that had killed Cimmie. Only Micky, who for years had hardly seen his father, regarding him as an amiable stranger, had no trouble in accepting another amiable stranger as his father's wife. But when the little boy referred to Diana as 'Mummy' he was surprised to hear his sweet-tempered nanny hiss furiously, 'Never call THAT WOMAN Mummy!'

Diana was unaware of this depth of feeling. Andrée, now the housekeeper at Savehay, was discreet enough to keep her opinion to herself.

Those of Diana's own family she cared for most had known of her marriage from the beginning and she was now completely reconciled with her parents, who liked and admired her husband and who both then shared his political views.

Altogether the first months of 1939 seemed to promise happiness. Perhaps the Munich Agreement really had removed the threat of war and with it the terror that Diana felt over Unity.

Diana, who saw things so clearly – albeit from one political point of view only – was convinced that 'reason would prevail' and that her husband's stance would be vindicated. The contract for the new radio station had been signed, the station itself was under construction and due to open on 1 October. With a large and regular flow of funds for the BU, its influence would increase and the prospect of power become more likely. When she and Mosley took their baby son from Grosvenor Road to Wootton, there to hand him over to the delighted Nanny Higgs, as she later wrote, 'My happiness was complete.'

True, there was a rift or two in the lute. Mosley and Bill Allen had quarrelled violently. Allen believed Mosley had swindled him over the money he had put into the Nordyke station. He withdrew from Air Time, telling Mosley that he was in danger of losing 'your one surviving friend'. Alienating Allen was a foolish move. Mosley was aware that Allen had links with MI5, but believed that anything Allen could report would show that British fascism was not a seditious force.

Without Allen, Air Time's subsidiary Radio Variety stood little chance of success. It was, as James Herd put it, 'a flop'. Nevertheless, details were posted to von Kaufmann at Kochstrasse 33, Berlin, while the construction of the station went ahead.

In March 1939 – the month that Hitler tore up the Munich Agreement and occupied Prague – Mosley's daughter Vivien made her debut. She was presented at court by her Aunt Baba – the rule was that only married women could effect a presentation, so Irene, who had been effectively her mother for the past six years, was unable to do so herself – though Irene gave her niece's coming-out ball at the actress Maud Allen's house in Regent's Park. Mosley had nothing to do with his daughter's coming out, neither escorting her to parties nor even enquiring how she was getting on, so furious was he still over the Official Solicitor's refusal to allow Leiter money to be used on Wootton. Only when a page of gossip about Vivien appeared on 4 June in the *Sunday Express* did he react. The paragraphs concerned, appeared under the headline 'The Perfect Debutante of 1939':

> Miss Mosley has the heritage of the perfect debutante: brains, beauty, elegance, riches, glamour and political fame.
>
> She inherits beauty from her grandmother, Mary Leiter, one of the loveliest women of her time, wealth from Levi Leiter, the American wheat king who left more than £6 million, elegance and brains from her grandfather, the late Marquess Curzon, the famous Foreign Secretary.
>
> By the end of this season, directly and indirectly, thousands of pounds will have been spent by family and friends, hostesses and guests in launching Miss Mosley into society. Florists, caterers, beauty parlours and dress houses, all will have benefited.
>
> Every precaution has been taken to prevent her becoming involved in politics and she has been protected from the publicity which normally surrounds a debutante of her standing.

There followed a resumé of Vivien's accomplishments and a mention of her mother's jewels, which she had inherited. The only reference to Mosley by name came at the end.

> Twenty years ago, when Lady Astor was winning her seat in Parliament, two of her most ardent supporters, 20-year-old Lady Cynthia Curzon and young Oswald Mosley used to meet by arrangement early each morning on Plymouth Hoe. They were married in May 1920.

Mosley chose to regard the article, with its flattering comments on his daughter, as an attack upon himself. He put out a statement in *Action*.

> As Miss Mosley is a Ward in Chancery whose every arrangement is in the charge of her aunt, Lady Ravensdale, her father has nothing to do with her present activities. Therefore the Beaverbrook malice shoots very wide of the mark, as an attack on a man who has neither the time nor the inclination for any social life at all. On the other hand, it should be stated in fairness to this pleasant but ordinary girl, whose appearance in the world would have attracted no extraordinary attention if the Press had not detested the person and politics of her father, that not one word written about her in the national Press should be believed without verification.

His statement was not only deeply wounding to his young daughter but breathtakingly inaccurate: he was still her legal guardian.

Diana continued her regular visits to Unity in Munich. Her two small Guinness boys were deeply envious of these trips and longed to be taken with her to meet this man whose photograph they saw in every newspaper;

like schoolboys everywhere, they ached for autographs to add to their collections. But Diana was adamant.

The second time Diana went to Munich in 1939 Unity had a flat of her own, found with Hitler's help – it had belonged to a Jewish couple who had 'decided to leave'. Not that it would have made any difference to Unity if she had known of the previous owners' forced eviction: for her, whatever Hitler did was supremely right. 'She believes Hitler to be more than a genius; those who know him well consider him as a God,' wrote Joseph Kennedy Jnr to his father, the American Ambassador in London. 'He can make no mistake and has made none. She has been afraid to go to England lately for fear there would be a war and she would be caught there.'

Diana came home in time for the biggest meeting Mosley had ever held, a 'demonstration for peace' in the Earls Court Exhibition Hall on 16 July. The backcloth to the platform was, as usual, a huge Union Jack; as Mosley marched up to the platform many of the 20,000-strong crowd gave the fascist salute. Among these was Tom Mitford, who had already joined a territorial regiment, the Queen's Westminsters. Later, he was reported for this action but his commanding officer took the view that, as he was there in a private capacity, there was no cause for complaint. Like many people, Tom believed war was inevitable but still hoped for peace.

Speaking for two hours, as usual without notes, Mosley gave a virtuoso performance that played on this despairing longing. His peroration roused the crowd to a delirium of hope as he urged that Hitler should be allowed unrestrainedly to go east, 'and then he would not want to fight Britain'.

Afterwards, Mosley took Diana, Sydney and Tom back to his mother Maud's house for supper.

As the build-up to war continued, so did Mosley's campaign for peace 'with the British Empire intact'. Apart from the fascist beliefs and the anti-semitism which he shared with Hitler, he was convinced that the German military machine was unstoppable. Once Hitler had got hold of the Ukraine, the granary of Russia, and the rich minerals and oilfields of the Caucasus, Germany would be able to ignore any British blockade. The British army was small, the Royal Air Force weak. Hitler's declared ambitions were to the east; he believed that the Slav peoples were born to be subject to the Reich; he often said that once he had regained the German colonies he would have no interest in a war with England.* Better, thought Mosley, that the British should negotiate peace early and

* The view of a number of revisionist historians today.

thus save the Empire. He claimed that he himself would be interested in power only after such a negotiated peace. But such was his reputation that few believed him.

In August, Diana flew out again to join Unity as Hitler's guest at the Bayreuth Festival, in what was to be her last pre-war visit to Germany. One evening Hitler invited her to his box at the opera, where she found him in a state of extreme depression. He told her that he thought England and Germany were set on a collision course. England, he feared, would not change her attitude over Danzig, while he could not abandon Germans placed under Polish rule by the Treaty of Versailles. Later, when Diana and Unity lunched with him at the Wahnfried on 2 August, the last day of the Festival, he told them he wanted to speak to them privately. Once apart from the others, he said sombrely that as far as he could see, England was determined on war – and war was therefore inevitable.

Diana told him that Mosley would continue his campaign for peace as long as it was legally possible to do so; Hitler warned her that if he did he might be assassinated.

When the sisters were alone, Unity told Diana again that she would not be alive to see the tragedy of war between England and Germany. That night they went to see *Götterdämmerung*. 'Never had the glorious music seemed to me so doomladen,' wrote Diana later. 'I knew well what Unity, sitting beside me, was thinking. Next day I left for England with death in my heart.'

On 21 August the Ribbentrop-Molotov pact was signed, guaranteeing non-aggression between Germany and Russia; and the last faint hope of peace disappeared. The countdown to war had begun. The British knew of the strength and power of the Luftwaffe and believed that bombs would rain down upon all large British cities, in particular London. On 27 August Irene, Nick, Vivien, Micky and his nanny arrived at Wootton – Mosley had suggested they leave Denham, so close to London, for the comparative safety of Staffordshire. 'A lonely block of a house,' wrote Irene, who found Diana's central heating 'suffocating. Only one small window in each bay opens.' Even worse, though, was to find a photograph of Goering with his baby in a silver frame on the mantelpiece. Once at Wootton, Irene, Nanny Hyslop and Nanny Higgs began to prepare for war, cutting out huge blackout curtains to cover the large and numerous windows. Diana and Mosley remained in their London flat.

With war – short of a miracle – inevitable, Mosley's activities came under closer scrutiny. On 1 September he issued to the 22,500 members of the BU a statement which narrowly escaped prosecution – only the fact that it was published two days before war was declared saved him. It read:

> The Government of Britain goes to war with the agreement of all Parliamentary parties. British Union stands for peace. Neither Britain nor her Empire is threatened, therefore the British government intervenes in an alien quarrel. In this situation, we of British Union will do our utmost to persuade our British people to make peace . . . the dope machine of Jewish finance deceived the people until Britain was involved in war in the interest of the money power which rules Britain through the Press and Parties . . .
>
> Our members should do all that the law requires of them and, if they are members of any of the forces or Services of the Crown, they should obey their orders and, in every particular, obey the rules of their Service. *But I ask all members who are free to carry on our work to take every opportunity within your power to awaken the people and to demand peace.*
>
> We have said a hundred times that if the life of Britain was threatened, we would fight. But I am not offering to fight in the quarrel of Jewish finance, in a war from which Britain could withdraw at any moment she likes, her Empire intact and her people safe. For the moment I am not concerned to argue about the incidents which preceded the outbreak of war. In time, we shall know the whole truth. I am now concerned with only two simple facts. This war is no quarrel of the British people. This war is a quarrel of Jewish finance. So to our people I give myself for the winning of peace.

This message was exhibited in the windows of the BU headquarters, distributed to various local officers and would have been published in the 9 September issue of *Action* had Scotland Yard not intervened. Special Branch, when considering whether to pass it to the DPP for prosecution, appended the note: 'When the Polish question assumed importance British Union took up the attitude that Poland was actually a Jewish-controlled state, a harsh dictatorship under which millions were oppressed, and did all in its power to influence public opinion against any pact between Poland and this country.'

Taken with Mosley's reaction to the invasion of Czechoslovakia in March ('Britons fight for Britain only') and his view that Russia had only entered into the non-aggression pact in order to incite war between Britain and Germany, Special Branch concluded that 'in effect, his followers are urged to do all they can short of breaking the law to propagate the need for a quick termination of the war without regard to the Government's commitments to Poland or France'. But they did not yet consider him dangerous.

When war was declared on 3 September Diana and Mosley listened miserably to the Prime Minister. Bryan Guinness, waiting for the announcement at eleven-fifteen, scribbled an affectionate and understanding note to his former wife. 'I am afraid this must be a time of more than ordinary difficulty and anxiety for you than for most people, with Bobo I imagine in Germany and your sympathy with the German régime conflicting with the great love I know you bear for your own country.'

Up at Wootton, Irene was shattered when, just after listening to Chamberlain, she went into Diana's bedroom to put up blackout curtains and there found a signed photograph of Hitler by the bed. 'I loathe it & long to smash it to atoms.'

War provoked an immediate reaction against the BU and Mosley's former office in the King's Road was wrecked. His response was a statement in *Action*:

> Our country is involved in war. Therefore I ask you to do nothing to injure our country, or to help the other power. Our members should do what the law requires of them; and, if they are members of any of the Forces or Services of the Crown, they should obey their orders and, in every particular, obey the rules of the Service.

At the end, he repeated his earlier message:

> But I ask all members who are free to carry on our work to take every opportunity within your power to awaken the people and to demand peace.

To anyone familiar with Mosley's campaign for a negotiated peace there was little unusual about these two apparently contradictory instructions. Mosley was a man who believed that in many ways what you said was more important than what you did. Words, to Mosley, often negated deeds: witness his marches through the East End when he would proclaim his respect for law and order and follow every directive of the police – yet ignore the fact that he himself had provoked the violence by marching through areas where a fascist presence invited confrontation.

In Germany, Unity kept her promise. She had ignored the frantic telegrams sent by her father in the last few days of peace, preferring to remain in her Munich flat, writing letters that were implicit farewells. 'This is my goodbye . . . I hope you will see the Führer often when it is over,' she wrote to her parents.

She wrote to Diana, she wrote to several friends and telephoned one of them, leaving the receiver off the hook. She took the small Walther pistol Hitler had given her from its drawer in her writing table and drove to see the Munich Gauleiter, Adolf Wagner, to ask him if she would be interned. Reassured that she would not be, she requested him 'if anything happened' to see that she was buried in Munich with her photograph of Hitler and her party badge, both of which she handed to him for safe keeping. After a few last errands, she walked to the Englischer Garten, a small park near the River Iser, sat down on a bench, took her pistol out of her handbag, fired a trial shot at the ground and then put the muzzle to her right temple and pulled the trigger. She fell unconscious, but she had not succeeded in killing herself.

She was picked up, her face so swollen that it was unrecognisable, almost at once: Gauleiter Wagner, already worried about her, had instructed two of his men to watch her. She was put in a private room in the Chirurgische Universitats-Klinik, paid for by Hitler. The bullet, it was discovered by the doctor who examined her the next day, had lodged in the back of her skull and was impossible to extract. When she regained consciousness she was unable to respond, simply staring vacantly in front of her even when her beloved Führer visited her.

Few people outside Hitler's inner circle knew what had happened. Her family heard wild rumours, most to the effect that she was ill. Only Diana suspected the truth. 'In my heart of hearts I knew what had happened,' she said long afterwards. 'She had always said that if there was war between England and Germany she would shoot herself, and she was very much a woman of her word. But at the time I didn't want to worry my mother by saying all this to her.'

Nancy's initial reaction was frivolous. 'Well, the family,' she wrote to Violet Hammersley on 15 September. 'Muv has gone finally off her head. She seems to regard Adolf as her favourite son-in-law (the kind of which people say he has been like a *son* to me) . . . Bobo we hear on fairly good authority is in a concentration camp for Czech women which much as I deplore it has a sort of poetic justice.'

On 18 September one of Unity's former lovers, Teddy Almasy, wrote from Budapest to tell the Redesdales that Unity was very ill in hospital in

Munich but improving – during the last three days she had been a little better. His letter did not arrive until the beginning of October.

Sydney wrote immediately to give Diana the news. They had become closer than ever before. For Diana, worshipping at Mosley's shrine, anyone who also believed that here was the political messiah was automatically 'one of our wonderful people'. Sydney's conviction of the rightness of fascism and increasing fondness for her son-in-law transformed Diana's previously lukewarm feelings for her mother into devoted love.

Within the family, however, there was schism. Unity, Diana and Sydney were united in their fascist beliefs and admiration for Nazi Germany and Hitler. 'Dined with Mama last night she is *impossible*,' wrote Nancy to Violet Hammersley. 'Hopes we shall lose the war & makes no bones about it.' David no longer shared these views. With the invasion of Czechoslovakia and the perception that war was inevitable, his glowing admiration for Hitler had vanished and he had, as Nancy put it, 'publicly recanted like Latimer, in the *Daily Mirror*, and said he was mistaken all along'. He and Sydney, who had always before presented a united front, were now bitterly divided.

When Hitler arrived in Munich on 8 November he went to see Unity and asked her whether she wanted to stay in Germany or go back to England. 'England,' she replied. Her belongings in Germany were put into storage at Hitler's expense* and she was despatched to a nursing home in Switzerland by train, in a reserved carriage paid for by Hitler. With her went a doctor, the nun who had been nursing her and Teddy Almasy. As soon as they arrived, Almasy telephoned the Redesdales, saying that she would be able to come home when well enough. It was Christmas Eve.

David, terrified that his daughter would be arrested on arrival, went to see Oliver Stanley, the Secretary of State for War, to explain that she was desperately ill and he would rather she stayed in Switzerland than suffer arrest. Stanley assured him that she could come home safely, and Sydney and Debo set off for Berne.

The sight of Unity shattered them. Gone was the handsome, vigorous young woman full of life, freshness and energy. In her place was someone, in Debo's description, with 'two huge dark blue eyes in a shrunken face, short matted hair and yellow teeth. Her sunken cheeks made the teeth seem bigger and more horrible and yellow and her skin too was dry and yellow. Her face was totally unrecognisable because it was totally different from her.'

* On Hitler's death in 1945 they were charged to the Redesdales, who had them brought back to England in 1949.

The journey back was horrendous, a saga of missed trains, faulty ambulances and pursuit by a ravening press. They left Berne on 31 December. At Calais Unity was taken off by stretcher, but her special carriage had missed its connection with the boat train and the three women had to put up for two nights in an hotel. At Folkestone David, pacing the quayside as the boat drew in, found to his horror that his wife and daughters were not on it. When they arrived two days later by the next boat there was a reception of military and other police and what seemed like an army of reporters. Five miles outside Folkestone the ambulance broke down and they had to return for the night while awaiting a replacement. Debo left first in a car with some of Unity's fourteen pieces of luggage; finally Unity emerged on the arm of her father – the hotel lift was too small for a stretcher – pale and hatless in a blue overcoat. They left with a comet-tail of the press behind them.

That winter saw the last Christmas at Wootton. After it there was an explosive family row. The Official Solicitor had already queried the use of Mosley's Trust income to finance the entire cost of the Denham household at Savehay Farm. The object of the Trust was not, he said, to relieve Mosley of all responsibility for his family so that his own considerable income could be devoted to the BU and Irene had then stepped into the breach to rescue Savehay. Now, the Official Solicitor raised a similar objection about Wootton: he would not consent to a house that was not the children's family home being partly maintained by their money.

Again, Mosley was reluctant to use his income to pay for his children's wartime home. 'Now that we've moved the nursery world [Mosley's name for his children's household] to Wootton,' he said to Irene, 'can you pay for it?' This time, Irene declined to come to the rescue. Bursting into tears she rushed out of the room. She had in any case decided to leave Wootton. She was chilled by Diana's indignant protests when in a broadcast Anthony Eden spoke of the iniquities of the Hitler regime. 'I felt very strongly I could not be with her for long because of her pro-German attitude,' her diary records.

With no one willing to pay, it was clear Wootton had to be given up; by 31 December the three months' notice required had been given. As the expected bombs were not falling on London, the children were brought back temporarily to 129 Grosvenor Road. There, while Mosley raged sarcastically at Irene, plans were made to return to Denham.

Mosley and Diana stayed briefly at Grosvenor Road. In the eerie calm of the phoney war they took steps to counter the disappearance of their servants. James had already joined the territorials and his wife had returned to Manchester; Dorothy had joined the WAAF. Diana and Mosley, joined by Debo, took cooking lessons. As their teacher was from the renowned Cordon Bleu cooking school, the dishes they learned were designed for grand dinner parties rather than as useful everyday standbys. The aim was to produce a three-course luncheon every day with each of them responsible for one course. Whenever a sauce curdled the two Mitfords would scream in unison. Mosley said, 'I'm glad I haven't got all six of you.'

As well as being the first time either Diana or Mosley had touched a cooking implement, it was also the first time the nineteen-year-old Debo had had more than a fleeting encounter with her brother-in-law, whom she now began to call 'Cyril' – short for a gabbled Sir Oswald.

Soon Grosvenor Road itself was given up – too big, with a cavernous interior which would produce huge heating bills. The Mosleys moved to Dolphin Square, London's most modern block of flats, which had been completed only two years before. Here they took two adjacent flats on the seventh floor of Hood House – two flats, because to Diana and Mosley a bathroom each seemed essential. The flats, snug and warm and needing only an occasional visit by a 'daily', were furnished with their smaller pieces, while the rest of their furniture was stored in the large garage at Grosvenor Road, where it remained for the rest of the war.

Meanwhile, at Denham, Diana, who found Cimmie's chintz-led decorative style intensely antipathetic, used the grey silk curtains from the enormous drawing room at Wootton to curtain as many Savehay windows as possible. Nanny Higgs (in charge of Alexander) and Nanny Hyslop (in charge of Mickey) were not on speaking terms, and passed the winter at opposite ends of the passage without exchanging a word. Nanny Hyslop addressed Diana as 'Lady Mosley' through tightly pursed lips whenever circumstances made some exchange unavoidable.

The investigation of the BU by the authorities rumbled on. They were aware that members were being urged to 'spread the gospel' by joining the special constabulary, Air Raid Defence volunteers, nursing reserve and other such bodies, and therein propagate the Mosley message. Scotland Yard reported:

> The verbal propaganda concerning the war which is spread by the movement is definitely defeatist and pro-German. The German standpoint is

> advocated even on such matters as the sinking of the *Athenia*,* air raids, francs tireurs in Poland. British efforts, as for example the raid on Kiel, are ridiculed and more than one senior official has gone so far as to state that this country would be better off under German rule than under the present system. In his talks with his headquarters and district officials, Mosley displays the usual optimism regarding his certainty that the BU will be within reach of power 'in under two years'. He also believes he may be arrested at any moment as a very dangerous man.

By now Mosley's involvement in the planned German radio station had become known to Special Branch. Herd, when interviewed, confirmed that Eckersley, front man for the station, was extremely pro-Nazi. 'I have heard him say many times that Hitler's ideas are his and that England wanted war to make money out of it. He agreed with Hitler's invasion of Poland and disagreed with Chamberlain's attitude . . . so pro-Nazi is he in sympathy that I would not put him in charge of anything that would bring him into contact with propaganda.' This was not, of course, direct evidence against Mosley but it was a damning choice of associate.

The authorities' view was darkened even further by a report of a closed meeting of London District BU officials on 30 January 1940. Mosley had told his audience: 'Reward and victory are in sight . . . You must bring in new members – not necessarily a large number but a moderate number of reliable men and women who would take their place in the ranks when the time came for the sweep forward, which the movement would make, as their brother parties in other countries had made when their hour of destiny struck.' After this, Special Branch reported to Sir Alexander Maxwell that the BU was 'not merely a party advocating an anti-war and anti-Government policy, but a movement whose aim it is to assist the enemy in every way it can'.

* The liner sunk, without warning and with massive loss of civilian life, by a U-boat eight hours after the declaration of war.

Chapter XXIII

By the spring of 1940 Diana's father could no longer bear living with her mother. Their political differences were such that they found themselves looking at each other over an unbridgeable chasm. In Sydney's eyes Hitler could do no wrong, while David, who had never been anything but deeply patriotic, had become totally disillusioned in March 1939 when Hitler had entered Prague. Remarks by Sydney like, 'WHEN the Germans have won, everything will be wonderful. They'll treat us very differently from those wretched beastly Poles', infuriated and appalled him. Their constant, acrimonious quarrelling reached its height during the news bulletins, to which both listened avidly.

David was also made miserable by the sight of Unity. The cottage was small, she was loud, careless and physically clumsy. The paralysis that had afflicted one side of her body was improving, but her hand still hung heavily and was not entirely manageable. Sometimes silent, sometimes chattering with outbursts of frenetic, excitable laughter, she had lost her memory and did not know why she was ill – she believed that the doctors had made a hole in her head. For all of them it was agonising to hear her say, 'I thought you all hated me – but I don't remember why.' For David, a man of fastidious personal habits who believed that women should be charming, dignified and feminine, it was more than he could bear. When, for instance, Unity reached across the table to help herself from the honey jar and brought the overflowing spoon back to her plate leaving a trail of honey across the tablecloth, all the while gabbling loudly, he would get up abruptly and leave the room.

Sydney's patience was angelic. As well as an hour of lessons, she would teach Unity to crochet, take her for walks and talk to her endlessly. 'The aeroplanes worry her rather as they make her think of the hateful war but she sleeps as well as ever and has a good appetite too.' Diana, by now heavily pregnant again, came to their cottage at Wycombe as often as she could, her pity for Unity untinged by remorse or guilt at having organised that first, fatal visit to the Nuremberg Rally in 1933. 'Unity was a revolutionary and passionately devoted to Hitler,' she said later. 'The end would have been the same whether I had taken her to Germany or not.'

David went first to Inch Kenneth and later to Redesdale Cottage in Northumberland. He was never to live with his wife or family again.

Sydney took Unity to Swinbrook, to the cottage next to the Swan Inn. Here she and Debo lived with the confused and incontinent Unity in a kind of grisly parody of the children's early years; 'here we lead a very regular life,' she wrote to Diana on 29 April:

> Bobo gets up at 8 to feed goat and chickens and then has breakfast and bath, and we do an hour of 'lessons'. She is reading and writing better and can read quite a long bit to herself, and tell it back to me, so I hope will soon be reading to herself for pleasure quite comfortably again . . . Bobo makes and lays the tea, and cooks very good buttered eggs, the extent of her cooking for the present. The chickens are positive marvels and the eight of them lay eight or seven eggs every day. Of course it won't go on much longer. One had to be killed, it was ill. Debo likes milking the goat but it is difficult for Bobo . . . Action is as good as ever.

Diana's fourth son, Max, was born on 13 April in a Sutton nursing home. After three boys, many women would have wanted a daughter, but Diana was delighted. She had grown up in a household of girls and women; boys were a refreshing change. Born into an era when marriage was virtually the only female career, she was very conscious of the effect of a woman's looks on her fate and she had a secret dread of giving birth to a plain daughter. Also, given her own height and Mosley's (five feet ten inches and six feet two inches respectively), she imagined that any daughter would be extremely tall. The sight of Unity, her height now exaggerated by increasing weight, her once handsome face distorted by ill health, was like a caricature of her fears.

Late in April she returned to Savehay. Wootton had been given up at the end of the March quarter and Savehay was again the Mosley headquarters. The three youngest Mosley children – Micky, Alexander and Max – lived there with their respective nannies; Viv used it as a base; and Nicholas Mosley and Jonathan and Desmond Guinness returned there in the holidays.

Since the outbreak of war Mosley had been actively promulgating a negotiated peace. 'I find his attitude quite dreadful,' wrote Baba Metcalfe in her diary on 16 February 1940:

> You can't make him admit Germany is to blame. He thinks we will never win the war, with the present Government. They commit folly after folly. He thinks Germany has more staying power than us: a long war of waiting is to their advantage and not to ours. . . . At the end of a devastating war he thinks the Empire will be lost and we may be on our knees.

In March the Germans occupied Denmark and invaded Norway. In the 9 May issue of *Action* (which went to press about a week earlier) Mosley wrote that in the event of invasion: 'every member of British Union would be at the disposal of the nation. Every one of us would resist the foreign invader with all that is in us. However rotten the existing government, and however much we detested its policies, we would throw ourselves into the effort of a united nation until the foreigner was driven from our soil.' There was no doubt that he meant this and in fact many members of the BU were already fighting.

It was a time of acute anxiety over fifth columnists. The Norwegian traitor Vidkun Quisling and a substantial number of his supporters had played a significant part in Hitler's conquest of Norway. Suspicion of Mosley was widespread, fuelled by his own pronouncements and by extreme right-wing cranks such as Admiral Barry Domvile, who declared themselves devoted to him. Mosley's name was synonymous with anti-semitism, while through Diana and Unity he was perceived as closely linked to Hitler personally as well as politically. Many of his associates were known Nazi sympathisers. William Joyce, who had left the BU only two years earlier after being one of their best-known speakers, was already broadcasting Nazi propaganda from Germany (earning himself the nickname 'Lord Haw Haw').

The government had always kept a close eye on the BU, but had not initially seen it as a proto-Nazi organisation and therefore had not closed it down on the outbreak of war. However, from now on events led to a rapid change of view.

On 9 May the Germans launched their *blitzkrieg* against France, Holland and Belgium. The next day Neville Chamberlain resigned and Winston Churchill became Prime Minister at the head of a Coalition Government. The last flimsy possibilities of the negotiated peace Mosley had been advocating disappeared. If Hitler, who had never yet kept his word on anything, succeeded in seizing not only most of Europe but all the Channel ports, any such 'negotiated peace' would mean a Britain that could be picked off by Germany almost at will. And in the event of a German invasion, Mosley was perceived as a potential Quisling.

Rumours of spies abounded. There were stories of farmers planting their crops so that the growing seedlings formed an arrow pointing to nearby aerodromes. Everyone was on the lookout for parachutists – said to have been dropped in Holland disguised as nuns – and all signposts in rural areas were removed so that they would not know where they had landed. Britain was plastered with posters warning that 'Careless Talk Costs Lives!' In this climate, British fascism was inevitably seen as a vehicle for future Nazi influence in Britain.

The feeling was widespread that, while the nation was preparing for a possible struggle to the death against a cruel and aggressive enemy, it was either cowardly or unpatriotic – and probably both – to advocate suing for peace.

Irene was sent for by the Home Secretary, Sir John Anderson, in mid-May, and asked if she had any evidence that Mosley would betray his country or if she thought that he was a fifth columnist. She did not – as she would have done a few years earlier – simply laugh. She said that while she had no evidence that he would be a traitor, quoting his statement that if Britain were ever attacked it would be the duty of every BU member to resist, she did believe that if he felt a British version of National Socialism, in conjunction with Hitler, was desirable for Britain 'he might do anything'.

Then came the Tyler Kent affair. MI5 learned that Anna Wolkoff, a member of the Right Club, and Tyler Kent, a cipher clerk at the American Embassy, had shown to Captain Maule Ramsay, a British MP secret correspondence between Churchill and Roosevelt, which was highly sensitive because Roosevelt was then pledged not to intervene in Europe. When MI5 further discovered that Maule Ramsay was linked with Mosley in an attempt to unite all the anti-war, anti-semitic fascist groups into a single movement, their fears were confirmed.

Suspicion of fascism was now intense. One woman BU member, Nellie Driver, who lived in Lancashire, recounts that the police had asked the man next door if they could take a few bricks out of his wall so they could listen to her conversation; he refused.

By 19 May the Germans had broken through the Allied lines, the French army was disintegrating and the British army was falling back on Dunkirk.

Invasion within the next few months now seemed a probability rather than a possibility. What the public thought of 'the Mosley line' was expressed by the results of a by-election at Middleton, where the BU candidate polled 418 votes and the Conservative 32,063. Mosley himself, who had gone up to speak for his man, was attacked by the crowd and had to be escorted to safety by the police.

On 22 May the government rushed through what became known as '18B' to an amendment (A1) of Regulation 18, the Emergency Powers Bill of September 1939. Regulation 18, as it previously stood, had given the government power to detain anyone whom the Home Secretary 'had reason to believe to be of hostile origin or associations' or to have committed 'acts prejudicial to public safety or to the defence of the realm'. The 22 May amendment now gave the Home Secretary the power to detain any members of an organisation which he believed to be 'subject to foreign influence or control' or of which the leaders 'have or have had associations with persons concerned in the government of, or sympathetic with the system of government of, any power with which His Majesty is at war'. As soon as these wide-ranging powers became law, they were employed.

On the morning of 23 May the offices of the BU were raided. Several of its officers were arrested, so quickly that no one was able to telephone the Mosleys at Denham. It was chance that Mosley himself was not there: his usual routine was to drive himself up to the office in the morning, returning home in the late afternoon or evening. That day, however, Diana wanted to visit Desmond, who was in a nursing home with a gland infection. After she had finished breast-feeding the new baby, Max, the Mosleys were driven up to London together by their chauffeur Perrott. They left Denham at about two-thirty, planning to visit the Dolphin Square flats first, after which they would separate, returning home together in time for Diana to feed Max again at six.

As their car pulled up at the entrance to Hood House, they saw four men standing there in silence, looking about them. Their expectant attitude was unmistakable. 'They must be coppers,' said Diana unsuspectingly. As the Mosleys got out of their car the men stepped forward. 'Sir Oswald Mosley?' asked the first. 'Yes?' said Mosley. 'We have a warrant for your arrest. Can we go upstairs to your flat?'

They crammed into the lift together and in the flat Mosley asked to see the warrant. 'But what is the charge?' he asked. 'There is no charge,' came the reply.

Diana, stunned at the suddenness of it all, asked where her husband would be going and when she could see him. 'He will be taken to Brixton and you can come and see him tomorrow,' she was told.

Within the protective walls of their flat they embraced. Then, downstairs again, she watched him get into the car with the policemen and be driven off.

At the nursing home, she told Desmond and his governess Miss Gillies, sitting with him, what had happened. Miss Gillies – understandably, in view of what Diana learned later – was much the more agitated of the two. Eight-year-old Desmond, taking his cue from his mother, was perfectly calm.

Diana's first reaction was that the authorities would discover they had made a mistake and then release her husband. She had not heard of the new provisions of 18B, and she knew Mosley was not alone in wanting to see the war brought to an early end. The Catholic MP Richard Stokes had relayed Nazi peace feelers through the pacifist Lord Tavistock; Liddell Hart had been advocating a policy of restraint 'until Germany showed herself willing to reach an agreed settlement' and Lord Beaverbrook had openly agreed with the Duke of Windsor (during the Duke's visit to London in January 1940) that the war should be ended at once by a peace offer to Germany. Mosley, of course, had been much more strident, but Diana believed that his known policy of instant co-operation with any police order or request, his statements that all BU members should loyally obey orders, would prove that he was law-abiding and, ultimately, patriotic. Even so, she felt she had better get back to Denham straight away. On the way she saw the words 'M.P. Arrested' on an evening paper placard. The MP was Maule Ramsay.

On the drive to Denham, Diana collected her thoughts. As soon as she arrived, she telephoned Nicholas's Eton housemaster, J.C. Butterwick, to tell him about Mosley's arrest and ask if he thought Nicky would be persecuted because of it. 'Oh no,' said Mr Butterwick. 'You see, he's grown up – he's seventeen – and he's got his friends and that's that. If I were you, I'd be more worried about your boy at private school.'

Fortunately Jonathan, who had just gone to Summer Fields, was largely protected from victimisation because his surname was Guinness. Nor, though upset by the news of his stepfather's arrest, was he particularly surprised. A highly intelligent child, he was well aware of Mosley's anti-war stance, especially in contrast to the fervent schoolboy patriotism around him. Jonathan thought that it was no wonder the government regarded him as an enemy.

Viv's reaction was both sympathetic and practical. Now on excellent terms with her stepmother, she realised that the complication of a large household was the last thing Diana needed at that moment. She would, she said, take Micky and Nanny Hyslop up to their Aunt Irene the next day.

That evening, as Diana and Viv sat at dinner, they saw the big wooden gate on the far side of the lawn open. Through it streamed the police. Diana got up from the table, went to meet them and asked if they had come to search the house. When they said they had, the Mitford urge to tease overcame her. 'You're going to have quite a job,' she said. 'It's such a funny old house, full of hiding places and corners, and there are several big barns, all full of things from Wootton. It'll take you *weeks*. And how do you know we haven't buried a lot of things?'

The police response was courteous and they began a cursory search. While they were looking over the drawing room, Diana ran upstairs and slipped her favourite photograph of Hitler, with its warmly affectionate inscription, under the mattress of her sleeping baby as he lay in his cot – the large, silver-framed one which Hitler had given the Mosleys as a wedding present was already safe in its leather box in the bank.

At five past nine the telephone rang. Diana picked it up in the library, where a policeman was desultorily pulling books out of the shelves and turning over pages. It was her mother, who had just heard of Mosley's arrest on the nine o'clock news. 'It's monstrous, isn't it,' said Diana. 'What was the charge?' asked Sydney. 'No charge,' replied Diana. 'How monstrous!' cried Sydney. 'Yes, MONSTROUS!' repeated Diana, delighted that the policeman was listening.

Soon afterwards the police left, taking with them Mosley's guns: his pair of Purdeys, a spare 12-bore shotgun kept largely for visitors, two 20-bore shotguns used by Nick and any friend he brought to stay, and a .22 rifle for shooting rabbits and rooks. Receipts were given for all of them.

Diana went straight to bed. Though anxious, she slept well. She believed that as soon as the police went through the papers and safes at the BU offices they would see that there was nothing with which Mosley could be charged. She had been worried at the beginning of the war that he might be arrested simply for being against it. Mosley had declared that BU members must do their duty, whereas pacifists were committedly and consistently anti-war. And yet pacifists remained free, thought Diana. So unable was she to see the difference between the conscientious objections of individuals and the effort of a highly organised group to work against the government of her country that she was unable to see how her husband's policies could be viewed as posing any kind of threat.

Any impartial vision she might once have had was subsumed in her absolute belief in Mosley and all he stood for. She simply did not comprehend the authorities' suspicion of a man known to sympathise with many of the ideals of Nazi Germany, a man who was a declared enemy of the Jews, who advocated handing her colonies back to Germany and allowing Germany to do what she would in Europe and, above all, who wanted to make peace with an enemy who might soon be poised just across the Channel.

Next day Vivien, Micky and Nanny left Denham. Diana drove to Brixton Prison, taking her husband a change of underclothes, soap, washing things and shaving tackle – he already had a day's growth of stubble. A prison officer was there while they talked, sitting in embarrassment as he tried to give the impression of one not really listening to a conversation between husband and wife. The first thing they needed, said Mosley, was a lawyer. He felt sure that the family lawyer, Mr Sweet, would refuse to act for him, but there was another solicitor who had acted in some of his lawsuits. Then he gave Diana a list of the clothes he needed and asked if she would also bring some food, as what was on offer in Brixton was so awful.

She went straight to see Mr Sweet, but before she could even bring up the subject he refused to act for Mosley. When she went to the lawyers whom Mosley had suggested, she received the same answer. Her own divorce lawyer refused regretfully, on the grounds that with all the younger staff called up he was alone in the office; he could not, he said, take on a case where every conference with his client – and many would be needed – meant an outside visit. Instead, he gave her some helpful advice. 'Go and see Oswald Hickson,' he told Diana. When she approached Hickson, he agreed immediately to act for Mosley.

Though Diana was only allowed to visit Mosley once a week, there was no restriction on leaving parcels for him at Brixton. The days passed in a rush, dashing up from Denham, between Max's feeds, to pay the staff at the BU office and sort out what queries she could for them, going to see Mr Hickson, writing to children, family and friends. She was wretched without Mosley but she managed not to cry when she saw him – she did not want to add to his suffering with her tears and she did not want to give 'them' the satisfaction of seeing her break down.

Alone in her room at Savehay, or holding her baby, she wept often.

She was lonely, isolated and miserable; she was desperately anxious about her four sons; and it seemed the stuff of nightmare that her husband could be kept in prison without a specific charge to answer. 'It seemed unbelievable that Englishmen could be held in prison indefinitely,' she wrote of that time.

Five days later, while addressing his Cabinet in his room in the House of Commons, Churchill spelled out his complete opposition to Mosley's views in words which explained the thinking behind Mosley's arrest.

Churchill began by telling the Cabinet that it was a question of saving as many men as possible through Dunkirk and also that there would doubtless be attempts to invade Britain:

> I have thought carefully in these last days whether it was part of my duty to consider entering into negotiations with That Man. But it is idle to think that if we try to make peace now we should get better terms than if we fought it out. The Germans would demand our fleet, our naval bases, and much else. We should become a slave state, though a British government which would be Hitler's puppet would be set up, under Mosley or some such person.

When he concluded with the words: 'If this long island story of ours is to end at last, let it end only when each one of us lies choking in his own blood upon the ground,' there were cheers and shouts and many of the Cabinet ran round to his chair to clap him on the back.

With Mosley publicly identified as a wrongdoer – for few saw him as a political martyr – the slow demonisation of Diana began. She became a pariah, isolated from the rest of the world. The first to break through this *cordon sanitaire* of obloquy and disgrace were the press, who had heard a story that Unity had returned from Germany pregnant, given birth to Max and asked her sister to cover up for her. But numerous witnesses, from doctors and nurses to those who had helped Diana clamber on to buses, quickly eliminated this promising story. Gerald Berners, defying those who told him to avoid Diana, came over from Oxford; Pam and Derek Jackson, Debo, Sydney and Unity, who could now just about walk, all came to see her. Not so Nancy, whose long estrangement from Diana continued, whose political views were deeply opposed to Diana's and who in any case hated her brother-in-law.

Chapter XXIV

'I am thankful Sir Oswald Quisling has been jugged, arne't [sic] you but think it quite useless if Lady Q is still at large,' wrote Nancy to Mark Ogilvie-Grant on 24 May 1940.

Nancy did not confine her misgivings to letters to friends. She also actively 'shopped' her sister, telephoning anyone she thought was influential. Among them was Gladwyn Jebb at the Foreign Office. 'I am disturbed about my sister, Diana Mosley,' she told him. 'She's madly pro-German and I think something should be done to restrain her activities.' Jebb, as he had promised on the telephone, passed on her views to the Home Office.

Letters from the public were also arriving. One was from a Mr Clarence Percy Hitchen, of Leeds, who reported that his stepdaughter Edith Mary Hitchen had sat next to Diana on a flight to Nuremberg in August 1939. At Nuremberg, reported Mr Hitchen, Lady Mosley was met by Julius Streicher and her sister Unity Mitford, who arranged for a motor car to take his stepdaughter to the railway station to continue her journey. As they said goodbye, Unity offered to show Miss Hitchen round Munich if she came there, but because of the international situation Edith Hitchen was summoned home by telegram before she could take up the offer. However, said Mr Hitchen, the girls corresponded briefly, and his stepdaughter had handed over her letters – letters which clearly showed how well the Mitford sisters knew various highly placed Nazi officials.

Jebb's report of Nancy's view to the Home Office, like the public's letters, were taken calmly by Sir Alexander Maxwell, the Permanent Secretary. On 30 May he wrote to Major-General Sir Vernon Kell:

> I have spoken to the Home Secretary about the case of Lady Mosley. He does not think it would be a wise course to intern her at present. He thinks there would be advantages in having her carefully watched as

> it is possible she may be a channel through which we can get further information. The Home Secretary would be glad if you would do all that was possible to have her correspondence and telephone calls and doings kept under observation.

The next day Lady Astor asked a question in the House. Ostensibly referring to a raid on the Anti-Vivisectionist Society, it was capable of another interpretation. 'Is it really wise to lock up the man and leave his wife free when the wife is more notorious than the man?' she asked, to a rumble of approval.

At Denham, Diana was notified that the authorities wished to requisition Savehay, and would need it by 1 July. They would also want the cottages in which lived the two gardeners, whom they would rehouse in Uxbridge. On her weekly visit to Mosley in Brixton, Diana suggested that the best thing would be for the families to split up, his children by Cimmie going to live with one of his sisters-in-law, while she, the two youngest Mosley children and the Guinness boys went to Pam at Rignell, in Oxfordshire. The house at Rignell was large; Pam's husband, Derek – a Mosley sympathiser – was away in the Royal Air Force; and Diana offered to pay a generous weekly sum for the keep of her family. Pam, delighted at the idea of company, jumped at the proposal.

Packing up Savehay began at once – there was little more than a fortnight before they had to leave. Andrée went to work for Irene and those servants who had not already joined up were paid off. Some of the furniture went up to Grosvenor Road, smaller items were put in high cupboards under the lofty vaulted ceilings, which could only be reached by ladder. Trunks were prepared for departure. Diana, husbandless and now homeless, planned to leave for Rignell on 30 June. Unsuspectingly, she continued to breast-feed Max, to collect and return her husband's laundry, to visit the Dolphin Square flats, to sit on the lawn by the river on sunny afternoons.

In the world outside, the net was drawing tighter. The fall of France on 22 June brought the prospect of invasion dramatically closer – and sharpened the terror of an enemy within the gates.

On 25 June it was the turn of Diana's former father-in-law, Lord Moyne, to wield the knife. He had been immensely fond of her, he had been devastated for his son's sake when she had walked out of her marriage and he had not forgiven her for what she had done. His retribution was painstaking and efficient: he had suborned her son's governess, Jean Gillies, to spy on her; now, he wrote to his friend Lord

Swinton, Chairman of the Security Executive, a committee so secret few knew of it.

> It has been on my conscience for some time to make sure that the authorities concerned are aware of the extremely dangerous character of my former daughter-in-law, now Lady Mosley [his letter began]. As the matter was mentioned in a Secret Session the other day, I have no doubt that those concerned are aware of the public comment on the arrest of Mrs Joe Beckett and the immunity of Lady Mosley and there are many who are glad to draw the conclusion that there is one law for the influential rich and another for the friendless poor.
>
> It has been suggested to me that the Authorities concerned have no detailed evidence as to the opinions which are freely expressed by Lady Mosley, so I enclose herewith a summary of conversations which my grandson's governess has had with her when taking him on visits to his mother.
>
> It is widely believed by those who are aware of Lady Mosley's movements that her frequent visits to Germany were concerned with bringing over funds from the Nazi government. I also enclose a list of the dates on which Lady Mosley went to Germany, which the governess has extracted from her diary. Owing to the governess being absent for many months in 1939, on account of a serious operation, she has no dates entered in her diary during the last year before the war. I feel it rather embarrassing to write you this letter and am sure you will understand that it would be very hard on the governess to call on her for any evidence on the subject, which would lead the Mosley family to know of the information which she is giving. For this reason, I do not mention the governess's name, though it could of course be readily supplied if required.
>
> I am not clear on the activities of your Committee. If this matter is outside your scope, perhaps you could have the letter passed on to whoever is responsible.
>
> Yours ever, Walter

The governess recalled that on 27 September 1938 Mosley had said that Hitler would be perfectly justified in invading Czechoslovakia and that when the *Athenia* was sunk both Mosleys described this as 'a good beginning'. The next day Diana had said of Poland, 'Surely now the goddess of wisdom has forsaken them and they will realise their colossal folly before it is too late.'

She had also said 'triumphantly' to the governess that when Belgium was overrun the British army could not be extricated. 'She added that it was perfectly obvious that the British army would be caught by a pincer movement, making no secret of her delight in what was happening.' More than five months earlier, said the governess, Lady Mosley had described exactly the strategy adopted by the Germans when they broke through in France, saying that they would cut off the British and French armies from each other and with a scissorlike movement take the Channel ports. The implication was obvious: where else would Diana have got this strategic prediction but from Hitler?

Lord Moyne's letter made such a deep impression on Swinton that he sent it on the following day (25 June) to Lieutenant-Colonel C. Alan Harker, acting head of MI5 and a member of Swinton's Security Committee, with a copy to Sir Alexander Maxwell at the Home Office. In the covering letter to Maxwell, Swinton wrote, '[This letter] confirms information given me very confidentially yesterday by another member of the family, who said that they regarded Lady Mosley as at least as dangerous as her husband; that they knew she had been kept out of his overt activities; but that she had been much closer to Hitler than Mosley had ever been.'

But Maxwell was not as convinced as Swinton or MI5 that Diana posed a threat to the safety of the realm. On 27 June he wrote to Swinton thanking him for his letter of the previous day with its 'disclosures' by Lord Moyne. The Home Secretary, continued Maxwell, still thought there would be an advantage in leaving Lady Mosley at liberty but under surveillance.

His letter, with its note of cautious moderation, arrived twenty-four hours too late. Swinton had acted without waiting for the Home Office response. The order for Diana's detention had been signed on 27 June by J. L. S. Hale, countersigned by Harker with the words 'I approve this recommendation', and despatched immediately. At the same time, Hale wrote to Maxwell:

> I have today forwarded to Rumbelow in the usual way a recommendation for the detention of Lady Mosley. I have not included in the details given under 'Remarks' the information which Lord Swinton forwarded to you today which he had received from Lord Moyne. I understand that the presence of Lady Mosley at large is the subject of wide and universal comment, and while I am not suggesting that this should influence any decision which may be taken in the matter, it is a point which it would be as well not to ignore completely.
>
> In view of present circumstances I do feel very strongly that this

> extremely dangerous and sinister young woman should be detained at the earliest possible moment.

The wheels for three years' imprisonment without trial had been set irrevocably in motion.

Saturday 29 June dawned a warm, sunny day. With all the packing done, Diana gave Max – eleven weeks old that day – his usual two p.m. feed, changed him, gave him to Nanny Higgs to put in his pram in the garden, then collected her book and went outside to sit near him.

A few minutes later, one of the maids appeared, to say that a man and a woman were asking for Lady Mosley. The maid's tone of voice made it clear that the two visitors were not friends who had called unexpectedly. Diana realised instantly that they must be police.

They walked across the lawn and told her that they had a warrant for her arrest.

'Now what happens?' she said. 'I've got a tiny baby and I'm feeding him.'

'You can take him with you,' replied the policeman – the rule was that a mother could take one child with her, which in practice usually meant the baby of the family. Diana asked which prison they would be taking her to, wondering if perhaps it would be Aylesbury, forty miles from London. When she heard it was Holloway, her reaction was the instinctive one that, with France now in German hands, intensive bombing of London was more likely than ever.

'I don't think I'd better take a baby to London,' she replied. 'It wouldn't be right. I'd better leave him with his brother.'

The policewoman took her upstairs to pack and Diana asked how much luggage to take. 'Oh, only enough for the weekend,' came the reply. 'You'll probably be out on Monday.' She threw some things into a case and went downstairs to find Nanny Higgs.

'They're going to take me to prison so I'm going to leave Max with you,' she said. 'I think it's better.' Nanny burst into tears. With difficulty, Diana remained calm, kissed her children and said goodbye.

On the way to London she began to think that if she was only going to be in prison for forty-eight hours, she ought to continue breast-feeding Max. She asked the police to stop at John Bell & Croyden, where she bought a breast pump, before continuing to Holloway.

Holloway prison was a menacing sight. Bleak and dark with its coating of London grime, its façade a replica of the main front of Warwick Castle, it was reminiscent of some medieval Castle Despair. The entrance to the central hall was flanked on either side by great stone pillars, surmounted by two ferocious-looking stone dragons, their huge fangs exposed. The long, curving talons of these guardians of the gate gripped the stonework; under the claws of one was an enormous stone key, a symbol to make the toughest shudder.

Inside, the first impressions were of high, smooth, impregnable walls, small windows, iron grilles and, above all, a harsh, constant noise. With nothing to muffle resonance, the sound of prisoners or staff moving from block to block of the cells that radiated outwards from the central courtyard or clattering up and down the iron spiral stairways leading to the upper floors echoed boomingly off the stone walls and floors. Over everything, like some despairing leitmotiv, could be heard the jangle of keys. Each prison officer carried a large bunch of them on a chain round her waist, a grisly echo of the châtelaines worn by the black-clad housekeepers of the great houses where Diana had once stayed.

For Diana, the greatest shock was the dirt, a filthiness unbelievable by ordinary standards. Years later, she was to write, 'I had no idea that dirt like that existed.'

Chapter XXV

Prison was to lie like a scar over the rest of Diana's life.

It was the realisation of all her most dreaded fantasies, the ones that as a child had fuelled her terror of boarding school – lack of privacy, constant forced proximity to others, the stripping away of natural dignity, physical discomfort and a harsh, unsympathetic ambience lacking all softness and beauty. She was accustomed to linen sheets, excellent food, servants, exquisite furniture and pictures; she was naturally fastidious and the comfort and charm of her surroundings had been of paramount importance to her since her childhood. More than anyone else in her family, she required space, mental, emotional and physical, in which to flourish. Prison, with its small, dark, airless, claustrophobic cells, its dirt, squalor and surpassing ugliness, was an experience drawn up from some dark well of nightmare. (Mosley, on the other hand, when asked after the war how he had managed to stand up so well to being jailed, replied, 'Oh, after Winchester, prison was nothing.')

For all prisoners, whether convicted of a crime or brought in, as most of the 18Bs had been told, 'for a few days' questioning', life in Holloway began the same way: hours of confinement in an enclosure so tiny that it was almost a straitjacket. This was a four-by-four metal cell, its wire mesh roof increasing its resemblance to a cage. Here the new prisoner awaited the admissions procedure, whiling away the time with a large mug of cocoa and two slices of bread and margarine. Diana remained shut in this coffinlike 'box' for almost four hours. Then came the routine of fingerprinting, delousing bath, weighing and examination by a doctor – during which prisoners were asked if they were pregnant – and inspection for venereal disease, headlice and scabies.

Finally prisoners were taken to their cells. These were twelve by seven feet and nine feet high, with whitewashed brick walls and floors of stone or wood. The heavy door was made of metal, with no interior handle; at eye level there was a two-and-a-half-inch peephole, through which a warder could survey the entire cell. Six feet above floor level on one wall was a solitary small window, consisting of fourteen panes, each six by eight inches, only two of which opened. The only other light came from a 25-watt bulb. Cells were furnished with a plain wooden chair and table, a bucket, washbowl and jug, a tin plate and either a knife, a fork or a spoon, a large handle-less mug of thick china stamped with GR and a crown, and a low bed of planks. The coarse sheets and blankets, so narrow they were almost impossible to tuck in, were stained and grey despite innumerable washings. A single 'clean' one was issued once a month. 'I was amazed at Holloway. I thought it would be like an Army camp – very spartan and regimented and very clean. But it was filthy,' wrote one prisoner, Louise Irvine.

Worst of all were the lavatories. In the middle of each landing was a recess, containing one or sometimes two lavatories to serve twenty to thirty cells, with washbasins beside them. Prisoners seldom used these lavatories, for the simple reason that when their cells were unlocked 'slopping out' (the emptying of chamber pots) made this impossible. The lavatories were often blocked and the floor around them covered in excrement. Most women preferred the row of outside lavatories that flanked one side of the paved exercise yard. Early in her imprisonment Diana, seeing one miraculously empty one morning, made a beeline for it; when she emerged, she was told that the red cross on its door meant it was for those with venereal disease (current mythology held that one way of catching this was from a lavatory seat). 'All the lavatory and washing facilities were archaic and absolutely filthy,' recalled another inmate.

Holloway was built on a radial plan, with half a dozen blocks, or wings, each with four floors, converging on a central point. Prisoners with skin diseases were usually sequestered in A wing, D was for first offenders and E for remand prisoners. Pregnant women and most of the prostitutes went to C wing. In 1940 B and D wings held a mixture of convicted prisoners and interned enemy aliens – Germans, Austrians and Italians; some were Jewish refugees and some married to Britons. Those imprisoned under 18B were at first placed among the convicted prisoners. Later, most were put in F wing.

Diana, who had arrived at Holloway at about three forty-five, was released from the admissions cage at about seven forty-five to see the 'unsympathetic' Doctor Rougvie. This was well past the time – six p.m. – that she usually fed Max and her breasts were swollen and painful.

Still believing that she was only in Holloway for a day or so, she told Doctor Rougvie that she had brought a breast pump with her and would prefer to deal with her condition herself. Without having been examined, she was taken to her cell, on the third floor of F wing. Here she began to prepare for the night, as lock-up time, when all lights were switched out, was at eight-thirty, but when the wardress came round she found there was no lock on the cell door. Diana's statement made to the authorities a month later describes what happened then.

> Thereupon I was taken to a completely dark, airless and very dirty cell on the ground floor. The tiny window was entirely blocked by sandbags. There was no light or air. Here they swilled the floor with water in an attempt to remove some of the dirt. There was no bed. They put a mattress on the floor, dumped some dirty blankets and grubby sheets on it and locked me in for the night. By then my breasts had become rather painful and I was terrified to touch them because everything was begrimed with dirt and, as everyone knows, cleanliness is of the highest importance when there is milk in the breasts.

This cell was in the notorious E wing, which had been unused for ten years and was later to be filled with internees waiting to be sent to the Isle of Man. The windows were so filthy that the dirt had to be scraped off with a knife before anything could be seen through them. The cell was so damp that the straw mattresses were soon soaked through. This wing had also contained the condemned cell and execution shed and its atmosphere of bitterness, fear and despair was almost tangible.

At the time Diana was too frozen with wretchedness and bewilderment to complain. Only a determination not to break down in front of her jailers prevented her from weeping with misery and outrage. Thus the warder who was on duty when Diana was admitted and had to transfer her to a different cell was correct when she wrote, 'Mosley made no comment on the manner in which it was cleaned.'

Diana was not alone in her disgust at the miserable, filthy conditions. The prison authorities, it was true, had done what they could to spruce the place up. The chapel had been renovated, with green and cream paint instead of the previous brown and orange, and the condemned cell – said to be haunted since the execution of Edith Thompson – had been turned into a storeroom. But the antiquated building, the primitive sanitation and the pervasive London grime meant that dirt appeared to have seeped into the very walls.

'It would be very difficult to describe in detail what my cell looked like to me as the pale morning light filtered in through the grimy window,' wrote Nellie Driver, who had also been detained as a keen fascist. 'The walls were running with damp and the mattress was filthy. My bed was a few planks on the floor, and my limbs were so stiff I had to raise myself, inch by inch, and roll off the mat.'

Diana could not bring herself to sleep on the damp, thin mattress, between sheets engrained with dirt. She sat up all night with her back against the wall of the cell, contemplating her position – bereft of her children, her husband out of reach, her home gone and the mere two days' detention the police had thought likely now looking like some delicious mirage. When her cell door was unlocked at 6.30 the following morning she felt ill. 'My breasts were very swollen and painful,' her statement continued:

> One of my fellow prisoners gave me some Epsom Salts. At about 9.30 a wardress came and told me to collect my belongings and bedding. She took me to E Wing and put me in a cell on the top floor. The cell I had in F wing had not been used before or since and had apparently been condemned as being unfit for occupation. There were not many people in E Wing at the time and they for the most part were Germans. During the afternoon Dr Rougvie came into the wing and I asked for Epsom Salts and cotton wool. These appeared 24 hours later but fortunately a fellow prisoner had already given me some . . . Owing to the dirt everywhere I was afraid I might get an abscess.

She remained in acute pain for several days. On Monday she was told to wash the stone landing with a bucket of cold water and an ancient rag. No one was allowed a kneeler; Diana, who could barely move her arms because of the pain of her swollen breasts, found it more difficult still. Eventually, a kindly fellow prisoner took her rag from her and did the rest of her portion of floor.

Almost as bad as the dirt was the cold. Even in summer an icy chill emanated from the walls. Most prisoners slept with all their clothes piled on their beds. During the day the limited exercise – half an hour in the prison yard – gave little chance of combating the cold with vigorous movement. Lack of fuel meant that, like much of Britain, prisoners could only have two hot showers a week.

The prison diet was sustaining but generally ruined by staleness, careless cooking or the poverty of the ingredients. Rations were eked out with bread, soup and potatoes, boiled in their skins and often with the soil still clinging to them. Breakfast was porridge, tea, bread and two pats of margarine. On Monday the main meal, bacon and haricot beans, was reasonably tolerable, but the fish pie which appeared on Tuesday and Friday, made from Icelandic stock fish, smelt so putrid that even the hungry prison cats refused it. On Wednesday and Thursday there was generally some form of stew ('quite decent', allowed the prisoners), on Saturday a soup full of vegetables. Sunday's corned beef was the treat of the week, the only prison meat that Diana was able to stomach. There was always cocoa for tea, 'unlike any cocoa known elsewhere', records one prisoner. 'It had a thin layer of grease on top – we would let the cocoa cool, skim off the grease with a bit of toilet paper, and use it as face cream.' No prisoner ever saw her butter ration.

Disgusted by the prison plates, their many cracks harbouring unknown germs which could not be eradicated in the cold water used for washing up, Diana asked Sydney to bring her in a plain china plate as soon as she realised she was not going to be released immediately. For the single utensil prisoners were allowed, Sydney brought her a fork, while the enormous prison mug was exchanged for a smaller one.

Meals were served from tin cauldrons so heavy it took two women to carry them. A warder escorted the prisoners who were fetching the meals to and from the kitchens. She locked and unlocked each gate or door in turn, adding to the general clangour as they walked through the central courtyard, through a garden where a few Borstal girls worked, past the nursery where prisoners' babies lay in wicker clothes baskets and finally arrived at the kitchens. Fetching meals was a sought-after task. In the kitchen, there were other women from different wings; messages and information were passed, gossip exchanged. Once back, the food was ladled out, sometimes in darkness by the light of a warder's torch – the dim, shaded blue lights on each landing were extinguished in air raids. Any pudding was put on the plates at the same time. Sometimes the bread was so mouldy it was thrown into the yard, where the pigeons and rats made short work of it. Diana, unlike many other women, had no desire to be a 'carrier' and never became one. When she first arrived, she was too ill to carry anything so was omitted from the rota; once overlooked, her name was forgotten by authority.

She was soon sent back to F wing, which housed most of the ninety-odd BU members. It was an immediate improvement. She preferred being high up – she had a corner cell on the third floor, with a tiny gothic window instead of the usual small four-pane square – and she was among her husband's sympathisers. In addition, the members of F wing, like the interned aliens, were treated in a more relaxed fashion than convicted prisoners. As the Home Office had pointed out to the prison authorities, 18Bs and internees had committed no crime and were not in jail for punitive reasons, merely to keep them out of the way until an Advisory Committee could decide on release or further detention.

Their life was easier in many small ways. They could wear their own clothes, instead of the blue and white striped prison uniform, and use cosmetics. Those who could afford it could have food sent in from shops outside and they could use the kitchen to boil a kettle for coffee – if they had any. Some fried up the tiny potatoes they had gleaned after the gardeners had been round. They could invite a friend to tea in their cell, the guest bringing her own chair, plate, mug and so forth and, if it was cold, blanket. Under prison regulations attendance at the service of the church to which, on admission, she had said she belonged was compulsory for the convicted prisoner, but for internees and those detained under 18B this was optional. They could also alleviate the darkness of their cells by making small lamps out of shallow tins, soaking the string wicks in sardine oil – Diana, who quickly discovered these lamps were useless for reading, did not bother to make one – and were allowed sweets and cigarettes. If, however, the 18Bs were caught passing these to the convicted prisoners, who craved them, they got the same punishment, the 'black cell', or solitary confinement. Otherwise, their paths and those of the convicted prisoners crossed most frequently in the chapel, where they were all allowed to listen to BBC news bulletins.

As it was generally the most outspokenly dedicated fascist women who had been arrested, Diana was warmly welcomed. Many wanted to show their devotion to the Leader by cleaning Diana's cell, making her cups of tea and scrubbing the passages when it was her turn to do so. Diana was determined to be treated like everyone else, although she did eventually let one particularly devoted older woman, Florence Hayes, clean her cell from time to time or fetch things for her.

Despite her hatred of communal life, Diana did her best to show solidarity with her fellow fascists. Two or three days a week she would join them at the long table on the ground floor where they met as a self-elected 'club' for lunch. She was invariably put at the head of the table.

But though her fellow prisoners remembered her as natural, unaffected and friendly, her clothes, her voice, her background, her preference for reading in the seclusion of her cell rather than, like the others, seeking company, set her apart. So did her beauty, which worked powerfully in her favour. The first time Louise Irvine saw her she was wearing a plain camelhair coat:

> a tall slender figure, her hair blonde and simply cut, and extremely beautiful. Quite involuntarily the description of Helen of Troy from Tennyson's 'A Dream of Fair Women' went through my mind: 'A daughter of the Gods, divinely tall and most divinely fair' . . . She looked thin and pale but her smile lit up her face. Later, when I got to know her, I was to find that her smile and her laughter were very much part of her.

These were also effective in the face of the warders' often sneering attitude. 'Letters for *Mosley*!' they would sing out, or remark sarcastically that prison must be very different from what she was used to. Smiling, she would agree. Most soon smiled back.

But alone in her cell, she wept bitterly. The realities of the situation were gradually becoming apparent. The 'few days' that she had been told her imprisonment would last had long passed; she was not a criminal yet no one could tell her when she would be released. She was not allowed to see her husband and she could have only one half-hour family visit a fortnight. Every time she thought of her own children she began to cry. The older ones were at school, but she missed the two babies with a physical agony. Nanny's letter from Rignell, a fortnight after her arrest, made her ache with longing.

> I was so pleased to hear about you and to know you were feeling better and no more discomfort from the milk. You will long to hear about your babies. Alexander is very well and happy but still a bit squeaky. He does want his own way but really has settled down, and the little girl [nursemaid] is very good at amusing him and keeping him happy when I am busy with baby's bath and feeding. We are all happy. Max had a milk bottle which seemed to agree with him but he was hungry before the next meal was due. I think he will be entirely on cow's milk this week. He is bonny and such a lovely colour and sleeps so well and is just a sweet pet, always a happy smile. He gained seven ounces last week, which is 4 lb. 3 oz. in three months, and is 26 inches long . . . We went to Banbury one day and got Alexander a red mackintosh. He looks the dearest pet in it, and loves to parade in front of the glass. He is very proud of it. I do wish you could see them . . .

Slowly Diana became accustomed to prison life, fighting the cold with hot-water bottles and warmer clothes. 'I am sending the highnecked grey jersey, and have asked Mabel to send the rug and green coat and skirt,' wrote Pam on 21 July, three weeks after her sister's arrest. Diana found the prison food largely uneatable, and existed mainly on the prison corned beef ('delicious') and Stilton sent to her by Mosley, which she would eat accompanied by a glass of port – detainees were allowed to have a little alcohol sent in, and Diana had ordered several half bottles of port.

The only organised entertainment was an amateur concert party from a charitable organisation which sang sentimental songs from the altar steps – reducing Diana to such agonies of stifled giggles that she had to bite her tongue not to burst out laughing. Far more enjoyable were the gramophone recitals given by a rich and generous *poule de luxe*, brought in on suspicion that her German birth would cause her to lure young officers into betraying their country. Furious at being arrested, this girl had somehow browbeaten police and warders into allowing her to bring both her jewellery and her large collection of classical records with her – these included Beethoven symphonies, works by Chopin, Wagner, Scarlatti, Grieg and Massenet. She would also send out for any piece of music not in her collection, if someone asked for it.

But nothing could make up for loss of liberty with no foreseeable end. As Nellie Driver later wrote:

> Nobody who has not lived in that endless-seeming imprisonment could realise the hopelessness, the cruelty and the mental agony of it all. To have committed no crime, to be branded a traitor . . . Years of imprisonment because 'in certain circumstances' you might commit an un-named, unspecified act against your own country? How many citizens could be imprisoned on the supposition that 'in certain circumstances' they might commit a theft or murder?

In the world outside, opinion was divided. Letters flowed into the Home Office. 'At last somebody has had the good sense to arrest Mrs Mosley,' ran a typical one. 'Ordinary people are getting a bit fed up with this protecting of monied people. Why should British food be used to feed traitors like Mosley, Mitford and Redesdale?'

In the House of Commons, the ethics of detention without trial were questioned constantly, and the uneasiness it provoked was reflected in the columns of national newspapers. 'Sir Oswald Mosley and his fellow fascists should be tried,' declaimed the liberal *News Chronicle*. 'If they are found guilty the British public will be able to stomach whatever penalty the State may inflict. If they are innocent they should be set free.' But this was not to be – not then, or for a very long time.

Chapter XXVI

Ten days after her arrest, Diana demanded through her solicitor, Oswald Hickson, to know on what grounds she had been detained, and applied to be seen by the Advisory Committee. But it was not for another two months that she received an answer.

The aim of everyone held under 18B was to be seen as soon as possible by the Advisory Committee. This was headed by Norman Birkett, KC, an advocate famous for his conduct of criminal defences and libel actions, who was to be the second British judge at the Nuremberg Trials after the war. The Committee had been set up by the Home Office to decide whether those caught in the wide net of 18B should remain in detention or be freed. In the event, the majority were released, but in the early months no one anticipated this.

For the women waiting either for a date or for the results of their hearing – often months later, such were the numbers to be assessed – it was a dreadful time. The stress was exacerbated by the poor diet, on which virtually everyone lost weight. Some became so anaemic they ceased to menstruate while, as the months went by, fatigue, inability to concentrate, lassitude and depression were commonplace. Worst affected were those with families and young children, who worried themselves to distraction. Some wrote to their MPs in the hope that they might be able to help or at least speed things up – Nellie Driver, still a convinced fascist, was surprised and delighted when her MP, the left-wing socialist Sidney Silverman, came to see her in prison.

Mosley's examination by the Advisory Committee had begun on Monday 15 July 1940, at 6 Burlington Gardens. It was immensely complex and long-lasting – the transcript is the size of a small book – and enlivened by shafts between Mosley and Birkett, who had faced each other before in one of the numerous libel actions brought by Mosley, which he always conducted himself. At the end of it the committee of seven were in no doubt that Mosley should remain in prison.

Neither Mosley nor Diana knew that he might have suffered the same fate as Napoleon: exile to St Helena. It had been suggested in the Home Office that the 600-odd members of the BU detained under 18B should be confined overseas. Jamaica, St Lucia (too many droughts), Ceylon, the Seychelles (too primitive) and Mauritius were considered. 'From a security point of view, St Helena would be the most satisfactory,' ran a Home Office memorandum dated 20 July 1940. 'But it was thought that, if Australia and New Zealand were willing, they would be more suitable.'

This idea eventually foundered on the distaste of all concerned for tampering with the Habeas Corpus Act. The problem was that if detainees were sent to Australia or New Zealand (self-governing dominions as opposed to colonies) they would be outside the jurisdiction of the Home Secretary and he would not be able to require their return; nor would the government have any control over the conditions in which British subjects were kept, or their release.

Such imprisonment was in any case unlawful under the Habeas Corpus Act of 1679. This provided:

> And for preventing illegall imprisonments in prisons beyond the seas bee it further enacted by the authoritie aforesaid that noe subject of this realme that now is or hereafter shall bee an inhabitant or residant of this kingdome of England dominion of Wales or towne of Berwicke upon Tweede shall or may be sent prisoner into Scotland Ireland Jersey Gaurnsey Tangeir or into any parts garrisons islands or places beyond the seas which are or at any time hereafter [shall be] within or without the dominions of His Majestie his heires or successors and that every such imprisonment is hereby enacted and adjudged to be illegall.

The only way round this was an Act of Parliament. The conclusion of a meeting on 22 July, at which the plan was decisively rejected, was that 'To send them abroad would require legislation and the Home Secretary thinks this would be controversial.' Even in the desperate circumstances of 1940, the National Government jibbed at amending the Habeas Corpus Act.

Mosley's experience on arrival at Brixton had been similar to Diana's, with the same small reception cell, carbolic bath, medical inspection and registering of property. But the prison governor, Captain Clayton, and the chief warder, Watson, described by Mosley as 'fair and honourable men', did their best to see that the 18Bs were treated like remand prisoners. Mosley had been allotted No 1 cell in F wing, previously condemned as unfit for use and seething with bedbugs; next door to him was his former follower John Beckett, furiously resentful that Mosley had been allowed a wireless and he had not. To add to Beckett's irritation, he could hear when Mosley's wireless was on but not what was being said. 'Nothing in the world was more frustrating than to hear fucking Tom Mosley's radio,' he would remark for years afterwards. 'To know that he could hear the news and I couldn't.'

Brixton was not built on Holloway's radial plan, but had blocks of three and four storeys surrounding a long narrow hall open to the air at each end, with iron staircases halfway along each wall. Facing the stairway on each of the floors was a recess between the cells, holding a lavatory with a half-door, a sink and a cold-water tap. Cells had the same furniture as in Holloway, with a plank bed about two inches from the floor. There were clean sheets once a month, only one pane in each of the four-pane windows could be opened, and throughout the blocks electricity was on a low-voltage system so that prisoners could not kill themselves.

At first, the 18Bs were treated like the other prisoners: let out at 6.45 for slopping-out, locked up again until breakfast was brought round at 7.15, let out at 9.00 for another slopping-out, locked up again until it was time for an hour's exercise on a narrow stone pavement round a grass plot. After this, they could be together until 11.30 when they were locked up again until 2.00. Dinner was brought round at 11.40. In the afternoon they went outside again for more exercise, until 3.30. The last meal of the day was at 4.00, after which they were locked up for the night except for a last slopping-out at 6.30. Lights-out was at 9.00. But it was not too long before the cells of male 18Bs were left unlocked except at night.

'Imagine trying to read in a genial babble of Mediterranean voices,' wrote Mosley cheerfully to Diana (many internees were the Italian waiters who had staffed the most famous restaurants in London). 'The final nightmare was permission to bring in a ping-pong table, when the echoing seashell of a building resounded with ping and pong and Latin laughter. The subsequent discomfort of being locked in cells while bombs were falling round the prison was nothing to it.'

As in the outside world, Mosley was a dominant figure. He would wander round the exercise yard surrounded by a dozen or so followers at a discreet and respectful distance and, as in the world outside, the scent of violence seemed to hang around him. Many of Mosley's former disciples were now, like John Beckett, bitterly opposed to him, and two hostile camps emerged. Beckett himself formed unexpected alliances, with IRA prisoners, and with Arnold Lees, a vet who thought the trouble with the BU was that it was not sufficiently anti-semitic. Viewing the scene in the exercise yard, Beckett would remark caustically, 'There goes Mosley with his kosher fascists.'

Where another internee might have been pathetically grateful for a visit, Mosley allowed neither the isolation of prison nor ties of friendship to affect his view of the loyalty due to him. He had refused, for instance, to see Harold Nicolson, who arrived at Brixton to visit him, on the grounds that in a broadcast the previous evening Nicolson had said something 'disobliging' about him. It weighed not at all with Mosley that Nicolson, as a junior Minister, had shown great moral courage in acknowledging the tie of friendship with a man assumed by many to be a traitor.

Mosley manifested the same stubborn sense of his position in his attitude to the three children of his first marriage. Less than two months after being jailed, he had more or less divested himself of them. They had earlier been made wards of court and for a number of years their aunt, Irene Ravensdale, had been virtually running their lives. With their father in jail, his role would be more distant than ever and she had been asked if she would be their joint guardian with Mosley. She replied that she thought 'the difficulties and divergences of views would make this unworkable'. Her solicitor, who had acted for her when Mosley had pushed her into paying his share of the Denham bills, strongly endorsed her view. Almost at once, the court hearing proved her point.

> The judge was willing to leave all question of guardianship and cooperate with Tom on important issues [records her diary]. Tom flatly refused it and wants entire and complete control for the boys – he was not interested in the girl. After discussion and deliberation the judge could only ask for outside guardianship as Tom had refused to have any assistance, and my name was put up. He sent me a lamentable message saying that I must understand he would never later have anything to do with the care or guardianship of any of his children.

From then on, Mosley and Irene hardly ever saw each other.

'The children today were duly removed,' he wrote to Diana. 'The judge took the line that I should remain guardian in name, but as it was clear that I should have no direction in their education or anything else, I declined responsibility without authority, and Lady R. was made guardian.' Or, as 'Lady R's' diary puts it, 'God in his illimitable mercy has handed me Cim's children.'

Diana's concerns for her own children were very different. Explaining her imprisonment to her two older sons had worried her deeply. Bryan, now stationed in Aldershot, had not seen Jonathan for some months but had managed to reassure Desmond, at home at Biddesden with glandular trouble in the neck:

> I don't think he is worrying over you. I told him about your imprisonment as being like an hotel for that very reason. I didn't suppose conditions were quite like that – but it seemed the best way of putting things to him . . . I shall be leaving here soon to go to a battalion but I don't know where that will be. I will give Desmond many messages from you and explain that though probably rather bored you aren't unhappy and that as soon as he is well enough he is to go and see you. Jonathan is being brought to me by Pam and Derek [Jackson] next Saturday. He evidently did very well last term and won lots of prizes.

At first prisoners were allowed only two letters a week. Diana's, naturally, were from Mosley; charming, loving, encouraging letters that did much to keep up her spirits: 'Darling Percher, I do miss you so – you are such a brave and wonderful Percher. I do hope you are not too depressed . . . the present hysteria must pass quite quickly and we might soon be together again.' Like all letters written to prisoners, these were censored but the warders soon gave up attempting to read Mosley's letters: to anyone except his family, his handwriting was virtually illegible.

Soon there was little restriction on incoming mail. 'Darling Honks,' wrote Debo from Inch Kenneth on 25 August, 'Syd writes saying one can write to you at last. Oh, I do so long to see your cell. I haven't seen you or your pigs for such ages that I have almost forgotten what you look like.' She went on to recount exactly the sort of chatty, day-to-day gossip Diana loved, from details of the weather ('never stops raining') to an improvement on the béarnaise sauce they had both learned to make in the cookery course that now seemed so long ago:

> I now put in equal amounts of wine, lemon juice and vinegar. I hear you cook like a mad thing too. I do hope you are given eatable ingredients? . . . If ever you were short of a visitor, I would come hurrying to Holloway.
>
> I wonder what Syd's form is now, I mean whether she's in a good

temper or not. Her favourite thing is going to see you. She always writes 'I'm going to see D' or 'I've just been to see D', usually from the tea room at Paddington. She will be the death of me. Much love, Debo.

On 7 September the bombing of London began. During the following twenty-two days and nights almost 7,000 tons of bombs were dropped on the capital and the Docks were on fire almost every night. Holloway was more crowded then than at any other point in the war – there were 682 18Bs and internees and 181 convicted or remand prisoners. 'We were horrified to see "Bombs drop outside London prison",' wrote Debo from Swinbrook on 14 September:

I hope all the well-known inmates were all right. I long to come and see you and I long for Cyril but what with one thing and another I shan't come to London this week because Syd announces her intention of living up in Scotland . . . Syd is learning to milk my goat but as she always either shuts her eyes or gazes into the middle distance the milk goes anywhere but in the bucket. She is the maddest lady.

Five days later Diana received a letter of a different kind: an answer to her solicitor's letter two months earlier asking why she had been detained. 'The Secretary of State has reasonable cause to believe that you have been active in the furtherance of the objects of the organisation now known as the British Union and that it is necessary to exercise control over you.'

It continued:

a) You have acted as a channel of communication between Oswald Ernald Mosley, leader of the said organisation, and leaders of the German Government.

b) After the detention of the said Oswald Ernald Mosley you gave instructions for the carrying on of the said organisation.

c) You have supported the said Oswald Ernald Mosley in his position as leader of the said organisation both publicly and privately.

d) You have publicly and privately given expression to pro-fascist and pro-German sentiments.

How true were these charges? There is little doubt of the accuracy of the first and most serious. As to the third and fourth, Diana would have proudly agreed. She had attended as many of his meetings as possible, sitting near the front, and she thought everything he did and said was right. She also believed that what Hitler had done for Germany was little short of a miracle. The answer to the second charge was less clear: on Mosley's behalf, she had continued to pay the wages of those staff still left in BU headquarters but she had not in any way concerned herself with BU strategy.

Diana viewed the charges as – to use one of her favourite words – monstrous. She thought that free speech was one of the things the country was fighting for, she believed totally in her husband's views and in his right to express them. She admired and liked Hitler and she was not about to deny a friendship. For many years, she had shut her eyes to the evils of Nazi Germany and to the reflection of Nazi attitudes in Mosley's fascism. She arrived for questioning on 2 October in a state of contemptuous, hostile defiance. She was also looking forward to her first decent meal for months.

It was a glorious day for her brief outing from prison. She was driven, with an escort, to Ascot – the house in Burlington Gardens where the committee had previously sat had been destroyed by a bomb. The committee was headed by Norman Birkett; its other two members were Sir Arthur Hazlerigg and Sir George Clerk – with whom Diana had lunched in 1931, when he was Ambassador in Athens.

She was questioned on everything from her visits to Rome ('Were they to meet Mussolini?' 'No, they were purely social') to her views on Hitler. 'Absence makes the heart grow fonder,' said Norman Birkett. 'Do you still entertain the same feeling for him?' 'As regards private and personal friendship, of course I do,' replied Diana. 'The history of Hitler in recent years has not affected your view about that?' continued Birkett. 'I do not know what his history has been,' Diana responded coolly. She told them that she liked Himmler 'very much' and that she did not believe what she had read of the Gestapo's actions. Even more indefensible was her statement that she agreed 'up to a point' with the German policy towards Jews. Finally she expressed the ludicrous opinion that England had been 'dying for a war' when, in the general view, the Chamberlain government had betrayed Czechoslovakia in order to avoid a war which was universally dreaded.

Her arrogant refusal to admit the slightest fault in the enemy of her country did for her. Nor can Mosley's testimony – when asked why he and Diana were allowed to marry secretly in the Goebbels' drawing room – that 'Frau Wagner, Frau Goebbels and my wife are the three women for whom Hitler has the greatest regard in the world' (in front of his earlier

Advisory Committee) have helped her cause.* When she answered 'Yes' to the question 'If you had power, you would displace the present form of government in this country with a fascist regime?' the committee must have felt they had little choice. 'You have a great contempt for democracy?' asked Sir Arthur Hazlerigg. 'Yes,' she replied shortly. Later, at lunch, a bottle of claret from Sir Arthur arrived at her table; Diana shared it with her escort.

Two days later, on 4 October, Birkett wrote to Sir Alexander Maxwell to say that the Advisory Committee recommended unanimously that Diana should continue to be detained. Indeed, in the prevailing climate, she might have been in some danger if released.

His letter said:

> They felt that there was no other possible course open to them. It would be quite impossible, having regard to her expressed attitude and her past activities with the leaders of Nazi Germany to allow her to remain at liberty in these critical days. Lady Mosley appeared to the Committee to be an attractive and forceful personality who stated her views with clearness and without ambiguity; and the Committee formed the opinion that the views which she entertained were held quite sincerely. In these circumstances, Lady Mosley could be extremely dangerous if she were at large and the Committee entertained no doubt that the proper decision was to recommend that the detention be continued.
>
> In their report the Committee commented on the fact that her second marriage was in Germany and kept secret by Hitler himself; this incident illustrates in the clearest way the terms of extreme intimacy which existed between the Mosleys and the leaders of the Nazi Party in Germany. The Committee entertain no doubt whatever that the friendship of Hitler was entirely due to the attitude of the Mosleys to the Nazi regime. Lady Mosley made no secret of the fact that she was a very great admirer of the Nazi regime and every fact that was investigated by the Committee confirmed this in every detail. They also say that it may very well be that the views of Hitler on the British statesmen have been to some extent coloured by the views put forward by Lady Mosley.

Though at the time Diana believed that Birkett was deeply hostile to

* Later Diana was to claim that this extract was a forgery, put in to discredit her 'by Blunt or one of the Communists. Kit would never have said anything like that'. In the opinion of the author, it is genuine and, when seen in context (see Sources for Chapter XXVI) perfectly understandable.

her, short of releasing her he had tried his best to help her. His letter continued:

> Lady Mosley would desire to have some opportunities of seeing her husband and the Committee felt this was a very natural wish that they would like to support. If it were possible that Sir Oswald and Lady Mosley could be detained in the same camp no possible harm could arise therefrom, and it would to some extent mitigate the detention, which may last some considerable time. Furthermore, Lady Mosley has two very young children by Sir Oswald Mosley, being babies of two years and five months respectively. Quite naturally, she feels very keenly not being allowed to see her very young babies and the Committee also felt that it was quite natural that she should desire some opportunity of seeing these young children. These are matters which strictly are outside the province of the Committee, but they felt that they were matters which, if they were brought to your notice, could be dealt with on humanitarian lines.
>
> You know better than anybody else that even in the case of convicted criminals opportunities for visits are allowed and the Committee felt that they would like to bring to your notice the situation of Lady Mosley with regard to her husband and particularly her very young children and to ask you if it is possible to see that her very human desire can be met.
>
> I would have sent this letter direct to the Home Secretary but in view of the very recent change [Herbert Morrison had succeeded Sir John Anderson the previous day, 3 October] I thought it would be more speedy to send it to you.

Diana was at the time allowed one half-hour visit a fortnight in the presence of a warder – the nicer ones pretended not to listen, though they had to make a report of all conversations. Only two people were permitted to see her at a time and she had not asked for her children because London was now being bombed more frequently.

Sir Alexander Maxwell replied to Birkett on 19 October, saying that there would be no difficulty about the children being taken to Holloway to see their mother – a special sitting room had been set aside for the visits of children to their mothers – and the only reason Diana's children had not been to see her before was that she had not requested it.

A meeting with Mosley was not to be allowed, continued Maxwell's letter, the reason being that other husbands and wives who were detained were miles apart from each other all over the country and such visits would set a precedent that the Home Office could not fulfil.

> It would look like favouritism if we allowed the Mosleys to exchange visits merely because they are both detained in London. Another difficulty is that, if such a visit were allowed, if expenses and transport were paid out of public funds, there would, I imagine, be public comment and criticism, and if it were not paid out of public funds, there would be inequality between rich and poor.

Birkett tried once more, saying that the Advisory Committee were suggesting that Lady Mosley be allowed out of prison on parole to visit her children. 'It is entirely natural that she does not want her children to come into the prison. The Committee are quite satisfied she could be entirely trusted if this course were possible. The Committee hope very much that arrangements of this kind, i.e., the children and parole, and being interned with her husband, can be made.'

He was turned down again, but there would soon be one much desired improvement in Diana's life.

On 11 December the MP Richard Stokes declared in the House of Commons that 'it was quite insufferable that people should be locked up for an indefinite period and not be allowed to have even an open trial'. His view chimed with that of Churchill, to whom the Habeas Corpus Act was at the heart of the liberty that Britain was defending, and on 22 December the Prime Minister asked the Home Secretary if conditions for detainees could be improved. Why were letters restricted to two a week, asked Churchill; were newspapers allowed, what were the facilities for exercise and, most important of all, what arrangements were permitted to husbands and wives to see each other?

As a result of the Churchill note, Mosley was to be allowed to visit Holloway once a month to see Diana.

Many years afterwards, Diana was to confide in a friend, 'I regarded my appearance in front of the Advisory Committee as an absurd and insulting farce. I was to regret this bitterly.'

Chapter XXVII

The Blitz was now at its height. During October 1940, according to German records, 6,500 tons of bombs were dropped on London. The rise-and-fall wail of the warning siren usually sounded at about five o'clock. In Holloway prison, where the glass roof made 'blacking out' difficult, electricity was immediately switched off at the mains, leaving the inmates in darkness until the following morning. All cell doors were unlocked and the women groped their way into the cells of friends, often using the flashes from anti-aircraft guns to light their way. Prisoners from the top two 'dangerous' floors were expected to come down to the lower floors. Surprisingly, there was only one terrible accident: a French girl who fell from a landing and broke her neck.

It was possible to get a good view from the top floor of F wing by standing on a table in one of the cells. Many of the fascist women came from the East End; from a top window they would peer, distraught with worry about their families, at the glow of the fires lighting up the sky as bombs rained down on the Docks. Without access to a telephone, they had to wait anxiously for the reassurance of the next weekly visit. The noise was incessant – there was an anti-aircraft battery just down the road – and the dark hours of night seemed endless. 'The crash of guns and the falling bombs nearly drove us mad as we crouched in our cells or tried to snatch a few hours' sleep on our planks. Our longing was for light, light, light!' wrote Nellie Driver.

On the night of 14 October, B wing, which held many interned aliens, was hit by a bomb. It tore through the roof and several landings, blew in most of the doors, cracked walls and started a number of fires. There was rubble and glass everywhere but fortunately no one was killed. Because of the pressure on space, prisoners were put back into cells which in some cases were barely habitable, with wind whistling through cracks or rain pouring through gaps in the roof. Gas and water were cut off and drinking water was strictly rationed.

Diana was not frightened by the air raids – she hardly cared whether she lived or died. The defiant bravado with which she had faced the Advisory Committee had evaporated in the face of her continued detention and she could see no prospect of release. She was physically below par, cold, hungry and wretched, and the winter lay ahead. She was reading Carlyle's *History of the French Revolution* and she thought that, like the prisoners he described, if she had to stay long in jail she would prefer to die.

She could not understand the basis on which those chosen for release were selected and gradually the feeling took hold that in some obscure way she had been made a scapegoat. Florence Hayes, the devoted adherent who had insisted on doing menial tasks for her, had been allowed to leave prison. 'I can't understand why they released her,' she said to her mother (in a conversation recorded by the attendant warder). 'She was such an ardent follower, a great speaker and a good organiser. Of course, like the rest [of the fascists] she wouldn't do anything against this country. But I'm positive nothing will change her views, so it's really amazing why they let her out.'

She asked Sydney to send Miss Hayes £5 – but she could not tell her mother her most secret thought: a growing conviction that she had been imprisoned in Unity's stead. In the public awareness Unity, rather than Diana, was the more notorious for her close friendship with Hitler and for her sympathy with the Nazi cause. Diana was well aware of the revulsion her sister inspired in a large section of the public. Nations, as much as individuals, looked for scapegoats. But Unity, by turning the gun on her own head, had pre-empted any idea of punishment or revenge. Poor, sick Unity, destroyed by the bullet, could no longer be the sacrifice demanded.

Diana had plenty of time to brood on such thoughts. Visits were now permitted every week, but only for fifteen minutes. Sydney came regularly, often bringing Debo. They would make the journey from their house in Kensington to Holloway by bus. On one occasion the conductor sang out, 'Holloway jail! Lady Mosley's suite! All change here!' Looking daggers, Sydney descended from the bus with Debo. She was forceful about conditions inside the prison, once writing in the visitors' lavatory 'This lavatory is a disgrace to H.M. Prisons'. She campaigned constantly for Diana, talking of the injustice done to her daughter every time she met someone she thought could help, writing to MPs and lobbying the influential. She also devoted a lot of time and energy to making Diana's life more comfortable.

> After I left you I went to Harrods and sent you a siren suit and a very warm jumper [she wrote on 20 October]. I only hope they are the colours you like, and the right size. They were rather expensive, suit £4 and jumper £2.10s. Nanny and I are going to look for another thick one among Bobo's masses of clothes. Nanny is here and I go tomorrow back to Swinbrook. Debo has gone to Andrew [Cavendish] for the weekend, he has leave. I hear there is no water at the Dorchester or Claridges. I don't know if it is true. I also sent off your Malvern water and your Elizabeth Arden. Nancy here last night. She is very tired, I fear, and couldn't sleep as she had hoped, but she is going away to the country for a week on Monday. I am sending a deposit for you to Harrods of £10. All these things I bought today I paid from my account and will repay myself from your money [Diana had arranged that Sydney could draw on her account at Drummond's Bank]. So now you have £10 clear at Harrods. I went to lunch at the Cordon Bleu [a restaurant in Sloane Street] only to find a notice on the door 'Shut – no gas'.

One November morning Diana's general wretchedness was pierced by an agonising worry. Jonathan, at school, had suddenly developed appendicitis and had been operated on a few hours later, at 5.30 p.m. His headmaster tried to telephone Diana at Holloway, but was not allowed to speak to her, so sent a telegram. 'What a fright Jonathan successfully operated on for acute appendix just seen him quite happy and as comfortable as possible writing Evans Summerfield.' This, however, was not given to her. Instead, the Governor told her that Jonathan had had an emergency appendectomy the night before and was very ill. She begged for permission to go to him, but this was refused by the Home Office. Distraught, she walked frenziedly up and down her cell all that day and the following night, unable to think of anything except her son, perhaps even then dying, but there was no word. It was not until the following morning that the Governor sent for her and told her that Jonathan had 'turned the corner'.

It was Sydney who told Diana that Mr Evans had not only telegrammed but sent a letter. As Sydney's constant efforts on behalf of Diana continued unflaggingly, as she arrived faithfully to visit, whatever the difficulty or, sometimes, the danger, so Diana's admiration and love for her mother increased. Once distant, resentful and rather scornful of Sydney, she had been changed by her time in prison into a loving and devoted daughter.

By now letters to the Hon. Lady Mosley, 5433, cell 23 F3, were flowing in. One of the earliest had been from Gerald Berners, who described taking Jonathan out from Summer Fields, to Fullers, a favourite Mitford haunt in Oxford.

> I gave him a good blowout – two maple ice cream sodas, three slices of chocolate layer cake and a walnut sundae. I deposited him at Summer Fields with his great coat bulging with sweets. I only hope he won't have a bilious attack. Luckily, most boys of his age have stomachs like the interior of a tank. Then when I got back I heard that you had been jugged. I was very upset. I do hope you are not too miserable. I read in the paper that you would not be allowed to see Tom but I can't believe it. I don't know if you will be allowed to write. If you are, please send me a line.

At first Nancy neither wrote nor visited, possibly feeling that any contact after her efforts to have her sister put in prison would smack of hypocrisy. She also refused to believe her mother's accounts of conditions at Holloway and of Diana's pale, depressed appearance. In November she wrote caustically to Mrs Hammersley, 'I would die of the lights out at 5.30 rule wouldn't you? I suppose she sits and thinks of Adolf.'

As the year wore on, the 18Bs tried to keep fit, clean and tidy, washing their underwear with scraps of soap in their cells, playing ball in the prison yard. They learned the tricks of prison life: how to make coffee of a sort, putting used coffee grounds from the kitchen (it was before the days of instant coffee) into a mug and adding water from the hot tap in the washing section. They hammered milk bottle tops into toast racks and other small gifts for each other, they climbed the wild cherry tree in the garden to reach the cherries before the birds got them. Many, desperate for affection, tried to adopt one of the cats that had found their way in from bombed houses or had been dropped as kittens over the prison wall into the Governor's garden. Others did their best to keep their spirits up with whist drives; tins of sardines, sent in from outside, were the prizes.

One woman threw a note out of her window, saying, 'I am dying of starvation' (no one took any notice), but few complained about the poor, scanty food. Everyone knew that it was the same 'outside', where making the rations go round was a constant preoccupation. What they minded much more was the dirt and the icy, bone-clenching Holloway chill. Sydney brought a rug or jersey almost every time she visited. Diana wore an old raccoon jacket day in and day out and knitted herself a two-piece suit, from thick grey unrationed wool so stiff it almost stood by itself. It was the first thing she had ever made and she was inordinately pleased with herself.

November brought Diana the first glimpse of her husband that she had had since her arrest. It took a libel case to make this possible.

Stories about the Mosleys had been appearing regularly in the press, mostly to the effect that for them prison was a bed of roses. The hardest hitting had been in the *Sunday Pictorial* of 4 August, which claimed that Mosley, unlike other 18Bs, was somehow managing to lead a life of luxury in jail. It was alleged that he played Bridge, drank champagne, wore silk underwear and had breakfast, lunch and dinner sent in by car. Mosley counter-claimed that this was libellous and sought an injunction to prevent the story's repetition. The judge, however, refused an injunction on the grounds that to say someone did this in prison even at a time of national emergency was not libellous although it might be untrue (as it was). 'There is no harm in playing Bridge and drinking "bubbly",' said the judge. 'Many of us do. I do.' It was not quite the answer wanted by Mosley, who in other circumstances would have agreed with him.

Emboldened by this decision, newspapers continued to print further damaging stories. Behind many of these fantasies was the truth that the authorities had decided that since the 18Bs were not criminals they should be treated as remand prisoners: consequently from 1 August permission was given for detainees to order food, sweets, flowers, cigarettes, cosmetics and half bottles of wine or bottles of beer from outside. Diana ordered cakes, honey, fudge and chocolate from Harrods (giving up her sweet ration) to send on to Jonathan and Desmond at school. It was the only thing, apart from her weekly letter, that she could do for them to show her love. Sometimes these parcels were simply sent on unopened (parcels that arrived from a shop and remained sealed did not have to be opened and examined by warders whereas anything coming from a relative was searched), sometimes she repacked them in wrapping paper from her numerous orders of books. The parcels and the port she ordered surfaced in the press as 'a life of luxury'.

Finally, and perhaps inevitably, one writer went too far. Under the headline 'Privation' the *Daily Mirror* unleashed a torrent of indignant sarcasm, a *tour de force* of imagination unimpeded by fact:

> All right-thinking folk will be angered by the fierce rigidity of our internment regulations that allows rich fascists to be reduced to a harsh diet of chicken washed down by magnums of champagne [the writer began]. I learn on good authority that the most prominent fascist is being victimised by a ruthless menu which includes the rigours of daily bottles of red and white wine attended by the merciless persecution of a private batman. Further indignities are inflicted on this gentleman by the despatch of his silken shirts to a Mayfair laundry.

Mosley, who enjoyed legal battles, pounced. As another article, in a paper in the same group, had implied that Diana had pulled strings to delay her arrest ('What strange delay has led to Lady Mosley being at large for weeks after her husband's arrest? It is perhaps significant that she has been allowed her liberty until after the birth of her baby son. Does the law act with such delicacy to ordinary women delinquents?') Mosley instituted proceedings in their joint names against Daily Mirror Newspapers Ltd. It was an extraordinarily bold action, as their unpopularity in the country and the prejudice against them was such that a jury might well have refused to find for them. Mosley usually represented himself in libel actions but, possibly fortunately, prison made this impossible and so demonstrably untrue were the stories and so powerful their legal team – their counsel was the noted KC, G.D. ('Khaki') Roberts, assisted by Gerald Gardiner* – that the opposition caved in, and settled. Diana spent her share of the damages on a fur coat, lined in her favourite pale grey rather than the usual brown, chosen and ordered for her by her mother. It was ready by March 1941, after which she seldom took it off her back.

When the Mosleys met in court, each was shocked by the other's thinness and pallor, Mosley's enhanced by the beard he had grown because of the difficulty of shaving in the icy prison water. His phlebitis, too, was recurring. After giving evidence, they were allowed a few moments together in a conference room in the presence of 'Khaki' Roberts. Soon afterwards, came the visit resulting from the Churchill note.

Back in Holloway, a grand Fancy-Dress Ball heralded Christmas. One woman went as the Mistletoe Bride; Nellie Driver, who had found an old handloom in a top-floor cell, went as a weaver, complete with shawl,

* Geoffrey Dorling Roberts was a well-known KC always known as 'Khaki' Roberts. He was later one of the prosecuting counsel at the Nuremberg Trials. Gerald Gardiner, a Quaker and a pacifist, was later a distinguished Lord Chancellor.

shuttle and reed. Some concocted costumes out of blackout material, including the winner, a Mrs Winfield, who went as Old Mother Hubbard in mobcap and shawl, walking on her knees for the entire parade to make herself look short enough to 'live in a cupboard'. Most striking was the girl who came as Hitler (some of the German women took this very badly so at the next Fancy Dress Ball she appeared as Anthony Eden). The motor-cyclist champion Fay Taylor, arrested for being a BU member, produced the Holloway Cabaret.

Gradually, Diana emerged from her despair. She had never lost her fighting spirit and, as she later told Mosley, even on her worst days, 'I always felt, "Isn't it lovely to be lovely *One*?" when I looked around me.' (Mosley found this remark so hilarious he retailed it to Nancy, who put it in the mouth of Cedric in her novel *Love in a Cold Climate*.) As Diana's natural good spirits began to reassert themselves she would sometimes drawl out a joke. 'Of course, they were expecting to see me in *chains*,' she said after her Guinness children first visited her. On a good day, she sometimes sang to herself or to friends, old childhood favourites like 'She Wore a Wreath of Roses' or 'My Grandfather's Clock', which she and her sisters had sung grouped round Sydney at the Asthall grand piano. The flow of letters cheered her and, thanks to the constant efforts of their solicitor, Oswald Hickson, permission had been given for Mosley to visit her once a fortnight rather than the monthly visit suggested by Churchill. When the raids slackened off at the end of December, she felt that it was at last safe for her to see her younger children. Sydney brought Alexander and Max to Holloway on 7 January 1941.

For this visit, the first time Diana had seen her babies for six months, she was given an hour instead of the usual fifteen minutes. Max stared at her solemnly; Alexander, aged two, had begun to talk. The warder in attendance reported that Diana asked her mother to beg for help from Lloyd George in gaining her release, as 'he is not in this filthy government'.

In January, too, came the first letter from Nancy ('I had no idea I was allowed to write'). Even then it was a response rather than an initiative: a letter of thanks for the Christmas present of money which Diana had sent her through Sydney. But at least it broke the ice.

More welcome were Nanny Higgs's comforting bulletins from Rignell.

> We had a most successful day. Alexander was very tired going home, but had a nice sleep on the way and quite recovered when we got back

> and ever so hungry. He enjoyed his day I am sure but the time was far too short. He keeps saying 'Mummy today'. I am sure he knew you. He was sweet with Sir Oswald and talked in his funny way but had a tumble but soon recovered. He really is brave over his falls. His Daddy thought him very goodlooking and quite fat enough . . . I am glad you are warmer and comfy.

In February she reported that Alexander

> says lots of funny things. You would eat him if you heard him. He often says 'a kiss for Mummy' and looks at your picture . . . Mrs Jackson went to Oxford yesterday and had the [Guinness] boys out for a short time. They say they write to you every week.

At the beginning of April there was a dramatic amelioration in the treatment accorded to the 18Bs. Orders were given that their cells should be left unlocked from ten a.m. until seven p.m., and they were allowed to leave their cells or walk in the exercise yard without a warder in attendance. This distant echo of a former life in which they had been able to move about freely made them all more cheerful. On a fine day Diana, wrapped in her fur coat – now that she was so much thinner she felt the cold more – would read her book in the garden. With the German-born Mrs Swan, another detainee, she had been reading the great German writers, including Goethe and Schiller, throughout the winter. She was still wretched, but she no longer felt that death would be welcome.

In April Debo was married to Lord Andrew Cavendish. There was a choir of girls because no choirboys were available, but Debo's wedding dress, a huge and exquisite affair of tulle by Stiebel, made up for any shortcoming, as did her going-away hat with its blue taffeta ruche. The Mitfords dressed up to the nines. 'Bobo wore a pink dress, navy blue coat and a pink hat made of veils and ribbon, Nancy had a tartan silk dress, Woman [the family nickname for Pamela] had a tiny little doll's hat, her dress and coat were brown and hat brown and pink,' wrote Sydney in her account to Diana on 20 April. 'Duchess Mowcher [the Duchess of Devonshire] had a magenta flowered dress and magenta coat, black hat with magenta feathers. She looked wonderful. I had a grey dress and navy blue coat and blue hat and feathers.'

As Diana's morale improved, so did her determination to fight what she saw as the cruel and inhuman circumstances in which she unjustly found herself. Urged on by Mosley, she made an official complaint about her initial treatment at Holloway. It was despatched by Oswald Hickson to Norman Birkett, who sent it on to Sir Alexander Maxwell on 17 April. Mosley had already stated his worries about Diana's health during his own Advisory Committee hearing the previous summer and Birkett, a humane man who was in any case uneasy about the morality of imprisonment without trial, had made immediate enquiries on the committee's behalf, but the Holloway authorities had replied that Lady Mosley was perfectly well and the matter had been dropped. Now Birkett caused a further letter to be written. It was tactful and polite but the hint of steel was unmistakable.

'If the facts in the statement of Lady Mosley are correct it would appear that the Committee was seriously misled. It is possibly a matter which you may think worthy of a little enquiry,' it began smoothly. 'I ought to add that we have been informed by Hickson that Lady Mosley is either bringing or has brought an action against the Home Secretary for damages for breach of statutory duty.'

The response was immediate. On 29 April the Governor of Holloway wrote to Maxwell, refuting Diana's claims of mistreatment; he said that she had expressed no dissatisfaction at the time and that her own doctor, who saw her soon afterwards, had said he was perfectly satisfied by her health. Of her chief complaints, the filth and the cold, there was no mention. On 9 May Birkett received a letter from the Home Office saying that there was no foundation for Diana's complaints. There was little more he could do.

The following day, 10 May 1941, London was battered by the worst raid of the war. There was a full moon, but its light was soon extinguished by the glare from the fires raging all over the city. The Law Courts, the Tower and the Mint were all hit and the Docks set ablaze. Westminster Hall was set on fire, Westminster Abbey badly damaged and the House of Commons gutted. In the British Museum, 250,000 books went up in flames. Every bridge across the Thames between the Tower and Lambeth was rendered impassable, and Marylebone was the only main-line station that did not have to be closed.

For the prisoners in Holloway, the most terrifying feature was the noise, an inferno of din that lasted all night long. The anti-aircraft battery down the road fired constantly, shaking the thick walls of the main hall. Shell splinters fell like heavy shot on roofs and tiles, dust filled the air and all the time there was the roar of guns, the whistle and crump of falling bombs and the terrifying sound of explosions. Some of the prisoners, frantically trying to judge if their neighbourhoods were ablaze, watched from their usual vantage point – a barred window which overlooked most of London in the direction of St Paul's. As the bombs rained down, they saw blazing incendiaries, firemen and ARP workers dashing to burning buildings. Diana, blocking her ears, remained in her cell as long as she could, then, with sleep out of the question, joined a group at the top-floor bathroom window, taking her turn at climbing on the only chair.

Holloway itself was hit again: one bomb fell on the roof of C wing and a 'Molotov breadbasket' of thirty-eight incendiaries scored a direct hit – one burned a hole in the wooden roof of the Catholic chapel, another set fire to the carpenter's shed. The prison officers accounted for thirty-seven of these bombs; the thirty-eighth was found later by a prisoner gathering dandelion leaves for a salad. The red blaze of burning London could be seen from Cuddesdon Hill near Oxford and – by the bombers of the Luftwaffe – from above Rouen. By the morning of the 11th, more than 3,000 Londoners were dead or seriously injured.

But the Blitz was over. Months would pass before there was another raid. Even so, 18Bs who went down to the kitchen after dark to brew tea still sat under the table until the kettle boiled and automatically kept away from the windows.

The war news was miserably depressing. At the end of May Diana asked if she could have a wireless sent in; the authorities refused on the ground that all wirelesses should be communal. In June Germany attacked Russia. It seemed as if the war would go on for ever. Diana, looking out at the sunshine, wondered how many summers would pass before she was free again. A large number of 18Bs had now been released; all the women with small babies apart from herself had been freed several months earlier. Many more were sent to the Isle of Man where, though interned, they would enjoy much more freedom, accommodated in some of the island's many boarding houses. Diana was terrified that she too would be sent there: family visits would be impossible and she would be further away from Mosley. Almost worst of all was the thought that she would probably have to share a room, her familiar early nightmare about boarding school. Unpleasant though her cell was, at least she could seclude herself in it at will.

She was reassured that she would remain in London by Sir Walter Monckton, another who had courageously called on Mosley several times

('Still a prison visitor?' asked Churchill one day). But when this kindly man asked her if there was anything she wanted, the brave face she was able to maintain to the prison world collapsed. She began to cry as she spoke of her children and of how not being allowed to write more than two letters a week cut her off from those she loved. 'I miss my freedom,' she sobbed. 'I never was in one place for more than a week before,' and she asked pathetically if the Prime Minister knew what it was like for her. 'He does,' replied Monckton. 'It is through him that I am here.'

Monckton spoke the truth. There had been frequent representations to Churchill on the Mosleys' behalf, all saying roughly the same thing: if they could not be freed, could they at least be imprisoned together? This question had been under discussion for several months. Sydney had been writing constantly, arguing for her daughter's release. In July Baba Metcalfe had approached Churchill. She had been to see her former lover in prison and horror at his deteriorating physical condition overcame her anger at the way he had behaved to her. 'I thought for my sister's sake I couldn't just not do something,' she said later, in explanation of this generous action.

Two months later (on 23 September) Baba was writing in her diary:

> A most entertaining dinner: the P.M., Mrs Churchill, David Margesson, Edward, Dorothy [Lord and Lady Halifax] and self. Winston started by coming and plumping himself down on the sofa at once and speaking of Tom [Mosley]. I had asked Edward to talk to him about the prison and efforts to get them moved to the country. This had been done and Winston was charming, most ready to listen and saw no disadvantages in putting the couples together but Herbert Morrison will be the stumbling block. He is hard, narrow-minded and far from human about a matter like this, and in this case he has a special dislike of Tom . . . One rather telling remark I thought was when I said it was awful to see someone like T in prison. Winston said: 'Yes, and it may be for years and years.'

Tom Mitford, now a serving officer in the Queen's Westminsters who went to see his sister whenever he had leave, was another campaigner. One day he told Diana that he was dining in Downing Street that night; was there anything she wanted him to say. 'Only the same as always,' she replied. 'That if we have to stay in prison, couldn't we at least be together?' At other times, Tom was able to drop into the ear of his best friend Randolph Churchill accounts of Diana's despair and her misery at the separation from her husband, knowing that Randolph would repeat these to his father and, as Diana's devoted admirer, argue robustly on her behalf. Meanwhile, a doctor sent down to check Mosley's physical condition reported that if he spent another winter in Brixton he might easily get pneumonia – the last thing the government wanted was to turn Mosley into a martyr.

For Diana, living on fifteen minutes a week of the company of those she loved and their letters, little seemed to change. There were letters from Nanny, giving longed-for news. 'Baby Max is a pet and does try and talk and so pretty.' 'Yesterday we all went over to tea with Lady Alexandra Metcalfe. Alexander behaved just nohow. He rushed back to the car as soon as he saw our hostess and children.' Others were distressing, such as the one from Jonathan that she sent to her solicitors: 'here is the thing that I really must tell you because although it might hurt your feelings it would greatly hurt the feelings of the maid who called you and me traitors and German spies. She is foul. I was caned yesterday.' There was a tart rejoinder from the Headmaster of Summer Fields:

> As a result of my enquiries I can state that anything that was said by the maid to whom you refer, to Jonathan Guinness, was more than justified by the insulting rudeness of the latter to this maid. To this he has frankly admitted and intends to apologise. Jonathan Guinness is a boy for whom I have the highest regard but I am afraid that his conduct behind my back is not always as pleasant and commendable as it is in my immediate presence.

There were reports on Unity from Sydney ('We walked round Selfridges and bought a pair of shoes and she loved it. Bumped into a friend in Piccadilly and Bobo gave such a loud scream that I had to quell her'), and visits from Unity herself, recorded by Doris Andrews, the Mitfords' favourite warder, whose accounts invariably began, 'Miss Mitford appeared to be very excited.' 'The conversation was very difficult to understand,' one such account began in October. 'Miss Mitford spoke at great speed and laughed almost incessantly . . . before leaving, Lady Redesdale told Lady Mosley that things would soon be better. "As Tom [Mitford] said in a letter yesterday, something is being done about Kit being allowed to live with you." I could not follow any more as Miss Mitford was very excited.'

The autumn cabaret came round again, its programme full of jokey references. 'Snorers, please close your doors, we want to hear the guns.'

'Dancing: Miss Jean Wallace, Holloway Mansions, with chorus The Holloway Hoofers.' 'Mrs Winfield's Registry: housemaids, parlourmaids supplied, all highly untrained.' 'Lady Mosley's rock buns, as good as their name and better than rotten eggs. Store for future use.'

But behind the scenes change was approaching. On 15 November 1941 Churchill wrote to the Home Secretary, Herbert Morrison:

> Feeling against 18B is very strong, and I should not be prepared to support the regulation indefinitely if it is to be administered in such an onerous manner. Internment rather than imprisonment is what was contemplated. Sir Oswald Mosley's wife has now been 18 months in prison without the slightest vestige of any charge against her, separated from her husband.

This was as far as the Prime Minister could go if he were to retain the support, essential to a united government, of the Labour Left. Morrison, for his part, felt able to take note only of the last phrase. He agreed that the fifteen 18B couples still remaining should be transferred to married accommodation.

For most of them this meant the Isle of Man, where many of the wives were already. Official opinion was that while there were certain couples whom the authorities did not wish to send to the comparative freedom of camp life on the Isle of Man, it would be wrong to keep these husbands and wives separated if the others were allowed to live together.

Into this category fell the Mosleys, Major and Mrs de Laessoe, and a consultant engineer and BU member called Swan, married to Diana's prison friend Mrs Swan. The Swans were to be released so soon it was not considered worth while sending them to the Isle of Man. Major de Laessoe, who had won the DSO and MC in the First World War, had, like his wife, been imprisoned for his BU membership. Mrs de Laessoe had proved herself 'very troublesome' in the Port Erin Camp on the Isle of Man and the authorities were glad to send her back to Holloway to be reunited with her husband.

On 11 December 1941 a press release was issued:

> It has been decided to allow certain married couples detained under Defence Regulation 18B to live together in a camp on the Isle of Man. Certain other couples detained under the Regulation whom it is not proposed to transfer to the Isle of Man may be accommodated,

> if they so desire, in a separate block of Holloway Prison which has been set apart for this purpose.

By 20 December the accommodation in Holloway was ready. The next day, Sunday 21 December, Mosley was brought over to Holloway from Brixton. 'We should like to give the Mosleys a day or two to settle down before any other couples come in,' wrote Carew Robinson, Chairman of the Prison Commission. For Diana, Mosley's arrival meant that 'one of the happiest days of my life was spent in prison'.

Chapter XXVIII

The house in which the Mosleys were installed was known as the Parcels House, as it had once been used as the prison's parcels office. It was separated from the rest of the prison by a high wall, behind which lay a yard – or as the press release put it 'space for adequate outdoor recreational facilities' – and a vegetable garden, largely abandoned and covered with grass.

The Parcels House contained two bathrooms, a sitting room, dining room, good-sized kitchen, and sleeping accommodation for about eighteen people, almost entirely in the form of single bedrooms. The floor the Mosleys were given, reached by an echoing stone staircase, consisted of the small dining room, a double bedroom and a small spare bedroom which Mosley sometimes used, a store room, the communal kitchen and a pantry. Like everywhere else in Holloway, the rooms were high ceilinged, dingy and felt cold even in summer.

As far as possible, the three married couples – the Swans, the Mosleys and the de Laessoes – were treated exactly the same as their fellow detainees. In the prison proper, all passages, lavatories and other common rooms were cleaned by prison labour, and the same principle was followed in the Parcels House. 'I'll give you sex offenders, they are clean and honest,' said Miss Davies, one of Diana's favourite warders.

Within their quarters, the three couples had complete freedom during the day. Only at night were the doors of their 'block' locked. They could also do their own cooking and were allowed to import some of their own furniture, on the grounds that there was 'civilian' furniture in the boarding houses in which lived their fellow detainees on the Isle of Man.

By the same reasoning, no contact was allowed with the other inmates of Holloway except by letter. If the Swans, the de Laessoes and the Mosleys had been sent to the Isle of Man they would only have been able to contact someone in Holloway by letter; ergo, this should be the rule even if they were in the same complex of buildings. In practice, as the other five had come from Brixton or the Isle of Man, Diana, the only one to have friends in Holloway, was the only person affected by this restriction.

She minded not being able to talk to her friends. If it had not been for the Mosley name, few would have noticed – or cared – if such a minor rule had been infringed. But where Mosley and Diana were concerned, the authorities, while anxious to be fair, had to be extraordinarily careful to avoid allegations of preferential treatment. There had been furious public resentment when it was disclosed that Mosley was to join Diana in Holloway. In Brixton a number of warders got up a petition to keep Mosley, not from spite – he was personally popular with most of them – but because they were convinced that Hitler had given orders to the Luftwaffe not to bomb Brixton prison because Mosley was there. If he went, who knew what would happen to them?

Other Brixton inmates felt straightforward envy. Mosley's former follower John Beckett, who already believed that Mosley enjoyed special privileges, was incensed at what he regarded as blatant favouritism. As soon as he got the opportunity (on 27 January 1942) he appeared before the Visiting Committee to ask if wives of detainees who were not themselves subject to detention could be allowed to be detained on a voluntary basis at Holloway. 'Or at least could married detainees who had been at Brixton for a long time be allowed one day a month parole for unrestricted access to their wives?'

Beckett spoke forcefully, and clearly felt that he and his fellow prisoners were being unfairly treated. 'They are under the impression that they are being penalised because their wives are *not* interned,' wrote the Chairman of the Visiting Committee to the Under-Secretary of State, Home Office. Both Beckett's requests were turned down flat, though the Home Office reasoning ('it would be unfair on the unmarried detenus') seemed curious.

In February, longer visits were allowed except on Saturdays. But when Diana asked if they could have their younger children to stay for the occasional weekend, this was refused.

'The argument that she can only see the children at rare intervals for an hour can be met by allowing longer visits,' conceded Carew Robinson, adding that even this was rather disapproved of because it might occasion further allegations in the press of 'special treatment for the Mosleys'. For the same reason, when the Mosleys' Dolphin Square flats were burgled a few days later, they were refused permission to go there (under escort) to see what had been stolen.

Diana next asked if she could keep six or eight chickens in a corner of the Parcels House yard, to be fed on scraps that would otherwise be thrown away. From her childhood poultry-keeping at Asthall she knew exactly how to look after hens, and their eggs would provide valuable protein at no extra cost. The Governor of Holloway told the Prison Commission that he was in favour of the idea, but the Commission response was a suspicious and humourless negative.

> The object is pretty certainly not to contribute to the national food supply but to produce fresh eggs for Lady Mosley and her husband. If we allow her to keep chickens we cannot logically refuse applications by other married couples [there were only two] in Holloway who may want to keep ducks or turkeys or guineafowl, or even a goat.

The sensitivity of the authorities was, however, well founded. In March a newspaper story that a fashion show had been laid on for Diana, with mannequins parading expensive creations solely for her benefit, produced uproar. Letters flooded in, filled with rage and vindictiveness. Fairey Aviation's Shop Stewards' Committee wrote angrily, condemning such 'luxury', as did the Letchworth Housewives' Union, the Women's Cooperative and countless individual members of the public, and a question was asked in the House.

Wearily the Home Office explained the true facts. Because internees, who could not shop in the normal way, had to order clothing sight-unseen, it often did not fit and had to be returned or exchanged, causing a flow of delivery vans from all over London. To save petrol, and to rationalise the whole process, a concession was given to Jones of Holloway, three hundred yards down the road, to supply clothing, so that only one van arrived at the prison, where the clothes were laid out on a counter in the E wing common room. Here the prisoners could try them on and make their selection. Diana had also been to look at these clothes but, because of the ban on contact with other prisoners, she had perforce been on her own. Her solitary visit had been transformed into 'an exclusive private showing'.

What made this particular outburst so noteworthy was its virulence. Hitherto, most of the bitterness and odium had been directed at Mosley – who was, after all, the leader and prime mover of British fascism, the man who had openly urged a negotiated peace, the figure viewed as a potential

Quisling if Hitler won the war. But from the moment of his arrival at Holloway to join Diana, she was bathed in the same aura of hatred and perceived to be every bit as wicked as her husband – consort rather than loyal wife, demon queen to his king, a Lady Macbeth equally capable of plunging in the dagger, and friend of the arch-fiend Hitler himself.

At school her Guinness children began to suffer persecution. 'I feel I must write and tell you Jonathan and Desmond went off to school this morning looking very forlorn but brave,' wrote Nanny Higgs at the beginning of the spring term. It was a contrast to the cheerful letter from the headmaster of Summer Fields at the end of the previous term:

> I am afraid I have not been very communicative of late about Jonathan and Desmond but at least the 'no news' has been 'good news' . . . Jonathan is so much better at getting on with everything . . . Desmond is full of shameless wickedness. I hope that good reports reach you from your family about them. I am myself quite pleased with them and particularly so with Jonathan.

Diana missed her children wretchedly, an ache intensified by a degree of guilt. If she had not been seen to be so friendly with Hitler during her constant visits over the commercial radio enterprise, would she have been imprisoned and so parted from them? 'If I had known how my visits to Germany would be viewed, I would never have gone. My duty was to my children,' she admitted later.

In every other way Mosley's arrival had transformed her imprisonment. In addition to the delight of being reunited with the man whom she loved besottedly, she could enjoy Mosley simply as a wonderful companion. In jail, his qualities showed to their best advantage. He was warm, loving, tender, full of jokes, never losing his natural buoyancy, cheering her, encouraging her, teasing her constantly. He would mimic her Mitford voice, with its long o's. 'If I haven't lorst the piece of paper, I'm orf to make an orffer to an orfficer in Orffer and Orffer,' he would intone (Offer and Offer were a firm of estate agents).

Their minds fitted perfectly: both grasped the chance to study and learn. Huge parcels of books would arrive, sent by Nancy from Heywood Hill's bookshop in Curzon Street – Nancy, whose relationship with Diana was slowly mending, had begun to work there in March 1942. In the evenings they listened to music on Diana's horn gramophone, or pursued some course of reading – history, economics, poetry, psychology, plays.

With Diana's help, Mosley, who had begun to teach himself German in Brixton, became so proficient that he was able to read her favourite authors in the original. 'Once it has gripped you it never lets you go,' she said of Goethe's *Faust*. 'What one loses first in translation is the poetry, which is what matters in Goethe.'

Though Diana did not realise it at the time, the most extraordinary, and unexpected, outcome of these days of close confinement was the effect on their marriage. Deeply in love though they were, being in each other's sole company all day and every day, added to frustration with their enforced inactivity, could have meant quarrels of an epic proportion. For two people normally so independently active, each with such distinctive and idiosyncratic personalities, both so personally fastidious that they had until then never so much as shared a bathroom, the irritations of close proximity and the general strain of their situation could have torn them apart. Instead, prison welded them together indivisibly.

It was, of course, the first time Diana had been the sole focus of Mosley's attention, with neither his political work nor other women to distract him. He was lavish with praise for everything she did for him, such as the meals she produced on the ancient gas stove in the shared kitchen. Mrs de Laessoe, terrified of appearing to intrude, would dash in and out 'like a little rabbit' said Diana, while Mosley and Major de Laessoe worked together in the garden (the Swans had left at the beginning of May). Mosley, always fascinated by something he could learn from someone else, copied de Laessoe, a natural gardener, thorough, neat and green-fingered. The gardening patch was largely given over to vegetables, from cabbage and kohlrabi to aubergines, but they also managed to raise a highly successful crop of *fraises des bois*, 'thanks to the soot', said Diana.

In Holloway, as in Brixton, Mosley was a centre of attention. The newspapers dubbed Holloway 'the Mosley harem' and it was not entirely untrue. He appeared as the solitary male figure, the stag in a herd of hinds – de Laessoe was quiet and elderly, the Governor an anonymous authority figure. As always, Mosley was consciously virile and fit from the exercises he did daily and he served as a focus for the dreams and desires of the enforcedly celibate women in the main prison. One was Nellie Driver, whose cell overlooked the garden of the Parcels House and who regularly watched him. 'If I stood on my bed I could see Sir Oswald's well-built frame, naked to the waist and very tanned. Sometimes he was walking round with his wife, or gardening.' Mosley shamelessly played up to the women, stripping off his shirt the moment the sun came out, doing press-ups in front of the window to an admiring audience, sunbathing in the garden with as little on as possible. 'So naughty,' said Diana. 'He *does* tease – he lies around sunbathing with all those poor girls all screaming out of their windows.'

At the end of May, the Mosleys were at last allowed to go and check what was missing from their burgled flats – the police believed they had caught the thief. He had taken all the wirelesses, clocks, watches, silver, cutlery and an eiderdown, wrapping everything in Diana's silver-embroidered white satin bedcover. Eventually, almost all the stolen property was recovered: the burglar, dumping the two items he did not want – the eiderdown and the bedcover – had inefficiently enclosed with them an unpaid electricity bill giving his address.

Diana was still striving to see more of her children and, finally, the bleak face of officialdom softened, in the shape of a representative of the Prison Commission, Miss Barker (later a notably humane Governor of Holloway). When Diana, almost weeping, said to her, 'Think what a joy it would be to me if I could bath my baby and put him to bed occasionally,' Miss Barker put her whole weight behind this plea. 'The Mosleys are behaving very well and causing no trouble to the staff,' she minuted reassuringly on 9 July 1942, before pleading Diana's cause. 'Lady Mosley can only see [her children] at rare intervals when they are brought from the country for a visit lasting an hour. She says it is so dreadful for her babies to grow up not recognising her as their mother but just as someone pleasant they see occasionally.'

Conscious of the perennial anxiety about 'special treatment', Miss Barker said that the children would be no charge upon the taxpayer. 'Lady Mosley would do everything for the children. She would also arrange for all their food to be provided so that they should be no expense.' (In the event, the Mosleys provided everything from cots and blankets to mugs.)

> On talking to Lady Mosley [continued Miss Barker], I really felt she was asking for something that on humane and reasonable grounds she might be granted. But I expressed no opinion to her. It is not likely to cause repercussions with the other 18Bs. No one else has young babies like these [Alexander and Max were three and two respectively]. They are too young to be used as message carriers in and out. I would suggest, if this could be allowed, the visits might take place from Friday afternoon until Tuesday morning. I would suggest that one such visit might be allowed and after this it could be decided whether it could be repeated and what length of period between visits could be decided.

'It is impossible not to feel sympathy with this plea,' Carew Robinson urged. Permission was finally given for an 'experimental visit' – but not without the usual agonising from the authorities. 'Up to what age should children be allowed to visit?' wondered the Home Office. 'Will this arrangement have to be discontinued in the case of the elder Mosley child, now said to be nearly four, if the War goes on for two or three more years?' But the visits began, with the first from 25 to 28 September and another a month later. This time, the usual disparaging stories were pre-empted by a press release. As the Prison Commission minuted to the Home Office: 'There does not seem much point in trying to conceal details which the Press can no doubt unearth if they give their minds to it; and there may be some advantage in giving correct details in order to damp down tendentious inventiveness.'

Relations with Nancy improved further when, in September 1942, she began the love affair that would henceforth dominate her life. The man with whom she fell in love was Gaston Palewski, General de Gaulle's Chef de Cabinet. He was not immediately prepossessing. The Colonel, or 'Col' as Nancy called him in her letters, was short, swarthy and stocky, with skin pitted by acne scars. But he was cultivated, had great taste, loved art, was amusing, witty and full of *joie de vivre*. Above all, he genuinely adored women and knew just how to please them. When he laid himself out to attract, most women found him irresistible.

When Nancy met him, he had just returned from commanding the Free French forces in Ethiopia and was enjoying the pleasures of a capital city, aided by the fact that he spoke excellent English. Though he never pretended to be in love with Nancy, he found her attractive, stylish and supremely entertaining – like Fabrice in the novel she later wrote,* he would encourage her to chat for hours on the telephone and never tired of hearing about the eccentricities of her family. She, for her part, introduced to true physical passion by this experienced seducer, fell deeply and irrevocably in love with him – and gradually, by a process of transference, with France and the entire French nation.

Her happiness spilled over and she was delighted to do all she could to help Diana. In the holidays, Jonathan and Desmond (now fourteen and twelve) would stay with her on the nights before and after the monthly day-long visit they were allowed to make to Holloway. Diana did her

* *The Pursuit of Love*

best to make these days a treat, saving her rations to concoct puddings designed for a schoolboy's sweet tooth. 'The gateau riche was *lovely*!' wrote Desmond after one of these visits.

Jonathan and Desmond got on extremely well with their stepfather. He was charming, amusing and affectionate with them. 'Ah, the Monthly Meal!' he would tease Diana. 'The only time I get a decent meal is when the Guinness boys come.' Jonathan would ask him question after question about his politics, to which Mosley would reply as to another adult, so that the boy soon understood exactly why his stepfather and mother were hated by the political establishment. Mosley would talk to him of the classics, of the German love of the ancient world, of the German writers he was reading (later, Jonathan learned German at Eton, becoming extremely fluent).

Prison conditions nevertheless took their toll on Diana that autumn. Until the Mosleys had been able to grow their own vegetables and do their own cooking, she had hardly been able to bring herself to eat prison food. For nearly two years, everyone who saw her had commented on her thinness and pallor. When, in October 1942, she caught the gastro-intestinal flu sweeping round the prison, she quickly became extremely ill. Her condition was aggravated by the coldness of the Parcels House.

Mosley, desperately worried, raged at the Governor, Dr Matheson – a man with whom he otherwise got on well – saying that when Diana got pneumonia he would hold him responsible. After a fortnight Diana was so ill that her own doctor, Dr Bevan of Gloucester Place, was sent for. When he saw her at the beginning of November he was shocked both by her appearance and the coldness of her room.

> I find Lady Mosley very low in general condition, very pale with subnormal temperature of 97.4 and with rather poor pulse [his report began]. The conditions in which she is necessarily living are unsuitable for an illness of this kind. The temperature of the room this morning when I saw her was only 45°, and since she has to get up and go to an adjoining room every time the bowels act, she runs a risk of chill, and there seems a possibility that her condition of colitis may become chronic. Her diet moreover is unsuitable. She should be on a milk diet but she has not been having more than half a pint a day. From a medical point of view, I feel that Lady Mosley should be removed to a nursing home, where in a warm room with a suitable diet, she should be well in about a week.

The prison doctor took a more robust attitude, saying that if Diana got another such attack, she should take a dose of castor oil, which would cure her in twenty-four hours.

The icy temperature of Diana's bedroom was yet another attempt by the authorities to counter stories of the 'luxury' in which the Mosleys lived. On 17 October the *Sunday Dispatch* had run a story that wardresses in Holloway were shivering while the 18Bs – notably, of course, Mosley and Diana – had plenty of coal and coke. The Governor of Holloway therefore decided that although the central heating should be turned on in the main prison on 1 November, that in the Parcels House should not be switched on until mid-January 1943. Even then, it did not make very much difference, as there were only four radiators in the entire block – one each in the de Laessoes' two single bedrooms, one in the Mosleys' double bedroom and a fourth, rather uselessly, in the store room. Otherwise, the only heating until mid-January was an open fire in the sitting room, which was on the de Laessoes' floor, and a small gas fire in the dining room, in the Mosleys' quarters.

'Here I never have a fire till teatime,' wrote Sydney consolingly on 5 November, 'but then I am out most of the time. The goats take a lot of looking after. I am so pleased, I have been given ten extra clothes coupons for my out of doors work. They really are needed as an overall takes seven, not to speak of rubber boots and gloves.'

The Mosleys had by now been in prison just over two and a half years. Diana's emaciated, run-down condition, the phlebitis on Mosley's leg, which now began to recur, the fear that yet another summer might pass behind bars, made Mosley determined to try once more for their release.

Accordingly, on 19 January 1943, when Sydney made one of her many visits to him, Mosley begged her to approach Clementine Churchill. He had written to the Prime Minister the previous autumn, through his solicitor, reminding Churchill of his promise to release the 18Bs when the war situation improved. The Battle of Alamein, for Britain at least the turning-point of the war, had been won two months earlier; would Sydney now see Clementine and ask her to press the Prime Minister for a reply? The warder listening to their conversation reported Mosley's words:

> The Prime Minister had promised to release the 18Bs when the war situation became better. Now, according to the Press, we were 'galloping to victory'. He told Lady Redesdale he did not retract one word of what

> he had spoken. He had tried to stop this war, but they had got it. Now, they could get out of it as best they could. If they had listened to him it need never have happened. He would forget politics for the duration, and if they found he did not keep his word, they could give him two years. Lady Redesdale promised to do what she could.

Although Mosley had more or less abandoned the children of his first marriage in the summer of 1940, he had not quite lost touch with his eldest son, Nicholas. On 21 February 1943, Nicholas, now commissioned into the Rifle Brigade, came at his father's invitation to see him in jail for the first time. Nicholas, with his intelligence and interest in politics, had always been his father's favourite: Mosley did not have much time for girls and Micky, a baby when his mother died, had been brought up entirely by his Aunt Irene and his nanny.

Nicholas, who was allowed to spend anything up to a day with his father and stepmother, smuggled in luxuries including champagne, brandy and pâté under his army greatcoat. The wardress led him to a door in the high inner wall and there his father was waiting, beside his kitchen garden. After a delicious lunch cooked by Diana, Mosley and Nicholas went into the garden and talked.

The after-effects of Diana's illness lasted a long time. She felt listless and lacked energy. Though between visits she longed to see the children, when they did arrive she could hardly cope with them. 'Lady Mosley does not appear to be very interested in the children, whom she finds rather exhausting; Sir Oswald Mosley is the more interested,' ran the report of a visit on 8 March 1943.

The year wore on. In mid-April Unity was confirmed by the Bishop of Dorchester. 'She is greatly delighted,' wrote Sydney. But of Diana's father there was bad news. He had developed cataracts and though at first Sydney, with her belief in the Good Body's ability to heal itself, thought that time might effect a cure, she had to admit a month later that this was not happening.

> I fear Farve seems very far from well. He is ever so much blinder and can't really see anything except the line of sky and mountain. He must have got much worse rather suddenly, as he was writing quite long and good letters to me up to ten days ago. Now he can't write at all. He is very much out of spirits, naturally, and everything worries him. I should like to bring him down straightaway to see the oculist but he refuses utterly to come until he has a new boatman and Heaven knows

how long that will take . . . he looks dreadfully tired and when indoors sleeps nearly all the time.

Unity, too, was an increasing problem. She was incontinent, required constant nursing, had forgotten how to write and suffered from loss of balance. She had aged immensely in physical appearance – Diana's friend Cela (now married to David McKenna), seeing her in Oxford, mistook her momentarily for a woman of fifty-five. Already obsessed with religion, Unity had now become obsessed with the ideas of marriage and motherhood. 'Lady Mosley was most anxious to hear what they [her children] did but Miss Mitford didn't wish to talk about it. Somebody had told her that she was very beautiful but she did not make the best of herself and she was anxious to have Lady Mosley's opinion,' ran the report of one of her visits that spring. It is easy to picture poor Unity, a lopsided giantess, clumsy, pathetically well-meaning, romping and laughing excitedly as she chattered about her adoration for the Führer, her favourite hymns and her wish for at least ten children. Even Sydney's angelic patience was tried. 'Bobo is so extremely difficult I don't know what to do with her,' she said sadly on one visit. 'And you, Diana darling, who could do something with her, locked in here . . .'

As the summer passed there was other family news. In May Cela went to Jonathan's confirmation, reporting how well he comported himself and how proud she was of her godson. Nanny Blor, Diana's old nanny, wrote faithfully from Swinbrook for Diana's birthday. Jessica remarried; in an ecstatic letter to her mother she described her new husband Bob Treuhaft, a left-wing lawyer, as 'small and dark'. Diana and Bryan corresponded about Jonathan, who had won a scholarship to Eton – Diana sent him £5 in congratulation. Should he, his parents wondered, go into College or become an Oppidan scholar'.* The matter was settled when Jonathan told his father that, as six of his best friends were going into College, he too would like to.

At Eton, the chivvying and persecution to which Jonathan had been subjected at Summer Fields through being the son of the notorious Lady Mosley largely disappeared. This was partly because of Eton's

* Sometimes boys who had been elected to scholarships preferred not to go into College, and went instead into houses. They were known as Oppidan Scholars rather than King's Scholars.

more tolerant, adult atmosphere and partly because Jonathan himself, larger and tougher, was prepared to stand his ground.

Although their mother was in prison, Jonathan and Desmond still continued the pre-war pattern of shared holidays, half at Biddesden with their father's growing second family and half with Pam or Sydney. Diana did her best to make them realise how much she thought of them, with constant letters and thoughtful and imaginative presents: a hammock, a mileometer, a handsome fountain pen, marvellous crackers for birthdays, and food ('the nicest shop cake I have ever had' wrote Desmond appreciatively from school). Sometimes friends took the boys out – Evelyn Waugh took his godson Jonathan to the circus. 'I was overjoyed to see your writing after so long and you can't imagine how wonderful I thought it of both of you to invite the boys out twice,' wrote Diana to Roy Harrod. 'We are reading the neo-Hellenists . . . Heaven is also a preoccupation but in spite of all I should like to live to be 80 or so.'

Diana had imposed the semblance of civilisation on the rooms she shared with Mosley. She played 'The Entry of the Gods into Valhalla' on her gramophone, she wrote to Nancy for yet more books 'a complete Moliere, also I crave La Mort de Pompée, by Corneille'. But the pleasures of books, music, jokes, the careful cooking, the conversation and self-education, the spirited rejoinders to the Governor, the smiling faces they presented to their children, had worn thin. The report of a visit by Sydney on 24 September shows something of their wretchedness: '[Sir Oswald] complained of the deadly monotony, and the confinement and sense of frustration. Lady Redesdale told him to get out in the garden more but he said he was sick and tired of the garden.' Both Mosley and Diana were sliding into despair. There had been no response from the Prime Minister, no hint of coming freedom although the tide of war had turned. Like a black tunnel, confinement stretched ahead, its end nowhere in sight.

Chapter XXIX

By the autumn of 1943 the Mosleys had been in prison for three and a quarter years. As the days shortened towards another winter, both felt as if they would never be free again. Ahead lay icy cold, the inevitable illnesses that would attack them in their run-down condition and claustrophobic boredom. All the efforts they had made, all the representations of families and friends, seemed like so many stones dropped in a well for all the result they produced.

But behind the scenes these efforts were stirring the authorities into activity. Baba Metcalfe was a close friend of Lord Halifax who, like her father, had been Viceroy of India and Foreign Secretary. Halifax was a man of prestige and lofty moral principles, on easy terms with Winston Churchill and so close to the King that he was allowed to walk through the gardens of Buckingham Palace on his way to the Foreign Office from his Eaton Square home. Since January 1941 he had been British Ambassador in Washington, and on one of his visits home Baba had been to see him on Mosley's behalf. Omitting all mention of Diana, who for Baba did not exist, she described Mosley's condition in the blackest terms and painted a moving picture of a father separated from his children, the eldest of whom was about to be posted abroad on active service. Her visit bore fruit: on 4 September Halifax wrote to Herbert Morrison, the Home Secretary:

> I happened to meet a relation of Oswald Mosley's last night who told me that his (O.M.'s) son, now having his embarkation leave, is very unhappy at the thought that he has said goodbye to his father for ever. Apparently a clot of blood that was in his leg has shifted up into his stomach and it sounded pretty bad. You will no doubt be seeing the medical reports but I was not sure whether the latest development may have reached you, and I felt a bit sorry for the boy, just going off, and so thought you wouldn't mind a private word. I would have

> telephoned but it perhaps is now not a very good telephone subject. Don't bother with any answer, but I don't suppose we want O.M. to die in our hands.

Lord Halifax was correct. The last thing the government wanted was a political martyr.

In July 1943 Lord Dawson of Penn (who had been George V's doctor) and Dr Geoffrey Evans had examined Mosley. They were alarmed by both his physical and mental condition and reported to the authorities accordingly.

The Home Office doctors, however, were not especially concerned. The previous month Mosley had been seen by the visiting Prison Medical Officer, Dr Jameson, who reported that he could find no evidence of organic disease, and that Mosley's heart and lungs were normal, as was his blood pressure at 120/90. 'I cannot say I found Sir Oswald looking haggard and drawn.' Dr Matheson agreed that there were insufficient medical grounds for release, although he admitted that Mosley's loss of weight was substantial. 'He is 46 years of age, 6ft 2in high, and the average weight for a man of his age and height is 197 lb. His present weight is 157 lb.' (His weight when he entered prison had been 203lb)

Mosley was a man who had always needed and taken a great deal of exercise. His vigour, energy and high spirits derived in large part from his exceptional fitness. He had the physique of an athlete, with an athlete's slow pulse – 48 when resting, 64 otherwise. In prison it had gone up to 70. Physical activity was the way he kept his phlebitis (a disease of the circulation) at bay. In Holloway he did what he could to exercise himself, but inevitably his level of fitness declined and his phlebitis, which had attacked his left, undamaged leg, worsened. Dr Jameson conceded that this condition was not helped by the lack of heating in the Mosleys' quarters.

Less easily perceived by Dr Jameson – perhaps because as a prison doctor he saw so much misery and despair – was Mosley's mental state. Mosley was, in fact, showing all the signs of someone about to plunge into a deep depression:

> He stated he had given up reading philosophy because he was unable to concentrate sufficiently to justify continuing it. He was reading Shakespeare and apparently quite unable to memorise passages. We had a long talk about his earlier history, when his memory seemed to be quite good for distant events. He said he had given up gardening for

> the past month because it made him feel unduly tired. He certainly did not speak in a low monotonous tone of voice to me. His nerves seemed to be under control and he spoke quietly and with perfect clearness. His condition caused neither Dr Matheson nor me any real anxiety.
>
> I cannot agree that if he is detained in Holloway Prison there is a definite risk of his not surviving the winter. I can find no evidence to support Dr Evans' statement that confinement and isolation are breaking Sir Oswald's nerve.
>
> I agree that the loss of weight is unsatisfactory and is due in my opinion to a mild anxiety state arising from a deep sense of frustration and consequent depression. This is quite commonly found among detained persons, particularly those who have been detained for some years.

On 5 October Lord Dawson and Dr Evans examined Mosley again. The following day Lord Dawson sent a long report to the Home Secretary, which took a much more serious view. 'The veins in his left thigh and leg are the kernel of the problem. They are . . . now extended from below the knee to the groin. These vein inflammations high up in the thigh are always more serious and open up risks of extension into the body with detachment of clots.' Lord Dawson went on to say that if the disease remained stationary, at the present level, there would be discomfort and pain but no great danger. However, he said, it was clearly progressive. 'When I examined him last July the red and swollen patches did not extend above the knee. Today the inflamed thrombo veins extend into the groin.' Dr Geoffrey Evans, who was an authority on diseases of the kidneys and heart, reported in similar terms.

With the Home Office and Harley Street doctors in opposite camps, Lord Dawson now deployed the weight of his reputation to break the deadlock. He began silkily enough by saying that if Home Office doctors disagreed with consultants of standing it would be reasonable for a conference to take place between all the medical men concerned and then for a report, or if necessary dissenting reports, to be sent to the Home Secretary.

Then came the broadside:

> To merely pass over the outside doctors cannot be accepted because in fact the clinical experience of Geoffrey Evans with his many years at St Barts and myself at the London Hospital represent authority on this vein question with which no doctor connected with either the Home Office or the Prison Service could compare. Alternatively

Diana was twenty when she became chatelaine of Biddesden, the country house near Andover where she spent much of her short married life with Bryan Guinness. Their London house was in Cheyne Walk.

(*Below*) Evelyn Waugh developed an intense friendship with Diana while she was pregnant with her first child. (*Middle*) Diana's gifted and eccentric friend, Lord Berners. (*Far right*) Dora Carrington in riding clothes with Lytton Strachey.

(*Above*) Diana in Venice, September 1930.

(*Above right*) Diana in fancy dress. Photographed by Cecil Beaton in 1932, the year she met Mosley.

(*Right*) Diana with her two Guinness sons. She loved to dress her small boys in frocks.

(*Above left*) Sir Oswald Mosley addressing a British Union of Fascists Rally in Hyde Park on 10 September 1934. (*Above right*) Diana on Mosley's boat in the Mediterranean, 1935. (*Below*) Support for the BUF: fascist rally in London.

(*Above*) Unity speaking at Hesselberg in 1935.
(*Below*) Mosley inspecting members of the British Union of Fascists in Royal Mint Street, London, October 1935.

Hitler at the Wagner family's Haus Wahnfried,
where he told Diana and Unity that war was inevitable.

(*Above*) Diana with Hitler, photographed in 1936.

(*Below*) Diana and Unity at the Nazi Rally in 1937.

(*Above left*) Magda Goebbels, who became Diana's close friend. (*Above right*) Josef Goebbels. Diana was secretly married to Mosley in the Goebbels' drawing room on 6 October 1936. (*Below left*) Hitler's wedding present to Diana was this photograph of himself in a silver frame. (*Below right*) Diana at a Nazi party Rally.

At a Nazi party Rally
in September 1937:
(*Left*) Tom Mitford
(*Below*) Unity Mitford
with a smiling Diana.

> you might wish to call in two [other] doctors of equal standing in the hospital world.

At the same time Winston Churchill, who, although he loathed the idea of imprisonment without trial, could not afford to alienate public opinion or the more left-wing Labour members of his Coalition Government, was grasping at any evidence that might further the Mosleys' release. Along with Halifax's plea for clemency and the medical opinions of two of the most eminent doctors in the country, the Home Secretary now found himself the recipient of a note from the Prime Minister.

'Let me know what is the report of the Medical Commissioners upon Sir Oswald Mosley's state of health,' wrote Churchill on 6 October. 'I have received privately some rather serious medical reports about him but they are, of course, unofficial.'

Hardly had the Home Secretary received this than a further signal of Churchill's concern arrived, a long handwritten letter dated 3 November and forwarded from Downing Street. On it was written, 'The Prime Minister wishes you to see this.'

It was from Baba:

> Brendan [Bracken] has promised to get this note to you direct and advised me to let you know the latest news of Tom. I have just come back from seeing him. The change in three weeks is *ghastly*. He looks like my stepbrother Hubert Duggan* did two months ago. Two phlebitis patches have come up on his leg and the pain in the danger area of his thigh is much worse. The cold and the damp are naturally aggravating all this. I think the fact that there was no heating in the annexe in January last year is the cause of his phlebitis getting so much worse. I feel so strongly there is no time to lose. In his present condition he is a prey to 'flu or pneumonia and then, as his doctor said, nothing can save him. I don't think myself it will need either. Today he looked like a dying man to me. Tom has been told that a meeting is taking place between the different medical advisers and I hope this is true. Although already the report of the H.O. Medical Officer is out of date as his condition is worsening daily. If you could see him you would be so shocked and you would, I know, understand how desperately I feel.

* Hubert Duggan was one of the two sons of Lord Curzon's second wife Grace by her first husband. He had died only days earlier (on 25 October 1943), aged 39.

The morning of 17 November was cold and dark. In the kitchen of the Parcels House, Diana was preparing oatmeal porridge (to Mosley's fury, the *Daily Express* had just alleged that he had a personal maid and lived a life of great luxury; among his refutations he pointed out that both he and Diana breakfasted on porridge). As Diana stood stirring, in rushed her favourite warder, Miss Baxter. She threw her arms round Diana, cried, 'You're released!' and burst into tears.

The Governor told them of the conditions. They must not live in London. They were under house arrest, they would not be allowed a car, their passports were to be removed and wherever they lived must be approved by the Home Office. Both of them thought of Pam in her large house at Rignell in the Oxfordshire countryside.

Diana felt diffident about approaching Pam directly. She knew that whoever received them would have to face not only press interest in two notorious ex-prisoners but possible hostility from political opponents. For Pam, a woman on her own most of the time, this might be too high a price. Diana knew that Pam lived for the infrequent weekends when her husband Derek Jackson came on leave – he was Chief Scientific Officer at Fighter Command Headquarters – and she also thought that Derek Jackson did not particularly care for Mosley. She asked her mother to telephone Pam. 'Of course they can come,' said Pam, 'as long as Derek doesn't mind.' 'Of course they must come,' said Derek, when she telephoned him.

It was all settled by the time their release was announced on that night's nine o'clock news. 'Mr Herbert Morrison, Home Secretary, has decided to authorise [it] subject to certain restrictions.' He had before him, said the announcer, reports on Sir Oswald's medical condition submitted by the prison authorities. Lady Mosley would also be released. One of the main reasons for her detention had been the risk that she might act as an agent for her husband; with his release this reason was no longer valid.

'Darling Mummy, What News!' wrote Desmond ecstatically:

> When is it? I couldn't believe my eyes. Unbelievable, terrific, superb, paramount. I couldn't believe it when Jonathan said 'they're released' Of course when I saw the papers I knew he should have used the future tense. Owing to certain victorious matches it is a half holiday today. I can't write for happiness. What date is it? How soon? Where to? Other questions revolve in my mind. Thank you terribly for EVERYTHING. I can't describe my feelings.

The authorities were well aware of the loathing in which the Mosleys were held. They were released at seven a.m. on the morning of Saturday 20 November, as furtively as if they had been convicted of some black, inhuman crime. It was still dark and there was a thick fog, muffling sound inside the prison and cutting vision down to a few feet. They were led through this eerily silent, clammy blackness across the stone prison compound to a small, unobtrusive side door – the infamous 'Murderers' Gate'. Three cars, two of which held police, were waiting outside, their engines running. In a few seconds the Mosleys were seated in the middle car and the cortège slipped away unnoticed.

Outside, despite the earliness of the hour, the darkness and the fog, a noisy hostile crowd had already gathered by the main gate. Someone had erected a scaffold, others were marching up and down in protest with communist banners and slogans, and the press were there in force. It was a foretaste of what was to come.

At Rignell, Pam had done her best to prepare for the Mosleys' arrival. The police had telephoned her from Holloway just before they left and, again, when they stopped *en route*, to say, 'Your guests are on the way.' She had managed to keep a little anthracite in reserve by cutting back when Derek was away and now piled this recklessly into the central heating boiler. To add to the festive atmosphere she decided to use the electric lighting – the generator at Rignell needed a gallon of petrol every three days and the allowance was one gallon a month, so she had got into the habit of using oil lamps and candles, saving the petrol for those weekends when Derek brought brother officers back. Food was not so difficult: thanks to the Mitford custom of poultry-keeping there were plenty of fresh eggs for breakfast, she had already killed a chicken in preparation for dinner that night and there were vegetables from the garden.

The cortège arrived at about nine. Although Pam had visited them in prison, she was shattered at the appearance of both Mosleys, in contrast to the healthy faces of those around them. In the open air, away from the setting of prison and fellow prisoners, they appeared frail and ill. Mosley in particular, she thought, 'looked a mere wraith of his former self'. She had greatly disliked him before; but she could feel no antipathy towards this wasted, pathetic figure.

She gave everyone breakfast, including the police, one of whom remained with them as protection (he became so much part of the family that he was once heard calling Pam by her Mitford nickname, 'Woman'). After breakfast Mosley, exhausted, retired to bed. In the afternoon Pam set off for Swinbrook to fetch Sydney and Unity, who were to stay the night. Derek arrived late in the evening and there was a celebration dinner.

The day after the Mosleys' release, as the Home Secretary was preparing his statement to the House of Commons explaining his action, he received from Churchill, who was in Cairo, a cable suggesting that he attribute his decision not just to Mosley's poor health but to principle. Its resounding phrases show, as nothing else could, the Prime Minister's continuing regard for the liberty of the individual even in the midst of war.

> You might consider whether you should not unfold as background the great privilege of Habeas Corpus and trial by jury, which are the supreme protection invented by the English people for ordinary individuals against the state. The power of the Executive to send a man into prison without formulating any charge known to the law, and particularly to deny him the judgment of his peers, is in the highest degree odious and is the foundation of all totalitarian governments whether Nazi or communist . . . Extraordinary power assumed by the Executive should be yielded up when the emergency declines. Nothing is more abhorrent than to imprison a person or keep him in prison because he is unpopular. This is really the test of civilisation.

When Herbert Morrison rose to address the House on 24 November he told the assembled MPs that he had acted with War Cabinet approval. 'While considerations of national security must come first, I am not prepared to let anyone die in detention unnecessarily.'

He went on to describe the restrictions under which the Mosleys would live. They must not travel more than seven miles from where they were living and they had to report every month to the nearest police station. They were not to associate directly or indirectly with anyone who had been a member of the BU 'other than members of their [own] families'; they could not promote or assist at any political activities, make public or private speeches, publish anything, give interviews or enter any employment 'unless directed thereto by the Ministry of Labour or National service'. They were also, absurdly, prohibited from going anywhere near anything to do with the manufacture or transport of arms, chemicals, petrol, photographic equipment or machine tools.

The day after Morrison's parliamentary statement, he received another cable from Churchill, this time urging the abolition of 18B. 'The national emergency no longer justifies abrogation of individual rights of Habeas Corpus and trial by jury on definite charges.'

Churchill's view was not the popular one. The General Executive Council of the 1,200,000-strong Transport and General Workers' Union passed a motion pressing for the Mosleys' release to be reconsidered because it was

> a grave reflection and an insult to the people in the fighting services who are making such a great sacrifice in the interests of freedom and democracy. In regard to the past record of Mosley the Council feel that there is no justification whatever for the release, which will be regarded by the people in this country who have made such great efforts to defeat fascism as an indication that the Government are wavering in their adherence to the principles for which we are fighting, at the same time playing down to those traitorous elements with which Mosley has been associated.

The prose style may have been a little weak but the sentiment was unmistakable.

On 28 November a huge crowd marched to Trafalgar Square carrying banners and placards demanding Mosley's reinternment, with a mock gibbet from which his effigy dangled. Two Labour MPs took a petition to the House of Commons demanding his return to prison and stating that 'his release is an insult to the men and women of the fighting forces'. Ninety delegates, representing 20,000 factory workers, travelled to London at their own expense, forfeiting a day's pay, to march to Downing Street with another petition. 'His release will be a stab in the back to the British people . . . and may lead to a weakening of the war effort.'

The TUC, representing six million workers, deplored the Home Secretary's failure to take into account public opinion in Britain and other countries, nor did it feel that the restrictions placed on the Mosleys were sufficient to prevent them engaging in 'subversive activities'. Mass observation polls showed 77 per cent against the Mosleys' release. Workers in Manchester, factories in north Acton, Staines, Hayes and Cricklewood, all passed condemnatory resolutions. The President of the Mineworkers' Federation protested and the Yorkshire miners sent a separate wire to the Home Office, saying, 'The release of Sir Oswald Mosley and his wife will be resented generally by all thinking men and women, by the whole trade union movement and by the community generally.' Members of Parliament received deputations from all over the country urging the

Mosleys' return to jail, letters poured in saying that phlebitis was not a serious disease but, if this really was the reason, the answer was to investigate hospital conditions in Holloway. The *Daily Worker* issued stickers saying 'Put MOSLEY back in GAOL' (Nancy put these on the back of her letters to Diana).

There was also reaction overseas. From Australia, the powerful New South Wales Railways Union cabled Morrison that his action was opposed to the principles for which the democratic countries were fighting. In San Francisco Jessica, tracked down by the four San Francisco newspapers, wrote an open letter to Churchill and sent a copy of it to the *San Francisco Chronicle*:

> Like millions of others in the United States and the occupied countries, I have all my life been an opponent of the fascist ideology in whatever form it appears. Because I do not believe that family ties should be allowed to influence a person's convictions I long ago ceased to have any contact with those members of my family who supported the fascist cause. The release of Sir Oswald and Lady Mosley is a slap in the face of anti-fascists in every country and a direct betrayal of those who have died for the cause of anti-fascism.

Her bitterness was perhaps understandable, after Esmond Romilly's death in action. Nevertheless, it was an extraordinary outpouring, not only for its lack of family feeling – anyone else might have thought a discreet silence preferable to a public avowal of bitter family disagreement – but also for its inaccuracy. Far from 'ceasing to have any contact with family members who supported fascism', Jessica frequently wrote affectionately to her mother (who was still violently pro-Hitler and believed fervently in the Nazi cause) and she had continued to see Unity until the moment she left for the United States in 1939.

On 1 December the House of Commons debated the matter. Harold Nicolson, who lunched with the Liberal MP Violet Bonham-Carter at the Ritz the day before, noted approvingly that she was taking the correct line about Mosley: 'It is wrong to imprison anybody because you dislike his opinions. The only excuse is national security and when that excuse is removed you must release him.'

Morrison made a brave, fighting speech, declaring that he would not 'bow to the dictates of the mob', that he knew his decision would be unpopular but that he would rather go through his recent 'misery' again than act dishonestly. 'If you say I am to keep this man in because I hate him or disagree with him but I am to let another man out because I have sympathy with him, then the House must frame a law under which that could be done.' And if the House expected him to use his power in that way, he added, it would have to find another Home Secretary. He told them he knew when he put his signature to the order he would be denounced, but if he had not signed he would be unfit to hold the office. 'Believe me, that although this is Mosley, and plenty of passion can be worked up about it, I should not only have failed to do my duty but I should also have struck a blow at civil liberty, Habeas Corpus, Magna Carta and what not.' Though Harold Nicolson found himself saddened by the number of 'hatey' speeches in the debate, the House approved the Home Secretary's action by 327 votes to 62.

Outside Westminster it was a different matter. When Herbert Morrison spoke at Wembley on 6 December he could hardly make himself heard through the heckling. 'This issue is a conflict between a dangerous emotionalism and mob rule and a reasoned respect for law and the constitution,' he shouted through a megaphone.

All the while, letters were arriving at the Home Office. Some said that the Mosleys should never have been released without a prior debate in Parliament, others complained of the effect on Russia. Most deplored 18B, but felt that Britain's 'Number One Traitor' should not have been released. There were several suggestions that Mosley should be tried with other war criminals. In virtually every letter there were hostile references to the Mosleys' wealth and social position. Before their release, anger, unpopularity and odium had been directed at Mosley; now, after three and a half years of imprisonment, both Mosleys emerged into the glare of universal detestation, both were seen as malignant and satanic.

What the Home Office postbag also showed was (to quote their exact words), 'Lady Mosley's release is an almost greater source of aggravation than her husband's.'

Chapter XXX

Their arrival at Rignell had been kept as quiet as possible. The house was not overlooked; it was fairly isolated and surrounded by an unkempt 100-acre park. The most serious threat was from the builders, just arrived after months of persuasion by Pam to do some much-needed work on the roof. One of these men became so suspicious that Pam warned the Mosleys not to open their bedroom curtains. Mosley, after the first few days of weakness, delighted in the chance of some outdoor exercise and waited until after dark to walk in the park. Only Pam's gardener, the entirely trustworthy Smith, was in on the secret.

It did not take long for the press to track them down. One evening a week after their arrival when, as usual, Smith returned at ten p.m. to attend to the central heating after spending the evening at the pub, he told Pam, 'The village is full of newspaper reporters, Madam. They are asking everyone if they have seen Sir Oswald and Lady Mosley.'

It was a siege. The first move by the army of reporters, who had rented every available bed in the village, was to invade an empty cottage at the end of the drive and keep the house under constant surveillance with binoculars. Others hid behind trees and shrubs. Pam told everyone to leave the telephone unanswered – it had begun to ring incessantly – and to keep the front door locked. Diana and Mosley were confined to their room, its curtains still drawn. When Pam left the house, she locked the door and pocketed the key.

She was most vulnerable when exercising her dogs. She would let herself out as unobtrusively as she could and make straight for the fields, now covered with a heavy fall of snow. With her own feet protected by gumboots, she would take a mischievous delight in heading for where the snow lay most thickly, followed by reporters in their thin London shoes. 'You be square with us and we'll be square with you,' they would cry.

Ignoring them, Pam would stomp off into the deepest ploughed field she could find until, damp and dispirited, her pursuers tailed off. Many of the reporters were recalled to London after a week.

Within days, the Mosleys were told they had to move again. Derek Jackson was privy to the most secret information about the tactics of the Royal Air Force and Mosley's presence in the same house was considered too risky – presumably he might use his sinister powers to extract vital intelligence from his host and then somehow pass it on to the enemy. It was an overt indication that the authorities believed Mosley capable of actively betraying his country. Pam was informed by the Home Office that either the Mosleys left Rignell or her husband would not be allowed home for his weekends. Finding anywhere to live quickly in wartime, let alone for a sizeable family, was difficult enough; for two exhausted ex-prisoners it was a daunting prospect. But it had to be done.

Sydney came to the rescue, telling Diana of a half-abandoned small hotel called the Shaven Crown in Shipton-under-Wychwood. Diana and Mosley were able to rent it for a few months, moving in mid-December, and were soon joined by Nanny Higgs, Alexander and Max. It was six miles from Swinbrook, and Sydney, who as a keeper of goats had a small extra allowance of petrol, could sometimes drive Unity over to see the Mosleys, returning via Burford, where she sold goat cheeses. Debo, who had a pony and cart, could also drive over (her husband Andrew Cavendish was on active service and she was living with her mother).

The rooms were filthy and Mosley, who again needed nursing, lay in bed most of the time. Diana summoned what was left of her strength to look after him and, with the aid of a sixteen-year-old girl from the village, did her best to clean the place before the arrival of the Guinness boys for the Christmas holidays. Just after Christmas (they had eaten family lunch in the cottage at Swinbrook with Sydney and Unity), all four of the children developed whooping cough. Alexander was the most seriously ill, but both Desmond and Jonathan had it so badly they had to go back to school late.

The move to the Shaven Crown sparked further press stories. One report, saying gleefully that the Mosleys were so disliked that no one would work for them and that Lady Mosley had to look after four children, cook and clean the house herself, produced a flood of applications. From all over England, letters arrived offering to work for them in any capacity. 'I have never been in service before. I have not the clothes habitual for such situations, nor the coupons for same. My husband died nearly two years ago and I am employed in a local factory. I am willing to fit into the exigency of the times.' 'Dear Lady Mosley, Would a good cook, an educated, sensible woman be of any use to you? If so, will you let me have a post card and I can send full particulars. She can come almost at once.'

As the Shaven Crown was only a temporary stopping place, Diana went on house-hunting as best she could. When she heard of a likely house at Crux Easton, near Newbury, she obtained permission to go and look at it, escorted by two policemen. On the way she was allowed to stop for lunch with Lord Berners.

She loved Crux Easton at sight. It was a large, south-facing manor house which had been abandoned by its owners as too big for wartime; for several years its only inhabitants had been a rapidly increasing population of rats. It had the ten or so bedrooms needed for the Mosleys' extended family – between them, they had seven children – and a sunny terrace with a wonderful view. Most appealing of all in those days, there was a large, well-maintained, productive kitchen garden and eight acres – which meant the Mosleys could keep a cow. Their offer of £3,000 was accepted.

Diana's first thought was to find servants. Never in her life did she seriously consider doing without them. Like her family and most of her friends she had grown up with servants (Nancy's efforts at living alone in London had always foundered because, with no one to pick them up, every flat became kneedeep in discarded clothes) and Diana throughout married life had employed them – there had been servants at Savehay up to the day of her arrest. To her, a civilised life, let along a well-run house, was unthinkable without domestic staff. It was not simply that she had been imprisoned so early in the war that, unlike her mother or her sister Pam, she had never found herself running a sizeable house with a singly daily replacing a staff or four or five, nor that prison was an experience so outside 'normal' life that it had not affected her attitude. Rather, it was a determination to hold on to what she considered a priority in the kind of life she had decided early on to fashion for herself, a life that whenever possible involved the constant entertaining that both Mosleys loved. Tom Mosley was a man who liked the machinery of his household to run smoothly and if possible invisibly at all times, a view that chimed exactly with Diana's; neither of them would ever have been able to envisage a wife running in and out of the kitchen or distracted by domestic minutiae. As she regained a little strength, her first thought was to establish a home for herself and Mosley and this meant engaging, at the least, a cook.

One of the letters she had received because of the press stories was therefore particularly welcome. It was from Mrs Nelson, a marvellous cook formerly employed by Lord Berners. When Diana had stayed with him in his flat in Halkin Street in pre-war days, Mrs Nelson used to carry up her breakfast and they would chat. Mrs Nelson, who had had to leave Lord Berners because he found her husband unsatisfactory, now wrote, 'Dear Madam, I would come to you if you were at the North Pole.' Diana replied at once offering both Nelsons employment and they arrived in time to help with the move to Crux Easton. Lord Berners was a trifle put out to learn that Diana had acquired the best cook he had ever had, though Diana felt she had been paid back when on her first day Mrs Nelson used the family's entire butter ration for a fortnight to make a superb apple tart. 'Oh, Mrs Nelson,' wailed Diana, 'how could you?' 'People like you can always get more,' replied the cook. There was nothing for it but dry bread.

Within days of the move, Mosley had bought a cow, which was named Wellson by Max. It was looked after by a man from the village who had been turned down for military service; he also worked with Mosley in the kitchen garden.

At last, their food supply was assured. With plenty of fresh fruit and vegetables, milk and homemade butter, and an outdoor life, Mosley soon regained his health. Diana was still weak and unable to eat much, but with Sydney she painted the kitchen and lavatories and much of the rest of the house. The rats were gradually exterminated until they could no longer be heard galloping nightly inside the old walls and ceilings.

The Nelsons did not last the course (after the butter incident they never really settled) and in April Diana wrote to her former Wootton chef, hoping that he might be able to come. He replied that he was in the army:

> I was called up early in 1941 so I have been in the Services just over three years so this will explain why I have regretfully to say how sorry I am that I can't accept your kind and generous invitation to join your domestic staff. Had I been at liberty I would have been pleased and willing to have done so. I have seen James and his wife and little girl twice before he went to North Africa. I believe he is a corporal now. I am in the Royal Artillery. I am the Major-General's servant or batman now as my category has been lowered and I have to do a lighter job for the present. It would be fun to make eclairs and meringues again . . . I have often thought of you all.

That spring, Sydney, Unity and Debo, leaving behind her two small children, went up to stay with David on Inch Kenneth. David had made over the island to Tom; Sydney, anxious to make a little money for her son, had begun farming there. 'I imagine one can hardly hope for profit for a year or two as the animals etc have to grow,' she wrote somewhat despondently to Diana. 'The wages are devastating and come to £450 a year – partly due to having to have a boatman.'

It was not a happy visit. David appeared to be under the domination of a strong-minded former parlourmaid called Margaret, who had originally worked for the Redesdales at Rutland Gate Mews and on whom he was now virtually dependent. The old, politically based antagonisms between the Redesdales were too strong to allow of any *rapprochement* and David's fearsome temper flared up at the slightest provocation.

'Farve looks very ill,' wrote Debo on 9 May 1944:

> He's terribly thin and everything worries him, even the smallest little thing about the boats. He never sits with us in the drawing room but helps Margaret in the kitchen. I haven't had a chance to chat to him at all yet . . . he is dreadfully difficult and cross and unapproachable . . . It was evidently due to her that the babies were put off coming, because I think he would have liked to have them. Of course it's perfect nonsense as she only has to cook and do nothing else as Syd does the housework. It seems to me that Farve is a sort of scullery maid. Then of course there is the news, which Farve has at 7, 8, 10, 1, 3, 6 and 9 and midnight, and Muv won't listen to it. So Farve and Margaret listen together and she makes maddening comments. Oh Honks, you don't know what it's like . . . Muv is furious about the babies not coming but can't say anything as Farve flies into a temper so easily.

The situation was not made any easier by the fact that Sydney herself could not have looked after her husband. She was sixty-two and she was entirely occupied in caring for Unity. Margaret was strong, hard-working and made David comfortable, and he was terrified of upsetting her.

With the opening of the Second Front in June 1944, the danger of a German invasion of Great Britain was clearly long past, but the restrictions placed on the Mosleys remained. Within the limits of house arrest, however, Diana had achieved a reasonable facsimile of the life she had led earlier. The Wootton furniture had been brought out of store and she had managed to find people to work in the house. Like everyone else, she would rush to the butcher if there was a rumour he had some bread-filled wartime sausages and every week David sent her a small food parcel from Inch

Kenneth. But they were never short of garden produce or, thanks to Wellson, milk and butter. As she had done with her older sons, Diana began to teach Alexander and Max to read, write and do simple sums. Her sisters came to stay; her brother Tom, home from Italy and at the Staff College, managed to come almost every weekend. Viv visited, for the first time finding her father approachable and 'positively genial – not a word one would have used about him before'. Micky, now a schoolboy, came in the holidays.

Other social life was confined to the visits of certain pre-war friends. Randolph Churchill was not among them. 'Why will they see Berners and not me?' he complained bitterly to Nancy. But Mosley, who had blamed Churchill for the continuance of the war, regarded the entire Churchill family as arch-enemies and would have nothing to do with his former friend. Osbert Sitwell's sentiments, however, were warmly appreciated when he wrote:

> Later in the year, when the world is thawing, I would love to come and see you and Diana, after this long and insufferable gap. I could easily come down for luncheon one day (hope to be in London in May). Yes, the inadequacy of time is appalling and to have been unjustly deprived, as you have been, of a period of time, is beyond bearing. The only comfort for you must be that it is impossible to blame you for anything that happened in those years.

It was not a view generally shared: few of the Mosleys' neighbours wished to know them.

Sydney and Unity stayed on Inch Kenneth until mid-November (Debo had left a fortnight earlier). Sydney had persuaded her shaky, irascible husband, whose sight was rapidly deteriorating, to come down to London with her, but she did not know what he planned to do thereafter. It was while he was staying at the Mews that Jonathan and Desmond, arriving at the end of term, saw one of their grandfather's famous rages. Used to 'Granny Muv's' gentle sweetness, they were taken by surprise when, after bathing, one of them forgot to pull out the bathplug. David exploded, shouting at them furiously, 'That's the sort of thing people have to resign from their clubs for!' In the same month, Diana learned that her former father-in-law Lord Moyne had been assassinated in Cairo.

For the Mitford family, the following spring was to bring the most tragic news yet. Tom Mitford had been worried for some time that he might be sent with his regiment to invade Germany itself. He could not bear the thought. Rather than fight on the very soil of the country he still loved deeply, he obtained a posting from the Staff College to the Far East. He refused a staff job in Burma and was made second in command of a battalion. 'Every letter says how glad he is to be there,' wrote Sydney to Diana. 'He likes the Indian soldiers so much, they are fine fellows. I saw in the Sunday papers his division was on the Irrawaddy.' At the beginning of March the Redesdales were told that he had been wounded. They heard nothing more until, in early April 1945, came the news that he had died of his wounds.

The family gathered at the Mews – Mosley had rung the Home Office for permission for Diana to travel – to comfort each other as far as possible.

'So ghastly about Tom, we had all hoped he was getting better as it was some time before we heard he had died and still hardly believe it,' wrote Pam to her friend Billa Harrod on 18 April. 'Bobo said it will be so lovely for him to meet Dr Johnson. So like her!' A few days later Diana received a letter from Randolph Churchill, last seen many years earlier; 'how heartbroken I was to hear of Tom's death and how deeply I feel for you all in his tragic loss. I had seen very little of him during the war but he was my dearest friend. I saw him last just a year ago, in North Italy.'

Of all the sisters, Diana, always the closest to Tom, felt his loss the most. 'We had always been more like twins than brother and sister,' she wrote later. 'A day never passes when I do not think of him and mourn my loss. He was clever, wise and beautiful; he loved women, and music and his family.'

Grief aggravated Diana's difficulty in recovering her health. She was still painfully thin, with a poor appetite that showed no signs of improving. Psychologically she was deeply scarred: for years her sleep was disturbed by 'prison' nightmares – fears about returning to jail, terrors at being torn away from her sons again. Prison affected her far more deeply, and for far longer, than Mosley. Although he had been released on medical grounds, so weak he could scarcely get out of bed, he had begun to get better quickly. After six months, he had recovered much of his old vigour – so much so that he was looking for occupation. With anything that could be remotely construed as political proscribed, he turned instead to the subject that preoccupied everyone throughout the war: food. The war was clearly in its final stages but shortages, he believed, would go on for a considerable time. Therefore the most useful thing he could do would be to farm.

He was not allowed to travel more than seven miles from Crux Easton, but when he heard of a house and farm with 1,000 acres he bought it sight unseen. It was at Crowood, near Ramsbury, in Wiltshire.

The end of the war in Europe, celebrated by VE Day on 8 May 1945, meant the immediate cessation of house arrest for the Mosleys, along with liberty once more to own a car. As there was a petrol ration for every car, Mosley immediately bought four little Austin Sevens that were extremely economical on fuel. The first sortie made in one of these was to see Crowood for the first time. Although the house had the eighteenth-century façade that Diana so admired, with a later addition at the back, she did not really like it, though she did her best with the large drawing room – pale blue wallpaper, a large Aubusson and her French furniture. She ran the house with a married couple, the cook and housekeeper, a housemaid and, outside, a single gardener.

Mosley, anxious to buy more land, needed to raise cash. One way was close to hand. When they were married, he had given Diana the Mosley family jewels. Now he asked for their return. Without demur she fetched them from the bank and they were sold at auction. They were exceptionally beautiful, set with first-quality Brazilian diamonds, the most important pieces being a rivière of huge stones and an exquisite late-eighteenth-century tiara of blackberry-like diamond leaves and flowers, their centres made from large, lustrous single stones. Fortunately, the long earrings that matched the tiara did not reach their reserve, so Mosley gave them back to Diana. She had four of the stones made by Cartier into a pair of fashionable clip earrings of one big stone and one small one each, which she wore constantly. The remaining diamonds were made into brooches.

That summer was spent by the combined families at Crowood. It was a curious ménage. Irene, who had come with Viv, Micky, Nanny Hyslop and Andrée, still refused to speak to Mosley. His mother had been invited to keep the peace. Baba refused to come even for a meal because Diana was present. Jonathan and Desmond, trailed by Alexander and Max, worked on the farm. When Nicholas, who had won the Military Cross, returned in August, he went straight to Crowood. Later in the year, Mosley engaged a black gamekeeper, whom he discovered living in the village. But few except his old friends would accept his invitations to shooting parties.

With a farm, entertaining became much easier, and gradually Diana renewed contact with her old friends. John and Penelope Betjeman were within reach, as were Gerald Berners and Robert Heber-Percy. One of the friends Diana saw most of was Daisy Fellowes, without whom no party in

Venice, Paris or the south of France would have been complete in pre-war days. She and her husband lived ten miles away, at Donnington Park, a huge Gothic house near Newbury. Daisy Fellowes, beautiful, enormously rich, and famed for her chic, was also noted for her teasing. She loved Diana, but this did not stop her using the Mosleys as the best tease of all. 'The Mosleys are coming to dinner, do come and meet them,' she would say to her stuffier neighbours, watching for their horrified reaction.

For, in general, the Mosleys were ostracised. More and more horrific truths about the death camps were emerging, and few people would willingly meet the notorious couple who had so publicly allied themselves with the creed behind such unspeakable atrocities. When Ralph Partridge, one of the Ham Spray ménage who had been Diana's friends and neighbours fifteen years earlier, bumped into her on Hungerford Station he told his wife of their meeting with the words, 'I shook the hand of Fascism!'

Chapter XXXI

In the first few years after the war most of the Mitfords moved house. Nancy, with no one to please except herself, settled in Paris to be near Gaston Palewski. Derek Jackson, enormously rich, for whom Rignell had been ideal – he could hunt with the Heythrop and it was close to his laboratory in Oxford – was so indignant when the Labour Government introduced a top rate of tax at 97.5 per cent that he and Pam decided to move to Ireland.*

Sydney and Unity moved 500 miles north. The island of Inch Kenneth had been part of Tom's inheritance. He had willed it to his sisters and there was general consensus that it should be sold. Then, on 21 May, Jessica announced that her one-sixth of the proceeds should go to the Communist Party. 'One way to look at it is that my share will go to undo some of the harm our family has done, particularly the Mosleys and Farve when he was in the House of Lords.'

It was too much. With the consent of all the sisters, including Jessica, Sydney now took it over and made her home there. David, who with his failing sight, increasing deafness and general frailty found island life too difficult, moved to Northumberland to a cottage in Redesdale, where Margaret continued to look after him.

Sydney had always loved Inch Kenneth and had visited it as often as she could during the war – perhaps its isolation and encirclement by the sea subconsciously reminded her of her early life on her father's boat. Unity was less enraptured: she did not want to leave her life of hymn-singing, churchgoing and the Oxford cinemas for a wet, remote island and the company of goats.

* Income tax stood at 9s in the £. In addition surtax was payable on incomes over £2,000 p.a., rising to 10s 6d in the £ for incomes over £20,000. The top rate of tax on income was thus 19s 6d in the £.

Sydney was sixty-six and it was a dramatic change in lifestyle. Even the journey to Inch Kenneth was difficult, sometimes hideous, if wind or weather made the two sea crossings impossible. The first stage was the night sleeper to Oban, followed by the notoriously rough crossing to Mull. If this was impassable, it meant waiting in Oban until the following day. Once on Mull, there was a fifteen-mile drive along winding roads to the little village of Gribun, followed by a second crossing, by motorboat, to Inch Kenneth a mile offshore. When Sydney had established herself, she garaged her ancient Morris at Gribun, whence it would be driven to meet the Oban steamer. At Gribun her boatman, with her motorboat, *Puffin*, would be waiting.

Contact with the outside world was fitful. There was no telephone and, when the sea was rough, no post. If a black spot, visible through binoculars, was fixed on her garage door on the Mull shoreline opposite, she knew there was a telegram for her and would send her boatman in *Puffin*. Supplies of coal arrived once every four years, brought by the coal boat, which came as close to the jetty as it could at high tide and then tipped the coal into the sea. It was shovelled up at low tide by the islanders, who gathered with ponies and carts.

Sydney surmounted every problem triumphantly, adapting quickly to the hard life of a small island farmer. She reared sheep, Shetland ponies, goats, hens and cattle (the bull had to be swum over from Mull, attached to *Puffin* by a rope through the ring in his nose). She had always made her own bread and she had become an expert cheese and butter maker. Her amusements were simple: the wireless, two-day-old newspapers and library books – chiefly biographies and diaries – sent by post. She would also look through her field glasses to see if there were people on the opposite shore and would often send her boatman to ask picnickers or those from boats anchored in the bay to come and have tea with her. Among those she met in this way were the politician R.A. Butler* and his wife Mollie, who later often came to visit her.

The house, with its white-painted, bow-fronted, eighteenth-century elegance, was itself an unlikely piece of architecture to find on a Scottish island; inside, as Mollie Butler said, 'It was impossible to imagine you were in the Highlands rather than an elegant London drawing room.' The delicate cornices were picked out, there were satin curtains in the pale oyster drawing room with Sydney's favourite French furniture and

* Shortly to become Chancellor of the Exchequer and, later, Home Secretary.

pictures from Swinbrook. David, devastated by the loss of his adored only son, had instinctively wished to disembarrass himself of everything that had once been so carefully guarded as Tom's inheritance and had told Sydney to take whatever she wanted. Sydney had, however, managed to dissuade him from the wholesale dispersal of family treasures:

> Farve . . . will not sell family pictures, books or the Crown Derby (white and gold). The Crown Derby has 90 dinner plates, four soup tureens, 15 sauce boats and everything else en suite, including 13 flower bowls. Warren Hastings (they call it Berlin china) is not quite so big, five dozen plates and two soup tureens, etc as well as several large round dishes and all the knives, forks and spoons with china handles.

Diana was not to move for some years. The last link with the old life had been severed when, on 1 January 1946, Savehay Farm was sold to Mr Frank Cakebread, to be followed three weeks later by a sale of the Savehay furniture. Everything went for enormous prices: Cimmie's bulbous-legged oak refectory table, abhorred by Diana, sold for £300, a carved oak coffer for £140. Even the drawing-room curtains fetched £70.

Mosley had flung himself into farming with enthusiasm but soon, unsurprisingly, it was not enough. He resumed his political life, focusing on the aspect that now most interested him, the economic situation. He believed there would soon be an economic crisis of catastrophic proportions; in the ensuing chaos there would be a chance for the voice which spoke with reason and authority on such matters to be heard. If he could only regain a seat in parliament, he believed, his ideas and oratorical skills would do the rest. He organised large meetings, in London chiefly at Porchester Hall and Kensington Town Hall, in Manchester at the Free Trade Hall; and when he could not hire a hall he held meetings in the street.

For success, he needed to reach an audience far larger than those at his meetings – and to publish his views, he needed paper, then on quota and available only to those whose business was printing. Accordingly, he and Diana became publishers. *My Answer*, Mosley's explanation and defence of his past political life and the policies which had caused him to be imprisoned, was published by Mosley Publications in 1946. The following year saw *The Alternative*, a proposal that Africa, with its vast wealth of raw materials, should serve as a kind of garden estate for the more civilised European races; it would be run, said Mosley, by a new and superior type of white man specifically bred to carry out this task in the most efficient and altruistic way. After this Mosley Publications became the Euphorian

Press, which – as well as a constant flow of Mosley's pamphlets – published classics, among them Nancy's translation of *La Princesse de Clèves* and two Balzac short stories translated by Diana. *Stuka Pilot*, an account of the air war on the Eastern Front by the German air ace Hans-Ulrich Rudel, with a foreword by Group-Captain Douglas Bader, became a best-seller.

Ostensibly, Mosley had given up fascism. The British Union, put down by the government at the beginning of the war, no longer existed, and the day the war ended Mosley himself had said, 'Fascism is dead. Now we must make Europe.' But few believed that this particular leopard had changed his political spots; once again, his meetings were disrupted, there were counter-demonstrations and street fights with communists.

For those on both sides of the House of Commons, his past record clung about his shoulders. When he tried yet again for the return of his passport, his application was one of five refused by the Home Secretary, Chuter Ede, on the grounds that it was undesirable that the Mosleys should be allowed to travel.

Lord Jowitt, the Lord Chancellor, disclosing this ban to the House of Lords on 12 December 1946, explained that the government believed there was 'a reasonable chance' that Sir Oswald Mosley might make mischief abroad. 'Fascism and Nazism have not been rooted out,' said Lord Jowitt, in the first official statement that ex-detainees could be stopped from going abroad under the New Emergency Laws (Transitional Provisions) Act of 1946. 'They are lying dormant but it is possible that they might once more be fanned into life. If I were a Jew living in Europe today I could imagine that I should feel deeply annoyed at the visit to the country where I was living of a man who had, rightly or wrongly, been identified as being very anti-Jew in this country.'

Ten days later, Lord Jowitt must have felt justified when an East End 'Welcome Home' party was given for Mosley by many of his old followers – former 18Bs, party workers, members of book clubs started by Mosley and other sympathisers. 'There is stirring again in England and in the world some of those things we felt so deeply together,' said Mosley in his speech of thanks. 'I have always regarded the East End as the birthplace of those ideas . . . They tried to shame us, they did their best to throw us in gaols, into concentration camps but they never shook the faith within us and they never will.' The party finished with 'Auld Lang Syne', followed by the fascist salute and 'God Save the King' in uneasy conjunction.

An episode in Paris showed how strongly fascism and the Mitford name were connected. Nancy was now comfortably installed in a charming ground-floor flat at 20 rue Bonaparte, where she completed her new novel, *The Pursuit of Love*, dedicating it, naturally, to the man who was the centre of her own life. For discretion's sake, she had wanted to use only the Colonel's initials but he had insisted on his full name appearing. Neither of them foresaw the book's immediate and lasting success on its publication in February 1947 – nor the furore the dedication would cause. 'Hitler's mistress's sister dedicates daring book to M. Palewski,' howled the headlines in the French papers. The Colonel was angry, upset and fearful for his political future; Nancy prudently left Paris for several months. Fortunately the French papers did not realise she had gone to stay with the Mosleys at Crowood.

Two years into peace, the Mosleys' unpopularity showed no signs of diminishing. Graffiti were scrawled on the door of Diana's Dolphin Square flat, pleas to some of her oldest friends to invite them to dinner met with smiling acquiescence – but no invitation followed. She wrote to her old admirer Evelyn Waugh and received a charming letter in reply ('Dearest Diana, how very nice to have a letter from you. I think of you often, and of my poor heathen godson') but no visit, although he told her that Debo had asked him to stay with her in Ireland. When Cynthia Jebb, whose husband, Gladwyn, Nancy had urged to imprison Diana, ran into the Mosleys in Heywood Hill's bookshop, she recorded in her diary that they had 'the look in the eye that people no doubt acquire when they are accustomed to being shunned – a kind of studied indifference'. When half a dozen pistols brought back by Nicholas from the war were stolen from Crowood, the thief claimed in his defence that he had discovered an 'arms cache' at the house of the notorious fascist leader Sir Oswald Mosley. And when Micky, in a fit of adolescent rebellion, rang up his father one day from Eton pretending to be a communist agitator and saying in a threatening voice, 'We're coming to get you!' Mosley was so accustomed to such threats that he contacted the police immediately on another line. They traced the call at once and the luckless Micky found the police outside the box as he rang off. Characteristically, far from being annoyed, Mosley was amused at this show of boldness and intrigued, as always, by any sign of political consciousness in his son.

Unable to go abroad, the Mosleys spent three weeks in August 1947 with Sydney on Inch Kenneth, a visit which cemented Mosley's place in Sydney's affections as her favourite son-in-law. It was extraordinarily hot and the island was seen at its best, an idyllic place of peace and warm sunshine, made even happier for Diana by a letter from Bryan telling her of Jonathan's 'glowing' Eton reports. He was top of his division in German

– interpretership standard, said his form master – had good Russian and an excellent history report.

Privately, it was a time for building bridges. At Crowood, relations were sufficiently restored for a visit by Irene, who had just returned from Africa. Although 'Auntie Ni', as the children called her, still loathed Diana, she was determined to be polite for the sake of her nephews and niece, a forbearance which bore fruit when Nicholas became engaged later in the year, necessitating family consultations about the wedding. Diana wrote to Winston Churchill at the instigation of Daisy Fellowes, who had told him of Diana's outburst against him in prison in the depths of her misery. Churchill, who had always been fond of Diana was so pleased to get her affectionate letter that he immediately wired his thanks.

That autumn Diana began to suffer from occasional migraines. The printer who printed the Euphorian Press books would often invite both Mosleys to a splendid lunch in London. At one of them, to celebrate the publication of Mosley's *The Alternative* – a short political treatise published in October 1947 – within minutes of sipping a glass of the delicious Pouilly Fuissé offered by their host, Diana had a headache so terrible she had to leave the table and lie down. Only after several hours did the pain abate, leaving her exhausted and shaking. After several such episodes, she realised that white wine was one trigger, though she could safely drink a glass of red wine.

Mosley was steadily returning to his earlier political format: a party outside mainstream politics with himself as its unquestioned head. At the end of the year he decided to formalise his remaining support. Despite the events of the past, his whole character inclined him towards a precisely outlined body of support rather than an amorphous band of followers. On 15 November 1947 he attended a conference of Mosley Book Club fans at the Memorial Hall in Farringdon Street – where, eighteen years earlier, the New Party had held its inaugural meeting – and announced his intention of founding a new political party. It was to be called the Union Movement.

A fortnight later he gave a press conference in Pimlico to fifty-odd largely hostile reporters. As one of them ironically recorded, 'Purely from "a sense of stern and very painful duty" he has decided to float a new party.' Its aim, Mosley told the assembled gentlemen of the press, was to bring about a union of the European peoples, with the exception of Russia, after which Europe and the British Dominions would be asked to join with Britain in developing Africa for the white man. As the *Daily Mail* reporter pointed out, 'In the Mosley Plan, there will be no nonsense about "trusteeship for the natives".' There would be a direct Yes or No vote every four years on a Mosley Government and its programme, the Communist Party would be suppressed, Jews who had not had their roots in Britain for 'about three generations' would be resettled outside the country, perhaps in Eritrea. 'Something more vehement and violent than we have known in British politics for a decade was reborn last night in Pimlico,' concluded the *Daily Mail.*

The Union Movement's first meeting, at Kensington Town Hall, was in February 1948. In effect, it was fascism under another name: its members were largely former BU members and its aim was 'the Greater Britain that shall be born of the National-Socialist and Fascist creed'. It was already a lost cause. There were disturbances, riots, marches and demonstrations, but even in local government elections not a single seat was won. This did not deter Mosley, who was not only brave and tenacious but convinced of his own rightness.

At Inch Kenneth, Unity's health suddenly deteriorated. For the first time Sydney's self-imposed isolation seemed a penance rather than a joy. 'Bobo is much less well and I feel greatly worried,' she wrote to Diana on 27 May 1948:

> It is her poor head. The doctor here is bringing out a man from Oban hospital to consult. It is really terrible being here where it is so impossible to move anyone. No aeroplane can land on Mull and there seem to be no seaplanes. I wanted a specialist from Edinburgh or Glasgow but the doctor said they would require X-rays and all sorts of things before they would do anything. She is very fast asleep now I am glad to say. There is not a terrible deal of pain, thank heavens, but she can't move her head at all. Neck completely stiff. I have a day nurse who has been here since Monday.

Hardly had Sydney written this letter than Unity's condition worsened rapidly. The following day, 28 May, Sydney took her to Oban. It was an appalling journey, first the usual motorboat to Gribun, then by ambulance across Mull to Salen, where they arrived long after the daily steamer had departed. They crossed to Oban in a small hired motorboat, a journey that took five hours, Unity sedated with morphia so that she was not in pain, and arrived at the West Highland Cottage Hospital at one a.m. The following day, as the long, light Highland evening wore on, it became clear that the desperate trip had been in vain. At ten o'clock Unity died, the cause of her death given as meningitis stemming from the bullet wound she had inflicted on herself almost nine years earlier.

Her funeral was on 1 June at Swinbrook. Diana, with Mosley, was at the graveside of the sister once the closest to her in the family. Afterwards Sydney came to Crowood for a fortnight.

The years at Crowood were among the most domestic and enclosed of Diana's life. Her horizons had shrunk to the affairs of her immediate family, and outings were largely confined to those friends still faithful. She sent clothes to Frau Wagner, poverty-stricken among the ruins of Germany ('that beautiful warm dressing gown fitted as if it was made for me!') and a last link with Unity. Once, one April 21, she said wistfully: 'It's the Führer's birthday today. He did love his birthday.' But for most of the time the past was behind her. She sent hams and scented soap to Sydney on Inch Kenneth; she visited her sons at school – when she went to Eton, boys would hang out of the windows to gaze at her and the bolder ones would ask Desmond for a photograph ('Darling Mummy, do send Christopher the photograph. He doesn't want to hang it up, only gloat').

All her energies were bent towards creating for Mosley a cherishing, efficient framework for his life and work. He was a man who expected his household to revolve round him, its timetable geared to his needs, pleasures and wishes. If he was reading, noise was frowned on; if he wanted entertaining, he liked sparkling people – he had a low boredom threshold and loved jokes. The house had to be well run, the food good, with plenty of agreeable society. Diana, who had always wanted to write, found herself kept busy editing her husband's pamphlets and books; his autobiography took priority over anything she might think of writing for herself.

The most dramatic events were changes within the family circle. Nicholas became engaged to Rosemary Salmond. Desmond left Eton for Gordonstoun, where he could keep his horse. The usual preliminary talk with the founder, Kurt Hahn, lasted two hours (said by the Gordonstoun boys to be a record) and, as Desmond later wrote to his mother,

> I could see that Bryan had given him the impression that I hated Kit and Kit me. What Hahn would have said if I had told him that I would go to Kit for advice and not Bryan I don't know. There is but one thing that makes the dread of Gordonstoun less persistent and that is that I shall be allowed to take that Tom Thumb, the apple of my eye, to ride. Darling Bryan is going to buy me a proper horse at last, possibly for my birthday, possibly to cancel out Jonathan's gun. The only remark I have to make

> about Gordonstoun is that it is not the place for me, but it is more the place for me than Eton.

Desmond rode Tom Thumb to such effect that they won the Open Jumping class at the nearby county show.

Intellectually, Diana may have been dedicated only to the furtherance of Mosley's happiness, but in practice she soon began to find the everyday routine of life at Crowood stifling. Apart from her editorial work, she had never been involved in Mosley's political activities and he was often away from home. Her own amusements had always been metropolitan and the slow, unvarying passage of days on the farm was uncomfortably reminiscent of the boredom she had suffered at Swinbrook. Her social life was confined to those who still allowed themselves to know the Mosleys. She had always travelled a lot and could hardly imagine a 'normal' life without going to her favourite parts of Europe whenever she wished. She felt this emotional claustrophobia most strongly in the summer, when she and Mosley were used to spending long holidays in Italy or France, followed by a fortnight in Venice. The absence abroad of friends such as Daisy Fellowes, now revisiting all the places that the Mosleys so loved, only served to underline the constriction of her own life. It seemed to her bitterly unfair that after three years in prison without trial and with no criminal charge against her, she was now denied the right of any ordinary citizen to 'travel abroad without let or hindrance'.

Mosley's ingenuity came to the rescue. He had discovered that a British subject had the right to leave his country and return to it at will – and a passport was needed only to gain entry to another country. The answer was to go to sea. He bought a sea-going motorboat and they visited the Channel Islands. It was not far, and they were still on British territory, but it gave them a feeling of liberation.

The trip was such a success that they bought a larger boat, a sixty-ton ketch called the *Alianora*, for which they engaged a captain and an engineer. Without passports, they were not allowed to take money abroad, so every detail down to the last tin of food had to be carefully thought out. They planned to leave in June 1949 but just before they set off, with Alexander and Max, aged ten and nine respectively, they were given passports. That summer, revelling in the freedom of going exactly where they wanted, they spent four months away. The first leg of the trip was across the Bay of Biscay to the Gironde estuary, then on to Corunna, Lisbon and Tangier, where they ran into a storm. The *Alianora* had to be repaired at Gibraltar so they went by train to Madrid, then to Formentor, Majorca, where they boarded the *Alianora* again. At Cassis they met Nancy, and Daisy Fellowes's yacht *Sister Anne*, which they followed down the Riviera. At Monte Carlo Nick and his wife Rosemary arrived and the boys went off to spend the rest of the summer with their grandmother on Inch Kenneth. When it was wet, they devised an impromptu skating rink on the parquet floor of Sydney's large drawing room. 'I was pleased as it polished the floor,' reported Sydney. 'It went on for several hours and they sang quite a lot, which your two really enjoy.' Alexander, she added, needed a man's company. 'He doesn't think much of women (his age!) and of course Nanny has less than no authority and makes no impression. He has been reading Henty, Moby Dick and Heart of Midlothian.'

Diana and Mosley, spending most of their time at Daisy Fellowes's villa on Cap Martin, bathing in its seawater pool, were picking up old friendships. It was the same story in Venice: people from the past were returning and seemed delighted to see them. It was an enormous contrast to England, where they were still deeply unpopular, often cut and almost invariably referred to in hostile tones.

They began to consider living in Europe. There was little to keep them in their own country. Inch Kenneth was no more difficult to reach from Paris than from London. Nancy, in Paris, was well established: as an intimate of Diana and Duff Cooper, who had settled in Chantilly after leaving the embassy in 1948, she was at the hub of Parisian society. Pam and Derek Jackson (who had been appointed Professor of Spectroscopy at Oxford in 1947) were living at Tullamaine Castle in County Tipperary. Jessica – though between Decca and Diana lay a frozen waste of silence – was in America. Only Debo, now the Marchioness of Hartington,* was in England. After a final cruise in 1950, the *Alianora* was sold, and the Mosleys decided to leave England.

* Lord Andrew Cavendish assumed the courtesy title of Hartington when his elder brother was killed in action in September 1944, and became Duke of Devonshire on the death of his father in 1950.

Chapter XXXII

From 1951, with one brief break, the Mosleys lived abroad for the rest of their lives. Inevitably the word 'exile' was attached to them, a description loathed by Diana, who would at once react by arguing that anyone who chose to settle near the most civilised of European capitals and who visited Britain at will could hardly be called an exile. In fact, it was to Diana that the term was mostly applied. For her, notorious for her friendship with Hitler, living abroad was seen as a form of banishment, albeit self-imposed. Mosley's constant travelling and his frequent visits to England to speak or to engage in some political activity somewhat weakened this perception. In some quarters it was also felt that, although his methods were to be deplored, he had been all too correct in one of his most constantly reiterated political themes – that the war would mean the end of the British Empire as a major world power.

The Mosleys' plan was to split their lives between Ireland and France. In Ireland, Mosley was a welcome figure thanks to his denunciations of the Black and Tans thirty years earlier, while Diana's friendship with Hitler was viewed more tolerantly in a country which had been neutral in the war; Max adored hunting; they would be near Pam and Derek – who had become one of Mosley's closest friends – and to the Devonshires when they visited Lismore. In France, they had friends, they knew the Riviera well and visited it most summers, and in Paris there would be Nancy to smooth their path.

It was not long before they had found the two homes they wanted. During a visit to Paris in December 1950 they had seen a beautiful, dilapidated building called the Temple de la Gloire, twenty miles from Paris near the little village of Orsay. It looked like a cross between a stage set and a classical palace in miniature – when the Duchess of Windsor first saw it she remarked, 'Charming! but where do you live?'

It had been built in 1800 by Vignon, architect of the Madeleine, for the wife and mother of General Moreau, to celebrate Moreau's victory at Hohenlinden. The two adoring women had seen it as a place where their victorious hero could spend enchanted rural afternoons. In the days when it was painted by Corot, who rode over from nearby Ville d'Avray, it stood alone, overlooking a wooded valley, surrounded by woodland and long grass threaded with daisies and blue scabious. Since the Mosleys first saw it, Paris has advanced remorselessly; the Temple is now surrounded by suburban houses, with a football stadium, screened from the house by trees, on one side. Half a mile away is a station which provides a quick train service to Paris.

The house, although rundown in the extreme, was architecturally exactly what Diana wanted. It had the Palladian façade of a Graeco-Roman temple, with four corinthian columns rising from the first-floor balcony to the roof, and two small wings overlooking a lawn that led down to a lake. The two main rooms were the ground-floor dining room and, on the first floor, the large salon. The dining room was flanked on one side by the kitchen and servants' rooms and on the other by two visitors' rooms and a bathroom. The salon, reached by a narrow, spiral wooden staircase, occupied the central part of the house, with deep windows on each side, overlooking respectively the gravelled front courtyard and the lake. At each end were two small suites of sitting room, bedroom and bathroom.

In March 1951 the Mosleys bought the Temple and the three acres of meadow, woodland and lake that surrounded it for £5,000. Three months later, they were able to sleep there for the first time. At first they camped, carting chairs from room to room and eating with the minimum of china and cutlery.

All their furniture had been sent to Ireland, where they had bought another house in which they planned to spend autumns and winters. This was a former Bishop's palace at Clonfert in County Galway, a charming south-facing house, set in flat country on the edge of a bog full of snipe and waterfowl. It was reached by a great avenue of ancient yews called the Nun's Walk, down which the Bishop had exercised his hunters on days when the ground was too hard for hunting. Here Jonathan Swift had stayed, no doubt admiring Clonfert Cathedral opposite, or perhaps walking to the River Shannon a mile away.

Diana and her sons stayed with Pam at Tullamaine while electric light, bathrooms and central heating were put in. She would drive over every day to supervise the restoration of the palace and its decoration. At auctions she added to the furniture sent over from Crowood – one purchase was her own fourposter bed with curved mahogany bedposts and a blue taffeta canopy. In the dining room, with its French dining table inlaid with roses and ribbons, she hung the pictures of Mosley ancestors. Alexander and Max, aged eleven and ten, who had been miserable at leaving Crowood, which they regarded as their first real home, soon settled in; they had brought their ponies to Ireland and Max was able to hunt with the East Galway. Diana's two Guinness sons were now at Oxford; both were married, Jonathan to Ingrid Wyndham, Desmond to Mariga von Urach.

Another frequent visitor was Sydney. The great devotion engendered in Diana by her mother's loving support during the prison years had been deepened and strengthened by Sydney's obvious love for Mosley. Diana's realisation during one of Sydney's visits that her mother had developed Parkinson's disease was the only cloud in otherwise blue skies until, on a freezing night in December 1953, disaster struck.

Both Mosleys had been in London, staying at their Grosvenor Road flat, and Diana had remained behind for an extra day to see her father, who was making one of his rare visits south. When she arrived at Dublin airport the following morning she was met by Mosley, unshaven and exhausted-looking. Taking her hand, he said, 'Sit down. Nobody is hurt. Everything is all right.' Then he told her that Clonfert had caught fire the previous night and though every living creature was safe – from himself and Alexander and the two French servants to the horses and dogs – all the furniture, the pictures, their papers and most of the house had been destroyed.

The fire had begun in the chimney of the maids' sitting room, where an ancient beam had started to smoulder. By the time it eventually burst into flames, everyone had gone to bed. The alarm was only raised when a neighbour, woken by the whinnying of her horses, looked outside and saw flames appearing through an upstairs window. She shouted to her son, who ran to the house and threw pebbles at Alexander's window until he woke. Because there was no telephone, it took a long time to get word to the fire brigade and the fire was out of control before its arrival. Mosley and his children brought out what they could, but much was lost.

Made homeless overnight, the Mosleys were fortunate in their relations – the Devonshires lent them Lismore for Christmas – and even more fortunate that within a few days they had heard of another house for sale. Ileclash, on the cliff above the Blackwater near Fermoy, was famous for its salmon fishing and was on the borders of several hunting countries. They bought it at once.

It was at Ileclash that the couple who were to be two of the most important people in the lives of both Mosley and Diana joined them. Jerry (Jeremiah) Lehane and his wife Emmy (Emily Reilly), from County Waterford, were to stay with the Mosleys all Emmy's life and most of Jerry's (first as butler and maid, then as butler and cook) and their devotion was in every way reciprocated.

During the Clonfert years, Diana had been gradually restoring and refurbishing the Temple de la Gloire. Her first step had been to give its classical façade full weight by grassing in the drive that had once swept round to it. With the back door converted into an unobtrusive front entrance, reached across a gravelled forecourt from the high iron gates, there was nothing but a stretch of lawn between house and lake.

Diana believed that classical architecture should be set off only by green foliage, of different shapes, shades and contours. Accordingly she and Mosley planted beeches, sycamores, weeping willows and hedges of different heights. One, to the side of the house, concealed a swimming pool and the enclosed flower garden where she grew roses, delphiniums and syringa for the house. Trees round the lake made a curtain of green with a narrow opening opposite the house through which could be seen a pair of swans gliding on the silvery water. In autumn, mist and falling leaves gave the prospect a romantic, melancholy charm. In winter the bare branches sparkled with frost and only then could it be seen that there was not deep forest but a road on the far side of the lake.

Inside, the rooms were treated with the same mixture of grandeur and intimacy that characterised the exterior. In the large dining room with its high French windows, walls of dull terracotta were set off by a carpet patterned in black and white squares like marble tiles. A large chandelier hung in the drawing room above the pale pinks, corals and green of the Aubusson carpet. A pair of exquisite gilt stick barometers, bought at one of the Swinbrook sales, hung against grey panelled walls picked out in white.

Diana and Mosley took one each of the small suites at either end of the drawing room – they both preferred a bedroom and a bathroom of their own. Diana's sitting room had soft duck-egg blue walls, blue cushions on the brown sofa and a log fire; Mosley's had deep blue walls and a paler blue sofa, a French chair and stool covered in leopardskin fabric, leatherbound books and a blue porcelain stove. The whole house smelt of

Guerlain's Fleurs des Alpes, which Diana burned in a spoon carried from room to room.

Diana's joy in her beautiful house was marred by physical misery. The migraines she had suffered from time to time at Crowood now began to strike in a virulent, regular form. From two to four times a week these devastating sick headaches would occur without warning, leaving her with no recourse but to lie still in a dark room until they passed. All through Mosley's pocket engagement diaries the letters HA (for Head Ache) in his minuscule handwriting appear with monotonous regularity. It became difficult for them to accept invitations – for, as Nancy wrote, 'The French never chuck, even if ill.' In face of this torture, Diana hardly noticed the first signs of the deafness that was a family trait.

Eventually, after much experimentation, her doctor found a powerful drug, which left her sick and groggy but took away the acute pain. Without it, she felt helpless – once, when she was struck down suddenly in Paris, Mosley had to telephone Jerry and ask him to drive up with it.

Although migraine was a torment that was to dominate Diana's existence for thirty years, it did not stop her making every effort to establish a social life. She had always depended heavily on her many friends. Mosley, on the other hand, had few men friends – the closest, at various times, had been Bill Allen, Mike Wardell, who was the former editor of the *Evening Standard*, John Strachey, Oliver Stanley, Bob Boothby and, after a sticky start, Diana's brother Tom. He enjoyed the company of women and liked the stimulus of other people and the chance to discuss ideas with intelligent listeners. Now that politics was no longer at the heart of his life, he needed something to fill the vacuum. As for Diana, from adolescence on she had sought the company of the witty, cultivated and artistic, and – foreign country or no – it was unthinkable to her not to entertain and be entertained.

She set to straight away. She quickly became fluent in French, soon speaking it with the greatest correctness but with the most English of accents and all the Mitford intonations. As soon as the Temple was barely presentable, friends from pre-war days were asked for luncheon or dinner. Emmy, who had originally joined the Mosleys as a housemaid, now found her true métier, evolving into a superb cook. One of the visitors was Diana's father. It was the first time he had seen Mosley since the day almost twenty years earlier when he and Lord Moyne had gone round furiously to Mosley's Ebury Street flat to try to warn him off Diana. Now, with Diana clearly blissful in a marriage of fifteen years, David's perspective on his son-in-law was different, while Mosley's intelligence, courtesy and charm ensured a genuine liking. When he returned home, David sent Diana a large cheque with which to curtain the enormous Temple windows.

Occasionally a familiar face from the past would reappear – one New Year's Eve there was a reunion with Randolph Churchill, who had been out of their lives for so long. Both the Mosleys were shattered by his appearance: he looked old and ill and was drinking copiously. But the renewal of past friendships was not always easy. Their position was ambiguous, even in post-war Parisian society, which was riven by gossip and accusation as to who had and who had not strayed across the narrow border separating what could be called a prudent safeguarding of one's own interests from active collaboration with the Nazis. Daisy Fellowes, for instance (born Marguerite Severine Philippine, daughter of the fourth Duc Decazes), got her Paris house back intact almost as soon as the war ended, but her daughter Jacqueline, married to an Austrian accused of betraying people to the Gestapo, had her head shaved as a 'collabo'.

The Mosleys were often snubbed by former friends, including the Rothschilds. They were *persona non grata* at the British Embassy and British diplomats were forbidden to visit them. Their old friends Duff and Diana Cooper might have stretched the rules, but the new Ambassador Sir Oliver Harvey and his wife did not. One Embassy diplomat recalls Churchill asking the Ambassador, 'Oliver, do you ever see Paulsy [Prince Paul of Yugoslavia] or that awful, awful Mosley? Though it's such a pity about Oswald Mosley because he would have been a very great politician.'

Nancy was no help. An intimate of the Coopers and through the Col, now one of de Gaulle's ministers, an habituée of French political and diplomatic salons, she was at the heart of Parisian society. Her literary reputation enhanced her popularity – her novels, from *The Pursuit of Love* to *The Blessing*, which had recently been published, painted France and Frenchmen in glowing colours; her translation from the French of Roussin's play *The Little Hut* had opened in the West End to ecstatic reviews.

Secure and well known as Nancy was, all the old ambivalence in her feelings surfaced when Diana came to live in France, and Nancy did her best to keep her friends and Diana apart. Her excuse was that she thought they would dislike the idea of meeting the notorious Mosleys (though in fact many would have been intrigued to do so), but the real reason was jealousy. Nancy was the best company in the world, immensely chic with her beautiful Dior dresses shown to perfection on her narrow-waisted, slim figure. A busy and successful writer, she was, nevertheless, always conscious that Diana, although she had never written anything, was probably cleverer than she; while above all loomed the fact of Diana's beauty, with its strange goddess-like effect which effortlessly attracted worshippers of both sexes. In short, Nancy was terrified that, as in those Buckingham Street days twenty years earlier, her own friends might be drawn away by her sister's charm and looks.

Diana in her early forties was more compelling than ever. 'I think she is the most flawlessly beautiful woman I have ever seen,' wrote her old admirer Jim Lees-Milne in August 1953. 'Clear, creamy complexion, straight nose, deep blue eyes and grey-gold hair dressed in a Grecian bun swept rather to one side of her nape. Her figure so slim. She is just as beautiful as she was at 17 and more so than when first married.' Mosley, he noted, was fatter, rather greyer, with manners that were almost too good except when he got talking and seemed on the verge of delivering a platform speech, 'but he is no longer so extreme'.

Despite Nancy's social double-dealing – which at times amounted almost to an anti-Diana campaign – she could not keep away from the Temple. She was jealous of Diana, but equally Diana was by far her favourite sister: they laughed at the same things, each knew the other's literary tastes perfectly (Diana was the only sister to whom Nancy was to dedicate a book*), they spoke to each other on the telephone every morning and wrote often. Nancy disliked Mosley, but was too scared of 'Sir Ogre' to be anything but pleasant to him. His wit was as quick as her own and he would not have hesitated to counter one of her barbs with a scathing rejoinder. She saved her derogatory remarks about him for letters to friends. Neither Mosley nor Diana knew about Nancy's letter to Gladwyn Jebb urging Diana's imprisonment. Had Mosley known, he would have banned her for good.

Nancy's presence at the Temple meant endless jokes and stories of people who did or said the wrong thing, amid cries of 'Are you shrieking?' as some gaffe or gossip was repeated with gales of laughter. Later to be recycled as U and non-U, these jokes provided much of the raw material for Nancy's essay, *Noblesse Oblige*. Mosley, who also loved laughter and gaiety, would tease Diana endlessly. 'There goes Percher, *meaming* again!' he would say (he always pronounced 'miming' as 'meaming') at Diana's

* *Frederick the Great*

extravagant expressions or gestures of horror or amusement.

The most significant of the Mosleys' new friends were the Duke and Duchess of Windsor, who early in 1952 had leased the charming old Moulin de la Tuilerie, once owned by the painter Drion; later they bought it for £30,000. It was about five miles away from the Temple and also in the valley of the Chevreuse.

Both the Mosleys were already acquainted with the Duke. Diana, who had met him several times in the 1930s, also had a familial link with him: her grandfather Redesdale had designed a large part of the gardens at Sandringham – later the Duke told her that every time Lord Redesdale came to Sandringham he would give the prince and each of his brothers a gold sovereign. She had also met the Duchess once, when the Windsors were staying on Daisy Fellowes's yacht. Mosley had known the Duke – then Prince of Wales – reasonably well when both were young men attending the parties of the London season. Nevertheless, Mosley at first made no approach, thinking that as he had become such a controversial figure the Duke might prefer not to be reminded of their former acquaintance.

The Duke felt no such inhibition. As soon as he arrived he got in touch with the Mosleys. Soon there were weekly invitations to the Moulin, which was like a twentieth-century Petit-Trianon: the Duchess in high heels walking across the lawn of the English garden made by the Duke, their perfectionism apparent everywhere, their elaborate food – melon with a tomato ice in it, eggs with crab sauce. Diana would always curtsey to the Duchess, though her real affection was for the Duke – 'funny, charming, lovable'.

Nancy, still slipping in the dagger under a pretence of jokiness, implied that the two couples saw each other to bewail the dear dead days of the Third Reich. 'I believe their wickedness knows no bounds,' she wrote to Violet Hammersley.

Gradually Diana's life settled into a pattern. In the winter the Mosleys lived in Ireland so that Max could hunt. They returned to the Temple in April. Every year Diana would visit her mother in Inch Kenneth, often taking her back there after Sydney's annual visit to London, too. Often she would return via Northumberland to see her father. With Mosley she would spend a few weeks in August or September in Venice.

They now had two flats in Paris, a grand and elegant one in the boulevard de Montparnasse where they would stay, bringing Jerry and Emmy, for a weekend's racing, and a pied-à-terre in the rue de la Chaise. Their arrival in Paris had coincided with a flowering of French theatre: there were plays by Anouilh, Sartre and Giraudoux, and they went constantly to the opera. Gradually they made new friends, including the decorator Jean de Baglion ('You *know* that you are my goddess and platonic love'). There was also the welcome reappearance of Derek Jackson, now living in France – after almost fifteen years of marriage he had left Pam. In England, his need for constant intellectual stimulation had been satisfied by his work in the spectroscopy laboratories of Oxford. At Tullamaine in Ireland, there was only hunting and occasional visits to the University of Columbus, Ohio. They were not enough for him and he soon bolted with the woman who became his third wife (after Pam, he went on to marry four more times). Mosley was delighted to see Derek again; when they met they would kiss, in the manner of Frenchmen, Derek standing on tiptoes to reach the much taller Mosley's cheek. The fact that most of the French people who witnessed this spectacle disapproved of it, believing they were being mocked, only delighted Derek the more.

In the autumn there was partridge shooting in Normandy on the estate of French friends. When Jerry and Emmy went on holiday to Ireland over Christmas and the New Year, Diana and Mosley would move into the Montparnasse apartment, spending Christmas Day with their friend Countess Mona Bismarck and New Year's Eve with the Windsors. Diana's sole contribution to domesticity was to prepare a light supper on the days they were not dining out; the gardener from the Temple came up twice a week to clean the apartment and they invariably lunched out. In 1955 they began to go to the festival in Bayreuth again, staying in an hotel surrounded by romantic woods.

Mosley had made political contact with other right-wing leaders in France, Spain and South Africa. As the 1950s passed he became busier than ever, meeting and talking to proto-fascist groups all over Europe – though not in France. 'I did my utmost to spread my ideas throughout Britain and later throughout Europe with the exception of France,' he wrote later. 'We had long ago resolved to permit ourselves one happy land where I was free from all involvement in politics.' In March 1953 he launched the *European*, a monthly 'journal of opposition'; Diana edited it, wrote a 'Diary' in it and reviewed political memoirs. Now the doors between Mosley's quarters and the drawing room were quite often shut, so that he could work undisturbed. If Diana wanted to talk to him she would walk along the balcony, which continued right round the house to his bedroom.

Diana's life revolved entirely round her husband. When Mosley was not travelling, she was seldom apart from him. Gone were the days of constantly lunching or dining with people other than each other which had characterised their pre-war existence. Mosley would write, read or sleep on the leopardskin chair in his study; Diana, in her quarters, would review for *Books and Bookmen* or work for the *European*. They walked in the woods and, after a light dinner, read or listened to music together. In the household he was, as always, god. His wishes were treated as law, his appointments took priority; above all the rightness of his views was accepted. At a luncheon party at the Temple, Diana would fall silent if Mosley spoke, her eyes fixed upon him, listening to a story she had heard again and again, adding the same coos of delight, expressions of surprise or peals of laughter.

Yet beneath the surface of this idyllic picture there were darker currents. The result of Diana's rigid belief in Mosley, of her deliberate subjugation of her own wishes and desires to what she saw as the greater importance of his, meant stifling her natural impulses to irritation or anger. Sometimes these surfaced in furious quarrels about their children – the one subject on which she felt she should have an equal if not greater say than Mosley. Both had their favourites: Alexander was Diana's, Max his father's. Often when the children fought, their parents did so too. Each would take the side of their favourite, in arguments which they saw as purely rational, but which, for Diana, served as a release for pent-up emotion. She knew that Mosley wanted to return to the mainstream of British politics, with all that implied in terms of absences, disappointed hopes and, perhaps, further estrangements from friends. Worse still, she suspected him of unfaithfulness.

Chapter XXXIII

By 1956, Mosley had found a focus for his political activity. Large numbers of immigrants from the West Indies were being encouraged to enter Britain by the offer of jobs and housing, and many settled in North Kensington. Mosley now chose this area as his political arena, once more seeking to whip up racial antagonism in an eerie echo of his pre-war anti-semitism. 'I say, let the Jamaicans have their country back and let us have ours,' he would declaim, views that involved him in passionate argument with his son Nicholas.

While Mosley was making inflammatory speeches, Diana was immersed in family affairs. She was deeply worried about her father's health and her mother's isolation. For Sydney, the loss of two of her seven children had become a pain that was increasingly difficult to bear. 'It was so kind of you to think of sending a word for Tom's birthday,' she wrote to Diana in January 1957. 'I fear the sorrow for him gets no less. One misses him more as time goes on. I know also that you are the same.'

Her daughters were anxious that she come south, and the Duke of Devonshire thought of buying the Swinbrook School House, ostensibly to have a stake in Swinbrook, but in reality to help his mother-in-law. 'Debo and I thought it might do for me to retire to when really too old for the island but I fear it has many drawbacks,' reported Sydney to Diana in March:

> No outlook, practically no garden, a sitting tenant, no water laid on, and so on. Everything to do, in fact, but correspondingly cheap, £1,200 . . . I remember when we got the 'lodge' ready for Grandfather Tap at great trouble and expense. When it was all set for him to go and live there, he said: 'Two voices there are. One is of the mountains, one the sea, each a mighty voice. The "lodge" has neither. I cannot live there.' I fear this applies to all Swinbrook village.

It was her family's last attempt to persuade her to lead a less arduous life.

She loved the island and was perfectly content there. She dealt with her illness by ignoring it ('the Good Body') and obediently it advanced but slowly. She was happy, busy and social: she loved her garden, her cows – except when they escaped on to her lawn – and she picked the fruit from her orchard, made hay and walked her little dachshund José. Every summer her grandchildren came to stay; there was bathing off the rocks, fishing and long spells of fine weather ('we had breakfast out of doors every day'). Other friends visited – 'we have been a party of eight every meal' – and she filled the role of local *grande dame* with unpretentious charm. 'On Saturday had the W.I. to tea, about 30 old dears.'

David's life was in sad contrast. He seemed detached not only from his family but from life. He had never read, nor was he interested in politics or world affairs; he could no longer practise the country sports of shooting and fishing which he had once so enjoyed, nor was he a gardener; and he had become extremely deaf. Sydney and Debo had arranged to go and see him for his eightieth birthday. At the last minute Diana, who was in London, had a premonition. She raced to King's Cross, running along the platform until she found the carriage in which her mother and sister were sitting. All three stayed at Redesdale for two days and then went on to Inch Kenneth. Diana returned to London and three days later, on 17 March 1958, her father died. He was buried at Swinbrook, his ashes brought down from Northumberland in the sort of parcel he once used to bring home from the Army & Navy – thick brown paper and neatly knotted string.

David had cut Jessica out of his will after her earlier declaration that she would give anything she inherited to the Communist Party. He had not seen her since her departure for the United States in the spring of 1939. Three years before his death, when it was apparent that he might not live much longer, Sydney had written to Jessica, asking if she would like to come and visit him. Jessica had replied that she would love to – if he would promise not to roar at Bob and their daughter Constancia (true to the Mitford passion for nicknames, known always as Dinky). Sydney responded crisply that 'since you have set impossible conditions, I shall not arrange a visit with Farve', and Jessica never saw her father again. Nancy, feeling that her sister had been unfairly treated, gave Jessica her share of Inch Kenneth; a little later, Jessica bought out her other three sisters – the market value put on the island by an outside agent was £7,000. 'Decca and Bob and Dinky all seem really to like it,' wrote Sydney to Diana. 'The idea is that Dinky will have it one day.' Meanwhile, Sydney was able to go on living there.

* * *

In February 1958 Mosley relaunched his Union Movement. It was really the BU under another name: it consisted largely of former members, who rallied to him with the same aggressive fervour as before. There was the same demagoguery, the same salute, the same phalanx of tough young men surrounding the Leader and the same racialist bias, albeit this time with a different target, 'coloured immigrants'. The TUC described the Union Movement as 'fanning the flames of racial violence', a judgement borne out by the riots in Notting Hill later that year. *The Times*, while pointing out that the Union Movement was exploiting rather than creating the disturbances, summed it up accurately, dispassionately and damningly.

> A clear distinction must be drawn between official Union Movement policy which has been formulated by Sir Oswald Mosley and the reasons most members have for joining. Those immediately below the Leader look to him for political guidance and seem to have a genuine desire to see his policies instituted. All this is incomprehensible to the majority of his followers, who understand the clichés in which the ideas are expressed rather than the ideas themselves. They have joined the Movement for a variety of reasons: some because they are anti-semitic but the largest number, it seems, because they like fighting Communists and painting slogans on railway bridges. They fight because they have an instinctive desire to do so. It is admitted by the leaders that these elements are out of the control of the party, and probably would not have joined it were it not for the sinister avocation of the word 'movement', the dramatic salute, the hero-worship implied in the word 'Leader' and the fanatical hatred of the word 'Communism'.

Diana, as always, supported the Leader's political stance wholeheartedly. The peaceful serenity of her manner gave an uncomfortable edge to the sharpness of views that had hardened through the years. The mild anti-semitic prejudice of early days had become a full-blown conviction that 'international Jewry' was an enemy of both her country and her husband; now, endorsing his view that coloured immigration was to be resisted at all costs, Diana would cite the many ills it would supposedly bring, from overcrowding to resentment bred by competition for scarce jobs, housing or benefits. Even more specious was her argument that if employers got black labour cheaply they would not invest in new machinery and thus Britain would fall behind the rest of the world.

Domestically, she was miserable. Her migraines, always exacerbated by stress, became more numerous, with one following hard on another. She was now certain that Mosley had returned to his former habits, and was being unfaithful to her. Many of these women were drawn from their immediate circle. When he had deceived Cimmie with her own women friends, he had used the pretext of fencing bouts; now there was the excuse of appointments in Paris. Instead of the bachelor flat in Ebury Street, there was now the third-floor apartment at 5 rue Valledo (bought when the Mosleys sold their Montparnasse flat in 1956).

Although Diana declared, 'I knew what to expect when I married him,' she had believed that their great, all-consuming love, their sense of joint destiny, the unique bonding of their prison experience, the cocoon of love and cherishing in which she had wrapped him since, would be enough. Since she met Mosley no other man had held any attraction for her other than as a friend; dismissing her previous marriage from her mind, she regarded herself, as she put it, 'like my sisters, a one-man girl'. Her beauty could easily have brought her consolation, but the idea of paying Mosley back in his own coin or even of distracting herself with an affair never once crossed her mind. 'In all my life, I never once wanted to be unfaithful to Kit,' she said in old age. 'It simply didn't interest me.'

She was, though, better able to handle his infidelity than Cimmie had been. She was older, she was more sophisticated by nature, she was far more certain that she was the great love of his life and she had an inner core of steel that poor Cimmie had not possessed. She used what weapons she had: freezing disdain; clever, catty, unanswerable asides; and wittily malicious comments on the women concerned, a technique that sometimes cowed even Mosley. Yet it was still a terrible shock. She had no real fear that any of these liaisons might cause him to contemplate leaving her – or even affect his love for her. Nevertheless they caused her acute pain. 'There is no jealousy like sexual jealousy,' she wrote to one confidant. She tried to excuse Mosley by telling herself how attractive to women he was – he had always far preferred them to men, not only as sexual conquests but also as friends – but it did not help. Long after Mosley's death she said more than once:

> I knew from the very start it was no good being possessive with Kit – he was just somebody who would not be possessed. He was often unfaithful. The leopard does not change his spots. I guessed some of them and knew others. It made me miserable. Although I knew he loved me best it was wounding and annoying. One would feel very resentful. I minded terribly, even though I knew it wasn't going to make any real difference.

In France, Mosley's conquests included Rita Luke and Lotsie Fabre-Luce, both wives of writers. In England, his most ardent pursuit was reserved for Jeannie Campbell, Lord Beaverbrook's granddaughter. She was tall, dark, vivacious and buxom, with round rosy cheeks, sparkling eyes and a strong feeling for the conspiratorial. She was fascinated by Mosley's views, the power of his personality and the frisson of intrigue involved in such an affair. They spoke constantly on the telephone and Mosley would bombard her with huge bunches of flowers. When he came to London they would use chauffeur-driven Daimlers and meet in various flats – sometimes one borrowed by Jeannie, sometimes one Mosley rented from Nicholas. Nicholas and his wife Rosemary occasionally bumped into his father and Jeannie, realised what was going on – and realised also that Diana knew. It made for a painful conversational 'black hole'.

For Mosley, part of Jeannie's appeal lay in her relationship to Beaverbrook. She was the press lord's favourite grandchild and had lived and travelled with him for several years. Mosley hoped that she would further his cause with her grandfather and persuade him to give favourable publicity to the Union Movement. But Beaverbrook was far too canny to be drawn in this way. He called a family conference and told Jeannie that either she stopped seeing Mosley or he would cut her off. He resolved the matter by despatching Jeannie to the United States, there to write for the *Evening Standard*.

Diana's unhappiness over Mosley's unfaithfulness was aggravated by their quarrels over money for Alexander and Max. Mosley had always resented spending money on his family; to Diana he said that if their sons were healthy, clever and given a good education they should be able to make a life of their own. 'They've had far more advantages than 90 per cent of the population,' he would tell her. 'They don't need money as well.' Diana had never minded any economy that Mosley had imposed on her, but she felt desperately bitter over what she saw as his meanness to their sons. He and she lived in a substantial household, entertained lavishly and travelled constantly. Mosley's older children were well off in their own right thanks to the Leiter money they had inherited from their mother. Her Guinness boys were the sons of a rich father. She thought it deeply unfair that Alexander and Max, alone of the family, should not have some kind of annual allowance. She had a little capital of her own which she could always use on their behalf in a crisis, but it was not enough to provide a sufficient regular income.

She was especially worried about Alexander. He and his father were thoroughly at odds. Mosley had refused to allow him to go to university, saying that as he could not write an essay there was no point. He dismissed his son's political ambitions with the words (to Diana), 'It is a terrible thing for a boy of twenty years and three months who is politically ambitious never to have addressed a big crowd or even had an article published.' Alexander's work for a travel firm (later surfacing in Nancy's novel *Don't Tell Alfred*) had come to an end and he was now talking of going to South America. The prospect of her son adrift there with no money, no friends and no knowledge of Spanish or Portuguese appalled Diana. 'It must be a torment – what can I say?' wrote one of her friends:

> I have always had the feeling that of all Kit's children he was the one who came nearest to appreciating what a great man Kit is, and was the most crushed by this appreciation. He is large like Kit, brilliantly clever in an intuitive, rather rebellious way like Kit and was thus faced with an awful problem, either to submit and just be a total imitation of the father he so much resembles, or to break free and be as unlike as possible . . . hence I appreciate the South American idea: a country Kit has never been to and a language he doesn't speak. Manual work? because Kit has never lived this way.

Nicholas too was deeply concerned about his half-brother. He had always written as regularly to Diana as to his father; throughout the 1950s they had conducted vigorous but amicable arguments by letter about Christianity (Nicholas was then a committed Christian) and later about the question of race. Emotion crept into the arguments with his father, not so much because Mosley was scathingly anti-Christian as because Nicholas could not bear to see his father make the same political mistake twice – once again using racial hatred to recruit followers.

He now tackled his father on behalf of his brother. A meeting in Mosley's office erupted into a blazing row. Nicholas accused Mosley of being a terrible father, of always shrugging off responsibility for his children and of never learning politically from his past mistakes. Diana, who must certainly have agreed with Nicholas on the first two points, if not on the third, stood quietly by her husband. When Mosley said to Nicholas, 'I will never speak to you again,' she remained silent. 'Well, I'll always speak to you,' replied Nicholas and left.

His arguments had no effect at all on his father. In the Notting Hill race riots of 1958 Mosley had seen a chance of returning to parliament and was determined to seize it. He planned to stand for this district of west London in the next general election. Accordingly, he closed the *European*, diverting funds and effort into building up his following in the Union Movement. A steady stream of propaganda, much of it in the form of leaflets on the theme of 'Keep Britain white', heightened racial tension.

Mosley fought Notting Hill as the Union Movement candidate in the general election of October 1959. His platform combined a programme for economic reform based on the future need for a United Europe, with the logical outcome of the racialist divisiveness he preached – the repatriation of coloured immigrants. He was convinced of victory ('people running out of their houses to shake him by the hand as he passed', wrote Diana to a friend). But, despite his campaign to appeal via racist hatred to the lowest common denominator, he polled a mere 2,821 votes out of a total of 34,912. This result was so abysmal that at first he could not believe it and even contemplated bringing a legal action on the basis of irregularities in the poll.

The spring of the following year brought a distraction. Jessica's book *Hons and Rebels*, much of it taken up with a description of the Mitford childhood, was published. It was very funny and read with glee by the public, but it upset her sisters greatly. They disliked what they saw as its hurtful disloyalty and its inaccuracies – all of them, for instance, denied Jessica's descriptions of shoplifting expeditions to Oxford. As Nancy wrote to Heywood Hill (on 16 March 1960), 'It is rather dishonest for an autobiography because she alters facts to suit herself in a way that I suppose is allowed in a novel . . . Diana is outraged for my mother – I had expected worse, to tell the truth – and of course minds being portrayed as a dumb society beauty.'

Friends wrote to Diana to commiserate. 'Whenever I pass Burford I think of you all at Asthall, and Swinbrook, and the *Elysium* of it in my memory, and in Decca's such hell,' wrote Jim Lees-Milne adding loyally, 'Of course she wrote rot.'

The glacial breach that had opened between Diana and Jessica, originally instigated by Esmond Romilly, had been of Jessica's choosing. Unlike Diana, she had always been open about their estrangement publicly as well as privately. When her friend Philip Toynbee asked her, 'But didn't sheer curiosity drive you to want to see her again?' she answered, 'It well might have, if I hadn't once, long ago, adored her so intensely. To meet her as an historical curiosity on a casual acquaintance level would be incredibly awkward; on a basis of sisterly fondness, unthinkable. Too much bitterness had set in, at least on my part.'

After *Hons and Rebels* was reviewed in the *Times Literary Supplement* – a review that focused on the deficiencies of the children's upbringing – Diana was for once stung to public retaliation, though largely on behalf of her parents. She adored her mother and could not bear the portrayal of her in Jessica's book. On 8 April she wrote to the editor of the *TLS*, an icy and scathing rebuttal of Jessica's picture of their childhood.

> Your reviewer of my sister Jessica's book, *Hons and Rebels*, says that she describes 'an early environment of almost incredible aridity – tasteless, stupid, wasteful and idiosyncratic only in its scorn of all intellectual and aesthetic values . . . Her portraits of Lord and Lady Redesdale show them as monsters of arrogance and dullness whose neglect, in all but a material sense, of their children might well have resulted not in that rebellious pattern of behaviour so prized by the author but in alcoholism or the analyst's couch.' Does Miss Mitford realise, he concludes, 'how supremely unpleasant her father and mother appear, as seen by the most brilliant of their daughters . . . if she does, she is too wise and loyal to stress these points'.
>
> Doubtless the author realises how 'supremely unpleasant' she makes her family appear. Perhaps the object of the exercise was to demonstrate her good fortune in escaping from them and their way of life. May I, however, be permitted to correct one or two matters of fact?
>
> As children we had access at home to an exceptionally well-chosen library; therefore scorn of intellectual values was a matter of choice for the individual child, not of necessity. My brother was a talented pianist, contemporaries will remember, and through him our childhood was filled with the music of Handel, Bach, Mozart and Beethoven. Jessica does not mention this; she is concerned to denigrate and belittle.
>
> Of my sisters' marriages your reviewer states, quoting from this book: 'one married a jockey'. The 'jockey' in question is a distinguished physicist, a Fellow of the Royal Society who occupied the Chair of Spectroscopy at Oxford University. He also rode in the Grand National.
>
> The portraits of my parents are equally grotesque. My sister's book was probably meant to amuse rather than to be 'wise' or 'loyal'. Or truthful.

After this, any possible reconciliation was out of the question.

Nancy's sympathy with Diana over Jessica's book did not preclude her writing (on 20 August 1960) to Violet Hammersley, the regular recipient of the acid drops about Diana, 'Diana says Sir O has never been so busy – it makes my flesh creep. No doubt we shall all be in camps soon.'

Nancy's familiar waspishness now overlaid real unhappiness – as usual, over Gaston Palewski. For some time she had been seeing less of him and she was miserable when she heard he was having an affair with a married woman, who had borne him a son – something she, Nancy, could never do. Three years earlier he had been sent as French Ambassador to Rome (where he quickly became known as l'Embrassadeur) and she hardly saw him. 'When things go badly you don't need me,' she wailed to him, 'when they go well you turn to other, prettier ladies. So I seem to have no function . . .' But it did not stop her stream of work: her novel *Don't Tell Alfred* came out in the same year, 1960.

It was quite true that Mosley had intensified his political activity. So inexplicable did he find the Notting Hill fiasco that he decided to make one more serious attempt to win a seat in parliament. In the autumn of 1960 he and Diana moved to England. They took a flat in London, in Lowndes Court, where they planned to spend winters. Emmy and Jerry and their small son John came with them.

For Diana London meant a chance to see more of her mother, as Sydney spent winters in the Rutland Mews house. They would meet most days, sometimes on the sofas in Harrods' Bank in the mornings, or for tea and Scrabble at the Mews. Sydney, whose little dog had just died, was now a free agent and she came often to Lowndes Street. When Mosley held a meeting she would always insist on being there, whatever the danger of disturbance, which was now becoming a regular feature – at one meeting she attended, in Kensington Town Hall, opponents had managed to put tear gas in the heating pipes. Mosley, high above the fumes on the platform, wondered why his audience of the faithful was coughing and sneezing. Finally, with streaming eyes, everyone left the building.

Another regular attender was Max, who had just left Oxford, where, at the age of twenty, he had married Jean Taylor. He was reading for the Bar, but quickly became his father's right-hand man, at one point saving Mosley from a severe beating-up. On 31 July, 1962, at a meeting in Ridley Road, Dalston, a number of opponents managed to break through the cordon of 200 police surrounding Mosley and he was knocked down and pummelled. Max, who hurled himself to his father's rescue, was arrested, to cries of 'Down with Mosley!' and 'Germany calling!' Five minutes later the police shut down the meeting and Mosley's car drove off amid a hail of apples, oranges, copper coins and horse manure. The next day Max defended himself ably before the magistrates and was acquitted.

Alexander, in South America, was supporting himself by teaching English and French and working for the British Council. Nicholas, who had always thought his half-brother should have the chance of going to a university, now suggested that he go to an American one. A family friend was a professor at Columbus University, in Ohio, and was prepared to act as an anchor for Alexander – so Columbus it was to be. When the question of payment arose, Mosley argued that by suggesting that Alexander go to a university in defiance of his, Mosley's, expressed wish, Nicholas had taken on the responsibility for his brother. In his eyes, this meant one thing. 'About paying for it,' wrote Diana to Nicholas in August 1961, 'Dad is saying "Nick started this. He took Aly away etc." This is half a joke but you know what he is.'

It was more than 'half a joke'. Diana had to keep silent about her own contribution of £200 a year to each of her Mosley sons (from her £12,000 share of her father's estate), so much did Mosley resent the idea that his own income, or hers, should be used to assist his children. 'I could give Aly another £100 a year while he is at university out of my very tiny capital, but Max doesn't get enough and I must help him. Dad knows nothing of either, of course.' Generously, Nicholas and Micky agreed to settle Alexander's £1,400-a-year fees at Columbus, each of them making him a seven-year covenant for £700 p.a. There were no thanks from his father, but Diana was enormously grateful.

That winter, Diana went to Morocco to stay with friends, travelling south of the Atlas mountains and then visiting Tangier. For the first time in years, her migraines lessened dramatically; seeking for a reason for this miraculous remission, she put it down to being away from all her worries.

Mosley was still in constant contact with extreme right-wing groups abroad, representing Britain at a congress of these groups in Venice in 1962. The post-war Labour Government's fear that he would encourage such German groups to think that fascism was alive and well in Britain was justified by the fact that fourteen years later he was making neo-fascist speeches in Germany. In Britain, disruption at his rallies was such a regular occurrence that eventually, in 1962, he was forbidden to hold any more meetings in Trafalgar Square or most of the major halls.

In the spring of 1963 the Mosleys' Irish house Ileclash was sold. Max could no longer spend winters hunting in Ireland and for Mosley politics was once more a priority.

After three years in England, the Mosleys returned to Paris. At the beginning of May Diana received a telegram: her mother, who had just returned to Inch Kenneth, was very ill. Diana flew at once to Scotland. At Oban she had to wait for the next day's steamer to Mull. She had arrived too late for dinner – after seven p.m. no food was available – but eventually persuaded the hotel to produce a cheese sandwich, which she ate while attempting to telephone for news. She finally arrived at Inch Kenneth at teatime the following day.

Nancy, Pam and Debo were already there. It was her mother's eighty-third birthday and her friend Madeau Stewart had come to stay. Also there were Desmond's children Patrick and Marina Guinness with their nanny, Sydney's cook and maid and her boatman and his family. They managed to get from the mainland two nurses who did not mind the idea of being cut off by violent storms or the lack of a telephone.

Although it was clear that Sydney was dying, her heart was so strong that she lingered for three weeks. For her daughters it was misery to watch her discomfort as, desperately ill with Parkinson's disease, she lay unable to read and barely able to swallow or talk. She asked to see Desmond, who came at once. After five days in a coma, she died on 25 May. She made her last journey in *Puffin* across to Mull, as a piper played a lament. She was buried next to David at Swinbrook. 'I shall miss her terribly,' wrote Diana in reply to a letter from her prison friend Louise Irvine. 'Always so marvellous and loyal to O.M.'

For Diana, 'loyalty to O.M.' was prized above all. When *The River Watcher*, by Hugo Charteris, was published in 1965, Mosley felt he had been libelled – one of the characters was a mysterious fascist landowner, described by the hero as 'Hitler's best man', who 'did a trick with his eyes, blinking and making something flash with the pupils'. Mosley further claimed that Nicholas too had been libelled and through his solicitors – he was still refusing to communicate with his son directly – asked Nicholas to join him in his action. The River Watcher himself, said Mosley, was an unflattering portrait of Nicholas. Diana, who had never ceased her stream of affectionate though often argumentative letters to Nicholas, now wrote immediately to urge him to join in his father's action 'as any loyal son would'.

Nicholas, however, did not want to sue Charteris, who was a cousin of his wife Rosemary, a family friend and a fellow novelist. Though Charteris was known for putting traits or characteristics of friends in his novels, Nicholas, as he explained to Diana, did not feel that he himself had been libelled, nor did he think that the landowner in the novel was particularly like his father. Any lawsuit, he felt, would simply draw attention to what would probably otherwise pass unnoticed. In any case, he added, it was difficult to feel family solidarity with someone who would not answer his letters and communicated only through his solicitors.

Nicholas's decision brought a furious riposte from Diana. 'You are so lucky to be the son of the cleverest, bravest, dearest person in the world. I would have thought you wouldn't be able to wait to squash a spiteful insect like Hugo after it tried to annoy him.' But they went on writing, Diana's letters becoming gradually less spiky. Mosley, she knew, wanted a reconciliation: he was anxious to see Nicholas's sons Shaun and Ivo, his two eldest grandsons. In February 1966 Diana signified that she too had finally forgiven Nicholas's 'disloyalty'. 'Let us leave Hugo – what does it all matter *really*.'

By then, there were other preoccupations. Mosley had decided to make one last attempt to enter parliament, choosing as his arena his old stamping ground of the East End. In the general election of 8 March 1966, he contested Shoreditch as the Union Movement candidate.

The result was devastating. He polled only 1,600 votes – 4.6 per cent of the total. 'The people we worked with were perfect and this made it bearable for me,' wrote Diana to a friend:

> People from East London are often brilliantly amusing and when you add the loyalty and love they feel for Kit it was an extraordinary experience to work with them day by day for a month or whatever. We did not canvass and the result was no surprise. There was a feeling of enormous apathy reflected in the high rate of abstentions.

They were brave words, but Mosley had finally accepted that the call would never come. He gave up the leadership of the Union Movement and, with it, active politics. In all the years he had spent promoting his beliefs, his movements had never won so much as a single local council seat.

Chapter XXXIV

A month after Mosley's decision to withdraw finally from British politics Diana heard of the death of one of her oldest friends. On Easter Day, 1966, Evelyn Waugh died of a heart attack.

After Waugh had abruptly dropped out of her life more than thirty years earlier, she had seen little of him, but lately they had begun to draw together again. Waugh had been anxious to clear up any misunderstanding that might remain. 'I must not leave you with the delusion that Work Suspended was a cruel portrait of you,' he had written ten days before his death. 'It was to some extent a portrait of me in love with you, but there is not a single point in common between you and the heroine except pregnancy. Yours was the first pregnancy I observed.' When Diana heard of his death she drove immediately to console Nancy, for years one of his closest friends.

With no political ties, the Mosleys' life was much freer. As well as their usual late-summer holiday in Venice, they rented a house in Johannesburg during January 1967, in Bath Avenue, Rosebank. 'There are lots of flowers and a good maid,' wrote Diana to Nicholas. Again, as in Morocco the previous winter, there was a sudden, startling alleviation of Diana's migraines. This time she realised that it was not only freedom from worry but the warm dry climate that helped. Henceforth, they went to South Africa most winters.

Mosley did not remain idle long. In 1968 he published his autobiography, *My Life*. It marked a change in the way he was perceived. Age, withdrawal from active politics and residence in a foreign country had blurred his satanic image. Even the Left paid tribute: Richard Crossman said that Mosley had been 'spurned simply and solely because he was right', while Malcolm Muggeridge called him 'The only living Englishman who could perfectly well have been either Conservative or Labour Prime Minister'.

A.J.P. Taylor wrote that he was 'a superb political thinker, the best of our age,' and the BBC devoted a whole *Panorama* programme to an interview with him by James Mossman – the first time he had been allowed on television for thirty-four years. There was a sense of expiation, a feeling that he had 'served his time' and, to some extent, a softening among those who had previously refused to meet him. In London, while there were still many people who would not sit at the same table as 'that man', others were intrigued rather than horrified when they heard he was to be a fellow dinner guest.

None of this rubbed off on Diana. She was still thought of as Hitler's intimate, and her own views remained as sharp edged and unyielding as ever. In Mosley's original draft of his autobiography, he had written of Harold Nicolson (*à propos* his resignation from the New Party), 'One doesn't take one's pet guinea pig with one on an adventure into the jungle.' When he replaced this caustic comment with a remark that Nicolson would have made a good Foreign Secretary, Diana objected scornfully. In her eyes, Nicolson had lacked the all-important 'loyalty' and the fact that he resigned because the New Party had changed direction was neither here nor there. Mosley, however, held firm to his modification.

Apart from Mosley, Nancy was the person of whom Diana saw most. Through her writing, Nancy was now a rich woman – in 1968 her income was £22,000 – with a charming small house, 4 rue d'Artois, in Versailles. Successful, established and popular, she was as desperately in love as ever with Col, now no longer French Ambassador in Rome. She lived for their meetings, now less frequent. Sometimes she would catch the train to Paris, sometimes Col would drive down to Versailles to lunch with her or sit chatting in the small drawing room that overlooked a garden full of wild flowers.

What Nancy had succeeded in putting out of her mind was the well-founded rumour of Col's love for a married woman, Violette de Talleyrand-Périgord, Duchesse de Sagan. As long as Col was single, even though having affairs, she could feel that part of him still belonged to her – indeed, she could even hope that one day, perhaps, he would realise where his true happiness lay. She did everything she could to strengthen the thousands of tiny, invisible fibres that hold one person to another, even down to making her maid Marie, a peasant woman from Normandy, concentrate her culinary efforts on puddings. 'I have a friend who likes puddings,' she told her editor, Joy Law. When Col arrived to join them for lunch Mrs Law noticed that he took three large helpings of pudding.

The equation changed dramatically when the Duchesse de Sagan's husband finally consented to divorce her. As Col had retired from politics, marriage to a divorced woman could no longer harm his prospects – one of the reasons he had always given Nancy for not marrying her.

When he came to tell her of his forthcoming marriage, she greeted him with the words, 'Colonel, I've got cancer!' He felt unable to break what he knew would be devastating news and had to return the following day to do so.

Nancy was shattered, but concealed her feelings under her usual mask of flippant mockery, quickly promoting the notion that it was merely another of the Colonel's whims. She was aided by the fact that during the week he lived in his flat in the rue Bonaparte, spending only the weekends at the Duchesse de Sagan's château, Le Marais, forty kilometres outside Paris. 'Gaston's marriage is for those of riper years. Viz, no nonsense about living in the same house in Paris . . . nothing changed whatever in other words.' As for her Paris friends, from them there was none of the pity that she would have loathed. Her friendship with Col was so well known and open that few thought anything of it; those who did know of the wartime affair regarded it as long past, and no one except the most intimate had any idea that Nancy had never ceased to love a man who blatantly pursued every pretty woman he saw.

Col's engagement in March 1969 coincided with the beginning of Nancy's fatal illness. Her family always felt that her cancer was the result of years of repressed longings and jealousy, followed by the deathblow of Col's marriage – and certainly its severity only became apparent after his engagement. Almost from that moment, she began four years of suffering. She was in constant agony and none of the doctors – in all, she saw thirty-seven – could relieve it, as the pain, which began in her left leg, shifted all over her body.

The burden of care and consolation fell most heavily on Diana. During the years of Nancy's illness, Diana visited her sister at least three times a week. She was very conscious that, as she later said, 'The awful thing is, she doesn't come first with anybody.' Mornings were Nancy's worst times; but occasionally, wild with pain, she would telephone Diana in the middle of the night. Diana, who could do little except comfort and soothe, would get up and drive over. She took Nancy to her agonising tests in the Rothschild Hospital and to doctor after doctor to try to find one who might alleviate the suffering.

Often, Debo came over. On 24 April 1969 a grapefruit-sized tumour was removed from Nancy's liver in the first of several operations. 'I sat with Naunce [Diana's name for her sister] until 8.30,' runs Diana's diary.

> . . . Sunday today and Debo has gone to the clinic without me. I had a frightful migraine in the night and felt I might cry in Naunce's room. We finally spoke to the surgeon on Friday night and he said 'Les renseignements ne sont bons'. Debo and I cried so much and on Saturday we sat with her all day and I had a deep talk with the nun and it all seems hopeless. Colonel has been each day. I telephoned him immediately I had spoken to the surgeon. Violette answered and I asked for him and he was rather strange (I thought) but I insisted he must visit continually and he said 'Mais naturellement'.

There was no specific follow-up treatment after Nancy had returned to her little house in Versailles. 'I am half relieved and half afraid,' runs Diana's diary. 'Does it mean they think there's no point in it? Apparently the advantages are not proven and the disadvantages are huge – one feels ill apart from the worry of getting to Paris each time.'

Both sisters realised how desperately ill Nancy was. Even Jessica, icily hostile to Diana for thirty years, paved the way for a visit by a letter to Debo. ('I will be friends with Honks if she is willing to be friends with me.')

There were occasional remissions, making the plunge back into pain even more traumatic. In October Nancy was well enough to fly to Germany as a guest of the government to research her life of Frederick the Great, accompanied by her sister Pam, her editor Joy Law and Joy's husband Richard; but immediately after Christmas she was pulled down by 'flu. Diana, to her distress, could do little for her sister as she was looking after Mosley, who was just out of hospital after a hernia operation and reluctant to let her leave the house on any pretext. In any case, he thoroughly disliked Nancy.

As Mosley grew older he had become extraordinarily possessive of Diana. Years of being at the centre of her existence had accustomed him to his position as a domestic autocrat. Shirley Conran, who as a friend of Alexander's often visited the Temple, was amazed by Diana's subservience. 'I was deeply impressed by the way she *danced* round him. She struck me as almost henpecked.' The Laws also noted Diana's deliberate subjection of her own personality to that of her husband. 'For someone who so obviously had a mind of her own, she was absolutely at one with him always.' If he said anything that could be construed as ill judged, she would intervene smoothly to deflect the conversation into another channel.

Mosley did his best to stop Diana visiting Nancy and she would set off, miserable at upsetting him and sometimes half-blinded by tears, to a sister who wrung her heart still further. Nancy, who would make a tremendous effort to appear serene and cheerful in front of someone she did not know well, could not keep up this brave pretence with Diana, who saw her at her weeping, tortured worst. Harrowed by Nancy's misery, aware that her sister was utterly dependent on her and wanted her there as much as possible, Diana would return to a husband who did not hesitate to show his displeasure at her absence.

Torn between the two of them, Diana found her migraines rising to a crescendo, sometimes following each other with barely an hour's grace. Pam, who often came from Zurich to stay with Nancy at weekends when Nancy's manservant Hassan was away, was a furious spectator of Mosley's behaviour towards Diana. But nothing shook Diana's conviction of his unassailable rightness. 'OM is bursting with ideas,' she wrote to her prison friend Louise Irvine in August 1969 just after the Mosleys' return from Venice. 'How wicked and *stupid* it is not to listen to him.'

Her loyalty to Mosley was complete; she reacted like a spitting cat to anyone who offered even the mildest criticism of him, and only adulation would serve from anyone writing about him. The historian Robert Skidelsky, a friend of Max's, was writing Mosley's biography. Her diary records, 'Read seven chapters. I think it spiteful,' yet her fondness for Skidelsky and her natural truthfulness ensured that she wrote him charming, undoubtedly sincere letters. 'You are incapable of writing a dull page. It is a wonderful gift in my opinion and no doubt you realise it. Hard writing, easy reading.'

Diana, who had seen more of Nancy than anyone else in the family, was convinced that Nancy's anguish over Col was exacerbated by the acid of her temperament and that in Nancy's own disposition lay the seeds of her illness. Some years later, she set out her theory, which went some way to explaining the value she herself had always put on surroundings that were beautiful, serene and harmonious

> Spitefulness turns inwards. The French have an expression, 'Il fait du mauvais sang'. Nancy was not 'to blame' for her illness, because a certain sort of person can find enough unhappiness to create his or her own ill health within almost any frame. The mind and body react on one

> another at least as much as crazy Christian Scientists think they do. Nancy, who always secreted a good deal of poison, ended by poisoning herself – though again, nobody's fault and certainly not hers.

In Diana's case, the emotional strains she was suffering resulted in an ulcer, diagnosed in December 1970, just after Christmas lunch with a miserable Nancy and a brightly social evening with the Windsors. Another strain on her was Mosley's attitude to their son Alexander. Although Mosley had finally helped Max financially, he still maintained that Alexander was Nicholas's responsibility. Generously, Nicholas continued his support to Alexander, with Diana adding what she could in a secrecy she hated.

One of Diana's most salient qualities was honesty, often to the point of doing herself a disservice. She scorned dissimulation, yet where Mosley was concerned her true feelings often had to be suppressed or altered to achieve the complete loyalty to him and his ideas that she demanded of herself. In the final outcome, she would range herself on his side, no matter what the emotional cost of having to deny a principle she knew was right. 'I do love Kit and his courage more than I can say,' she wrote to Nicholas, 'but of course I see the harm he does to Aly and I suffer for it.'

Mosley, as Diana once admitted, was 'the strangest mixture of generosity and miserliness'. To the cynic, the generosity seemed to appear only in anything that involved himself – he kept an excellent cellar and was a generous host, he would never dream of taking Diana, or his friends, to any but the best restaurants, his clothes were beautifully tailored, his shirts handmade and he was extremely good to the servants who looked after him. He expected Diana to look elegant and liked her to dress at the great Paris couture houses – but he did not give her much to achieve this. Once a year she would buy a dress, or a coat, always in grey, black or beige, worn with simple white polo-neck jerseys; for dinner parties she invariably wore the same long black skirt. But neither short commons nor internal stress affected her beauty. Secure in both her looks and her style, she seemed completely without vanity ('the last person you could imagine looking at herself in a mirror', said Shirley Conran). She wore her grey-gold hair drawn back and twisted into a simple bun. This style, adopted to avoid hours in the salon of the fashionable hairdresser, Alexandre, suited her classic looks so well she kept it for years.

The terrible months dragged on. When Diana was in Venice, Debo and Pam replaced her at Nancy's side. Nancy's courage was immense. She never lost her teasing, acerbic wit and the semblance of gaiety. To one cherished correspondent, Raymond Mortimer, she wrote (in October

1970), 'I've always felt the great importance of getting into the right set at once in heaven.'

In December 1970 Mosley was invited to address a Round Table at the Centre of Advanced International Policy Research in Washington on his European ideas. He was delighted and planned, with Diana, to stay with friends in Nassau first. Just at that point Nancy's condition suddenly deteriorated. Unable to bear the thought of leaving her sister alone in France, Diana spent the morning of 12 January telephoning doctors in London to find a nursing home there for her sister.

> All this in the teeth of opposition from K, who doesn't want me to be bringing Naunce over to London on the first leg of my American journey, on the 19 [records her diary]. He thinks it will be an extra burden. Too true. Well, then I got a headache and went to bed for the middle of the day. Then K went down every hour or so to look for a letter which hasn't arrived, the proof of one of his 'broadsheets', short articles he writes and sends to a long list of journalists and M.P.s. The wretched thing didn't appear. It's been in the post five days. He got more and more cross and finally very cross with me, because I'm supposed to have advised him not to have letters expressed. I probably said why not have only *important* letters expressed. At the Temple they are sometimes slower than ordinary post because they wait for a special delivery man. Anyway it was one of those days when K is in such a state that anyone living with him must feel their nerves biting and it was exactly the atmosphere conducive to a stomach ulcer. It's nobody's 'fault'. K is angelic kindness itself about my headaches or any other ailment but he gets into a state of nervous gloom – about small things – which reacts on everyone near him. It is useless to sympathise and I just go into my room and read until the mood has changed.

Once in Nassau, as usual in the heat, Diana had a welcome cessation of her migraines – though they began to recur when a long telegram from Debo told her that Nancy had to have an operation on her spine. Gradually, the worst of the pain had begun to focus on Nancy's back; now the doctors had found that a vertebra was pressing on a nerve, which might account for the excruciating torment. 'Cry bitterly and long to be with her,' records Diana's diary; it was two days before she was able to stop tears rolling uncontrollably down her face.

Nancy spent eleven weeks in hospital, visited constantly by family and friends – among them Joy Law, who would call in to see Nancy and another friend there on the way home from work, taking with her trays of caramel creams, Nancy's favourite toffees. But the operation did nothing to alleviate the pain Nancy was suffering.

Back in France, Diana acted as Mosley's assistant, despatching and ordering books, sorting his papers, organising his letters, editing what he wrote; she was so useful that he was crosser than ever when she tore herself away to see Nancy. In April Mosley had a minor operation and in July, to speed his recovery, they went to Venice. Alexander came for the weekend. 'He is brilliant and happy, more than I ever saw him,' wrote Diana delightedly to Nicholas. And on 6 October, there was another 'brilliantly happy' day when she and Mosley celebrated their thirty-fifth wedding anniversary with lunch at Maxim's.

But there was always Nancy. 'One has completely run out of comforting words,' wrote Diana to Nicholas. 'They just sound too silly in face of the reality of pain.' Her own headaches increased as the saga of misery and the strain of being torn between husband and sister continued. 'I found Naunce in utter despair,' wrote Diana of one visit to London. 'She felt deathly ill and was in tears. I longed to stay but it was impossible. K had given up his room at the Ritz and was on the way to the air terminal. If I had stayed he would have had his birthday completely alone.'

Though Nancy returned home again, it was clear the end was near. When Debo telephoned on 19 May, Diana begged her to come over; Debo came, ostensibly to see her horse run at Longchamps. Together, Debo and Diana visited Nancy, who was being given increasing amounts of morphine. Then, on 21 May, Jessica came from California. 'Naunce v. unkind to Decca,' noted Diana's diary. 'Our meeting after all those years seemed completely easy and natural because we were both thinking only of Naunce. Decca has kept her childlike face but her voice has changed, not the accent but the tone of voice. I felt an unexpected sympathy, even affection, for her, and was surprised.' Pam came at the beginning of June.

Apart from the hour or two four years earlier, it was the first time Diana had seen Jessica since the mid-1930s. Their meeting at Nancy's deathbed was that of two strangers, though by the end Diana was writing (on 18 June), 'I felt very drawn to Decca. Her communist past makes her less influential perhaps than a more politically neutral person would be. I felt all my old love for her come flooding back and have quite forgotten her bitter public attacks on K and me, or at least quite forgiven them. She struggles away on behalf of suffering humanity.' With the conversation confined to Nancy's illness Jessica, though less able to overlook the past, was able to say afterwards: 'It wasn't dreadful – we actually got along very well on that level.'

By the end, Nancy was torn between longing to die, as she often said to Debo in letters, and clinging to life – or, as Diana's diary records, 'at least some vague hope of recovery. I detect in her a quite strong desire to live, which in a way makes the whole horrible business even harder to bear. I can't think of her without crying.'

Fittingly, Nancy wrote her last letter, on 8 June 1973, to Col. The pain, she wrote, was so terrible that 'I hope and believe I am dying . . . the torture is too great. You cannot imagine. I'm very weak and would love to see you now.'

He was the last of those close to her to see her alive. Driving into Paris on 30 June, the feeling suddenly came to him that he must go to her. He ordered his chauffeur to take him to her house in Versailles. She was lying in her bed apparently unconscious but when he gently took her hand she smiled. She died a few hours later. She was cremated at Père Lachaise, with speeches from the Mayor, who read in French from her books, and an address by V.S. Pritchett – much of which Diana missed because of her increasing deafness. Nancy's ashes were to be buried at Swinbrook, next to Unity's; adding to the complications of funeral arrangements in two countries was the threat of a strike by airline employees. Diana changed all the tickets from air to sea, but until the last moment they did not know if they could get Nancy's ashes to Swinbrook in time for the arranged burial.

One result of Nancy's death was the growth of a real friendship between Col and the Mosleys. Diana had always liked him and the days she spent in his company after Nancy's death – he helped with the notice for the *Figaro* and the arrangements for the cremation – served to turn liking into deep affection. Before Nancy's death, Diana had refrained from accepting the invitations of the new Mme Palewski to Le Marais (the chateau was close to Orsay) for fear that Nancy's feelings would be hurt. It was a delicacy that annoyed Mosley, who pointed out that Nancy would have had no such scruples in similar circumstances. Soon the Mosleys were lunching and dining almost every Sunday at Le Marais. Mosley too became very fond of Col. ('Kit and Col. as one,' wrote Diana in her diary.)

Nancy left her little house in Versailles to Joy Law. Joy had become a friend and by the time of Nancy's death was virtually indispensable to her. She had worked extraordinarily hard on *The Sun King*, correcting, fact-checking, suggesting, amending, but had been paid only £500. The book had brought Nancy many thousands of pounds – the reason, thought Mrs

Law, for Nancy's bequest, since *The Sun King*'s accuracy and presentation had much to do with its success.

Nancy's death brought no alleviation of Diana's migraines. Diana was also rapidly losing one of her chief pleasures, listening to music. Her hereditary deafness had increased to the point where she could hardly hear music any more. The last time she had really enjoyed opera was when Robert Skidelsky took her to hear Callas sing Norma in Paris in 1965; she finally stopped opera- and concert-going after a Mozart concert, to which she had been taken one evening by her friend Jackie Gilmour, when she could no longer hear the clarinet.

On 1 March 1974, the Mosleys went through another marriage ceremony. In order to make their wills, they needed to produce their marriage certificate, which had disappeared somewhere in the ruins of the Third Reich. It seemed simpler all round to marry again. Weddings were in the air: Nicholas told them he was remarrying and nine months later Alexander telephoned to say that he too was getting married, to the beautiful Charlotte Marten, who became a devoted and much-loved daughter-in-law.

As well as the busy social life he shared with Diana, Mosley was still writing, appearing on television from time to time, and lecturing – at the Frankfurt Book Fair he spoke in German to a large audience without a single note. Though his main emphasis was on the desirability of a united Europe, his anti-semitism had not abated. The Nietzschean Superman might have been transmuted into the Hero, but the villain was still 'the Judate [sic] money race'. When Frank Cakebread telephoned from Savehay to say that Cimmie's tomb had been vandalised – local rumour had it that her jewels had been buried with her – Mosley's immediate response was, 'It's the Jews!'

In 1975 Robert Skidelsky's biography of Mosley was published, attracting much interest and many favourable reviews. From Mosley's point of view, the attention of a respected historian set him firmly in context as a serious player on the political stage, many of whose forecasts – the loss of the Empire, the post-war decline of Britain, the need for integration into Europe – had now been shown to be correct. Skidelsky did not gloss over the more unpleasant manifestations of British fascism. But the reverence of the British for age (Mosley was now almost eighty), combined with the fact that he had long retired from active politics in Britain, had given him the status if not of an institution at least of someone who with hindsight had perhaps been judged too harshly. It was not exactly rehabilitation – but it certainly added a patina of respectability. When Lord Longford took him to lunch at the Gay Hussar, a favourite literary and political haunt of the 1970s, they found themselves next to Michael Foot, sitting at his usual table. Nervously, Longford introduced them as Foot was leaving. 'A pleasure to see you over here, Sir Oswald,' said Foot courteously. Mosley was delighted. 'It couldn't happen anywhere but England,' he said to Longford. 'Two old political enemies fraternising like that.'

In old age, Mosley's personality had mellowed dramatically. 'He has acquired a tolerance and wisdom which, had he only cultivated them forty years ago, might have made him a great moral leader,' wrote Jim Lees-Milne. Earlier, at the Chatsworth ball for the coming of age of the Devonshires' son Lord Hartington, Cynthia Gladwyn had described him as looking 'fat, smiling and benign' as he and Harold Macmillan chatted together – two old men with the exquisite manners of a previous age. Though there were flashes of the old arrogance and the same desire to command the table as he had once held crowds, everyone who visited the Temple spoke of his charm, his wide-ranging intelligence, the fascination of his company and his courtliness. He would refer to Hitler as 'a terrible little man' – words unthinkable from Diana. When Joy Law and her daughter were asked to luncheon at the Temple, Joy, who had made no secret of the fact that she was Jewish, was apprehensive if not hostile. Her mood soon changed:

> I sat next to Kit and my daughter was on the other side. He made quite sure that my daughter was being looked after by the young man next to her and then spent the entire lunch talking to me as if I were the only person in the world. I was totally mesmerised. I'd been very leery about him before because a lot of my relations had perished during the war and so I had an inbuilt antipathy to what he stood for if not to him. But I found that this belief was just suspended while I listened. I found him an amazing, wonderful, charismatic man.

A luncheon with Baba Metcalfe was not so successful. For years, Baba had refused to come to the Temple. Neither her rage over Mosley's treatment of her nor her hatred of Diana had lessened over the years. But Mosley, growing old, wanted reconciliation and Diana wanted this for him. Eventually, reluctantly, Baba was persuaded to come in 1976. The atmosphere was glacial and the experiment was not repeated.

In November 1976 *Unity Mitford, A Quest*, by David Pryce-Jones, was published. Like Jessica's book *Hons and Rebels*, it made her sisters furious – and it was, they felt, again largely Jessica's fault. 'I'm afraid Decca behaved in an extraordinary way,' wrote Diana to one friend. 'She took D. Pryce Jones to see cousins and old servants who chattered merrily away. She is very much to blame because she is by way of having loved Bobo, her presence made them trust Jones.' On 18 November the sisters wrote to *The Times*.

> Sir,
>
> A book has been published about our sister, Unity Mitford, which we do not accept as a true picture of her or our family.
>
> We hold letters from a number of people quoted in the book, saying that they have been misquoted. Some of these letters were sent to the publisher before publication but to little avail.
>
> Yours faithfully,
> Pamela Jackson,
> Diana Mosley,
> Deborah Devonshire

Diana, too, had been writing. Her first book was her autobiography, *A Life of Contrasts*, published in the spring of 1977. It was witty, entertaining, vivid, moving, elegant – and contained no hint of recantation. 'Jamie* comes Sunday morning, to ask me to put more condemnation of German atrocities. Write him furious letter which K persuades me not to send,' runs a diary entry of the previous October.

The book had a great success. Friends including Harold Acton and Alastair Forbes gave it predictably good reviews – but so did those who did not know her. Mary Warnock called it 'This fascinating book' in the *Listener*, and from the *Sunday Times* came, 'Lady Mosley writes extremely well. She is constitutionally incapable of being dull. An autobiography of real distinction.' 'Wit and unsentimental warmth . . . To all those not averse to a little powdered glass in their bombe surprise: enjoy,' said *The Times*. There were admiring letters from strangers. 'We do not know each other but I wanted to say how immensely fascinating I found your book – gay, evocative and moving,' wrote Alec Guinness. Diana appeared on the *Russell Harty Show* and was guest of honour at a Foyle's Literary Luncheon at the

* Hamish (Jamie) Hamilton, her publisher.

Dorchester on 12 May, where she expected trouble. 'I made sure of my sons coming so if the Chosen did decide to come and shout they would see them off the premises,' she wrote to a friend in September. 'In the end it was peaceful. A brave old General, a V.C. [Sir John Smyth] took the chair and Sir Arthur Bryant made a speech in which he said my book was about the two things that matter most in life, love and courage.' The only person to be discouraging was Mosley. 'Kit has made me promise not to write another [book],' she wrote to Robert Skidelsky. 'What he wants is for me not to write anything at all (though I always do it when he's asleep).'

Mosley was becoming much less physically active, though he retained his regular weight of thirteen stone two pounds. He had Parkinson's disease but responded so well to the drugs prescribed that the only giveaway was the occasional fall. 'We are going to Paris for Xmas as we always do,' wrote Diana to Louise Irvine on 15 December 1978, 'and he loves the Palais Royal to potter in, just near our flat.' He was no longer able to go for the long walks he had so enjoyed but he went out into the Temple garden as often as possible. Once, walking round the lake with his son Micky, he said, 'I've had an extraordinary life and I've blown it all. But I have had extraordinary good fortune in the two women I've married. I adored your mother and it was a great tragedy when she died. I was equally fortunate to marry Diana, whom I love so deeply.'

Chapter xxxv

Mosley's last years were sunny. Diana no longer liked him going to Paris by train, as she was frightened his slowness would cause him to be trapped in the automatic doors, but although at eighty-three he could only walk a few steps and was having treatment for his leg, he was still swimming twice a day. His ability to savour the moment enhanced his delight in the Temple. Almost every day he would say to Diana, 'We've had a lot of bad luck in our lives but what LUCK we had when we found this.'

He enjoyed social life as much as ever. At luncheon parties he was the leading actor. At a pause in the conversation, he would make his entrance, emerging from his small sitting room into the large salon, where the party was gathered for drinks. He would beam at all the guests in turn, then slowly and dramatically descend the winding staircase to the dining room. Sonorous phrases, many rooted in past attitudes, would roll out over Emmy's *boeuf à la mode* or *poulet aux poireaux*. 'The weak must be rooted out in order to ensure the survival of the strong,' he would intone when discussing a damaged plant in the garden. Jewish guests would look uncomfortably at their plates.

One such Jewish guest was Maître Blum, the Windsors' lawyer, who had met the Mosleys through Gaston Palewski (her husband, a French general, was a friend of Palewski), but she was in any case a useful acquaintance. Lord Longford, as chairman of Sidgwick & Jackson, had commissioned a book about the Windsors from her. It was published in 1980 and drew mixed reviews. The well-disposed called it a tribute to a friend, other critics complained that she had glossed over many of the more discreditable episodes in the Windsors' lives.

Soon afterwards, on 13 September 1980, Diana wrote to Nicholas: 'I am a bit worried about Kit. He would LOVE a visit if you are not too busy.' When Nicholas came over the following month he brought up a subject that had been on his mind for some time. 'Someone should write about you,' he said to his father during the course of one of their long conversations. 'People either think you are God or the Devil, but what's interesting about you is the truth.' Mosley was at first non-committal, then, in the middle of Nicholas's last lunch at the Temple, suddenly turned to Diana and said, 'Diana, I have made up my mind. When I die I want Nicky to have all my papers.'

As the autumn drew on Mosley became weaker. After his eighty-fourth birthday (on 16 November) he spent most of his days sitting in his big armchair by the blue porcelain stove in his study. Diana kept a daily chart of his temperature, watching him lynx-eyed for any change in his condition. Often at night he almost fell out of bed and she had the exhausting task of heaving him back into position.

On 2 December, after two restless days, he seemed better, sitting in his chair and eating his supper with appetite. At about 9.30 Diana helped him to bed, undressing him, as she had for months past, and putting on the bedsocks he needed for warmth. As her diary records: 'We said to each other "You're all the world to me" and over and over again "I love you". I went to bed myself, because that's where the bell would ring. I read for a while and then went to sleep and almost at once the bell woke me, at 11 p.m., and I flew to his room. He was uncomfortable and I pulled him down in the bed and rearranged the pillows and he said "You are so kind" and I said "You're the whole world to me" and kissed him.' The same thing happened at one a.m. and at two. When Mosley tried to apologise for waking her Diana reassured him lovingly that he must ring whenever he wanted her.

At four a.m. she awoke with a violent start. She ran towards her husband's room and, seeing the light was on, tiptoed to the door and peeped in. He was lying on his back on the floor and there was blood on the carpet. 'I hugged him and covered him with blankets and cried and begged him "Darling, darling, come back to your Percher" but I think I knew he had died.'

She rushed out and banged on Jerry's window. He came at once and telephoned Aly and the doctor, who confirmed death. Aly and Cha arrived at 5.30.

> We all sat in despair. I blamed myself terribly for not staying with him. But if I had been in his chair and dozed off I probably wouldn't have heard anything because I'm so deaf – the bell was so loud there was no question of not hearing it.
>
> But I couldn't and can't and never shall be able to get over the fact that he died alone. My only comfort is that it was not that he couldn't find the bell – I mean that he was groping for it – because he'd got his light on.

Next day the tributes and the visitors began. Diana, sitting in the salon that suddenly seemed so empty, 'told everyone who would listen how terrible it was that I couldn't hold his beloved hand at that great and unique moment of death'.

Mosley's funeral, elaborate, secular and moving, was at Père Lachaise on a freezing December day. The music included the March from *Tannhauser* and passages from Mozart's Requiem, Handel's 'Hallelujah Chorus' and Bach's *St Matthew Passion*. Nicholas read Swinburne's 'Before the Beginning of Years'; Micky read 'Say Not the Struggle Naught Availeth' by Arthur Clough. Aly read stanzas from Paul Valery's 'Le cimetière marin', beginning 'Et Vous, Grande Ame . . .' In place of Max, who could not trust himself to speak, Jonathan read a poem by Goethe. There was no nonsense about stiff upper lips: as Diana wrote later, 'I must say one was in floods the entire time.'

There was one incident of black farce which Mosley himself would have appreciated. Following normal French practice, the cremation, which took around twenty-five minutes, took place beneath the funeral room during the service, so timed that as the last piece of music was played curtains at the front of the room drew back to reveal the urn, ready to be collected by the deceased's widow or family. But at the end of Mosley's funeral it was clear something had gone wrong. The final piece of music was repeated, to be followed by a deathly silence.

Eventually a sacristan crept up to Diana, wringing his hands. For twenty minutes the congregation waited, until the sacristan reappeared, twisting his hands more furiously than ever. Mosley's corpse, it transpired, was taking an unprecedently long time to be consumed. 'Even in death he's made his last V-sign to the world,' thought his son Micky admiringly. Diana, glancing around at elderly Parisian friends, by now chilled to the marrow in the icy chapel, rose to her feet with a murmured, 'Well, darlings, I think we'd better go back to Orsay otherwise he won't be the only one . . .'

Two days later, the urn was delivered and Mosley's ashes were scattered by the lake he had so loved.

Mosley left everything to Diana for life and then to Alexander and Max, with the exception of family pictures, which went to Nicholas and some *objets* to his older children. The Temple was valued at £300,000 and the flat in the rue Villedo at £60,000. Diana wrote at once to Nicholas to explain why his father had not remembered his older children in his will. 'He intended to write to you and Viv and Micky telling you that his reason

(*Above*) Wootton Lodge in Staffordshire where Diana and Mosley lived from 1936 to Christmas 1939. Desmond Guinness and Diana's dog Rebel are on the lawn.
(*Below left*) This formal portrait of Diana with her two Guinness sons, Jonathan and Desmond, is dated December 1937. Diana had been married to Mosley since October 1936, but the secret was kept until November 1938, when their son Alexander was born.
(*Below right*) Mosley at Wootton, playing with Alexander. Oswald Mosley was known to his friends as Tom; to Diana he was Kit, as her brother was called Tom.

Kit & Alexander

Diana, April 1939.

Holloway, where Diana was imprisoned in June 1940.

VISITING ORDER

(VALID FOR 28 DAYS ONLY)

H.M. Prison Holloway.

13-11-1943.

Reg. No. 5433 *Name* Lady Mosley

has permission to be visited by Lady Redesdale,

Swinbrook, Oxford.

1 *The visit to last only* 30 *minutes* 15 in Sat 10.30 to 11.30 a.

2 *Visitors admitted only between the hours of 1-30 p.m. and 3-30 p.m.*

3 *No visit allowed on Sundays, Christmas Day, or Good Friday*

4 *Such of the above-named friends as wish to visit, must all attend at the same time, and produce this order*

GOVERNOR

Permit to visit Diana in Prison

Tom Mitford (Diana's beloved brother, seen here with Derek Jackson, Pam's husband) was killed in Burma just before the end of the war.

Mosley and Diana, with their sons Max and Alexander, at Crowood where they farmed after the war.

After their release from prison, the Mosleys were not permitted to travel abroad. They spent August 1947 at Inch Kenneth, the Redesdale's island off Mull in Scotland.
(*Above*) Diana with her mother, Lady Redesdale, and her youngest son, Max.

(*Above*) They were denied passports, but could sail in their ketch, *Alianora*.

(*Right*) With Alexander, Formentor, July 1949.

Diana at forty.

Diana on the Lido, 1954.

BUF reunion: Mosley meeting supporters in an East End pub in the 1950s.

Lady Redesdale with her goats on Inch Kenneth.

Mosley and Diana at a literary luncheon to celebrate the publication in 1977 of her autobiography, *A Life of Contrasts*. Christina Foyle on left.

Diana coming out of her first floor drawing room on to its balcony, at the Temple de la Gloire, her home in France.

Diana and Mosley in the Temple garden, overlooking the lake.

Dining with the Duchess of Windsor: Mosley at seventy-eight.

Diana in her nineties.

for doing so was not that he loved you less than them but because he thought it equitable, as you inherited from your mother, whereas I have nothing to leave to my sons. When he told me about this I urged him to write to you three but he obviously never did so.' Under French law, the three older children would have been entitled to a portion of his property unless they renounced it. All three did so. Mosley's beautiful clothes – too small for Aly, too big for Max – went to Emmy's brother.

Diana's grief was wild, overwhelming, endless, impenetrable. For months her sister Debo thought she might commit suicide, so total was the blackness in which she was shrouded, so complete and utter her despair. Weeping, she told Nicholas, 'I sometimes feel that the day that ruined my life and your father's life was the day I met Hitler.' Nicholas, immensely admiring of this truthfulness, attempted to comfort her. 'It wasn't YOU who ruined his life,' he said. 'You brought him great happiness. In any case, he was fifteen years older than you – what he did he did himself.' Good sense, the love of her family and the constant support of friends finally pulled her back from the brink of killing herself.

At the beginning of 1981 she faced a task she had been dreading: the sorting of Mosley's papers. Most were stored in the cellars of Lismore, where they had lain since the burning of Clonfert. She invited Nicholas to come with her and Alexander, so that he could look at them and take what he required for his book. Sack after sack was brought up and spread out on the billiard table; Nicholas took most of them, with Diana's blessing. 'Darling Nicky, I know you'll do the book wonderfully,' she said. To Jonathan she wrote, 'Nick is very fond of Kit. Although his book will not be quite what I'd have written, it will nevertheless squash some of the stupider rumours and ideas. And the fact that it's written by someone who doesn't agree with him politically will make it more effective.'

That spring and summer, Nicholas went again and again to Orsay, staying in his father's old room, to talk to Diana about every aspect of his father's life. One day she put a package of letters in his hands. They were his mother's letters to his father. Diana had discovered, after his death, that Mosley had always kept Cimmie's letters by his bed. 'I don't know what to do about these,' said Diana, 'but I think you ought to have them. They're awfully sad – but they do show I wasn't the first.'

At the end of June, Diana felt able to go and stay with an old friend, Lord Lambton, in Tuscany. She took the train to Florence, but unthinkingly did not lock the door of her sleeping compartment. For safe keeping, she had put all her jewellery in her handbag; during the night, this was rifled and her jewellery taken along with all her money. Penniless, she took a taxi to the Excelsior in Florence, where Lord Lambton was well known, and breakfasted there as she awaited his arrival, brooding sorrowfully on the loss of her diamond brooch, the last of the Mosley family jewels and a precious remembrance of her husband. 'Then I heard Kit's voice,' she wrote in her diary, 'saying as he always did: "Those things don't matter in the least. What matters is people, and love. You must never let the loss of material things get you down."'

Hearing the beloved voice affected her so deeply that out sightseeing later, when she felt faint, she mistakenly dismissed this as the after-effects of grief.

In August Diana, Alexander and Charlotte went to stay with Max and Jean in the south of France. Walking back to the car from dinner at a restaurant in the village, Diana fell flat on her face. Spells of giddiness followed. More alarming still, she found that her body would not obey the commands of her brain. She had volunteered, as her daily holiday 'chore', to clear away the simple lunch while the others went off for a siesta, but she found that washing up the few plates and glasses and putting them away took her two hours instead of ten minutes.

Worse was to come: at Orly, somehow becoming detached from Alexander and Charlotte, she was found wandering dazed and confused at the other end of the airport.

'I am in a very poor way,' she wrote on 23 August in handwriting that was barely recognisable as hers,

> lost all balance, can't do my hair it just falls down in a tangle. It takes me three hours to dress, I can't find the sleeve to put my hand in. Can't find my mouth with food but try to stick the food in my eye. If I read on the sofa I find myself sitting on the floor. Charlotte found me a doctor – there aren't any in August – and I am having injections, and about 20 different pills. Results so far nil. Do be sorry for me . . . I still can't believe Kit has gone.

Her family realised it was an emergency. Max, insisting that his mother be flown to London to his friend Professor Watkins, a brain surgeon, chartered a private plane. From Biggin Hill she was taken by ambulance to the London Hospital. She was operated on at the beginning of October by Professor Watkins, who removed a large brain tumour, which proved to be benign. She was kept icy cold, without food or drink, for two days. As she swam slowly back to consciousness, she heard the hospital chaplain ask if she would like communion, followed by Debo's voice saying indignantly, 'She's not even allowed Ovaltine so there's no question of wine.'

Physically, she recovered amazingly quickly. Visitors, from Joy Law to Lord Longford ('He thinks I'm Myra Hindley,' muttered Diana), were soon allowed. Emotionally, she felt unbearably weak. 'I feel I am not fit for human company,' she wrote to Nicholas. 'I am *so* depressed. Longing for the Temple and yet what is in the Temple now that he isn't there? I have completely lost the will or desire to live.'

Arrival at Chatsworth proved the turning point. 'Within minutes I felt completely different. The terrible stale air and general hideousness of the hospital made me so miserable the nights seemed endless. I now feel as if I've come back from the dead. Just looking at beautiful things makes me cry.'

After ten weeks she went home.

Diana's illness altered her appearance. Her weight fell from her lifelong nine stone nine pounds to seven and a half stone; when her hair, which had been shaved off for the operation, grew back, she kept it short. The most dramatic change of all was also the most welcome. After the operation, the migraines that had been torturing her for years vanished completely, never to return.

She would have suffered them again willingly if it would have brought Mosley back. 'Nothing seems worthwhile to me now and I just grieve over Kit and wonder and wonder whether if I'd been cleverer I could have kept him going,' she wrote in her diary on 5 September 1981. And to Robert Skidelsky: 'A saying. L'homme est un animal inconsolable mais gai. That exactly describes me. I can be gay at moments but I shall never never get over the loss of Kit. Somehow when it comes the end is so terrible in its finality.' It was particularly so for Diana who, as an atheist, was denied the consolation of any belief that they might ever be together again.

The Temple had become like Verlaine's 'vieux parc, solitaire et glacé'. The male swan had died; putting his head beneath his wing, he was found like a great white egg on the edge of the lake. The melancholy sight of the female floating solitarily on the lake only served to reflect Diana's own loneliness. On 3 December 1981 she wrote to Nicholas, 'Darling Nicky, One year today but for me it really seemed like the night before last, because that was the terrible night.'

Nicholas was still the one to whom she wrote most intimately. 'I cry much less now,' she told him on 29 September 1981. 'I can't help it at times. It's so awful not to hear him say: "Percher, you're all the world to me."' And at the end of January 1982, a short note said simply, 'Darling Nicky, I miss you terribly.' It was the last time she was to write him a loving letter.

In 1982, before publication of the first volume of Nicholas's biography of his father, *Rules of the Game*, he sent the typescript to Diana with a note saying, 'Everything here has come out of our conversations over the last year. I don't think there's anything in it you'll say isn't true, even though you may not like everything.'

Diana's first response was enthusiastic. 'Darling Nicky, I have just finished your excellent, fascinating, funny, at times unbearably tragic book. Though if you don't mind my saying so, I don't think you've quite got the whole story of Kit's wonderful optimism . . . I realise he had faults but they are puny in relation to his virtues and talents. All fond love, do come over.' There followed eight pages of minor corrections, mostly factual.

'I am so moved by your letter, I don't know what to say,' replied Nicholas, adding that he had made all the corrections 'and as the book only takes us to 1933, the "wonderful optimism" comes later and will be in the second volume.'

Four days later, on 17 March 1982, another letter from Diana arrived. It was wild, bitter and miserable:

> Now for a more considered opinion [it began. The book] is a bitter attack on the man, his politics, his whole being, such as no enemy could have written, because naturally something from 'inside' is so much more damaging. My own feeling is intense disappointment. When I think that . . . his most intimate letters to your mother and hers to him should be published less than two years after his death! How terribly she would mind. I am sure you remember her well enough to know. All their rows, her jealousies and naggings, his duplicities (all inseparable from infidelities but which, because people want to hate him and because she died, become a sad prelude to tragedy) exposed and by their own son! . . . You have painted the portrait of a brute and a fool and not of the amazing person your father was. In any case there's little to be done because it's the whole bitter tone throughout the book that is so painful. IT IS BRILLIANTLY DONE WHICH MEANS IT WILL BE READ FOR ITS EXCELLENCE as well as for the scandal and the tragedy . . . 'Taste' is a matter of opinion, I suppose, but it does seem distasteful to the last degree to publish their poor little letters, for the whole world to gape

> at . . . I started this letter 'darling' and to me you have been a darling but my poor wonderful real darling – you have kicked him below the belt over and over again.

Scarcely had Nicholas recovered from the shock and pain of this attack than another letter arrived on 21 March. This one had no opening endearment but simply started:

> Thank you very much for taking Bryan out of your book. He would have minded so dreadfully. As to me, I used to quite like a fight. All my fights, all my rude letters to people, were about Kit. Frank [Longford] was remembering last night I wrote one to him . . . If I fight you now, it is with one hand tied behind my back because it will be thought it's because of a 50-year-old scandal. You know, and I know, that I don't mind a bit what is said about me. But I've lost the will to fight. I long and long not to. I thought Kit was absolutely safe in your hands. You gave no hint. Unluckily it is a brilliant book with a dramatic and tragic story which will give unending joy to all our enemies. The whole family will suffer. Those poor little letters! How could anyone, least of all their son, give them to be gloated over. Oh Nicky! How I wish I'd died when they operated on my brain. Any other biographer would of course have had no access to such private and painful things because I should have read everything and put safely away what was impossible to print now.

It is easy to understand Diana's feelings about the Mosley–Cimmie letters ('they are FAR the worst part' she wrote in a second letter on 17 March). They are loving and intimate, often written in baby talk, sometimes using various animals as pet names. Embarrassing though they may be to read after the passage of years they are, in many ways, typical of their era, with its somewhat whimsical sentimentality and its passion for animal nicknames such as 'Bunny', 'Piggy' or 'Kitten'.

It was useless for Nicholas to point out that Diana herself had handed over these letters freely – as he thought – for him to draw on. Or that both of them had always known that, as Robert Skidelsky had written Mosley's official, political biography, Nicholas would write something more personal and intimate, showing his father as a human being as well as a politician. Or that, as a biographer, Nicholas's instinct was to set down everything that could contribute to the portrait of his father as he was.

For Diana, none of these considerations clouded the narrow, piercing focus of her judgement. Her tunnel vision allowed her to see only what she thought of as Mosley's supreme virtues. No compromise was possible. Her violent shift of opinion from praise to venom was, in aggravated form, an example of the split in her personality that had taken shape over the years. At her core she was essentially honest, but over-riding even her deepest feelings was her worship of Mosley. Denying this would be to deny the huge emotional and psychological investment of the years of increasing unpopularity, the suffering of prison and of isolation in a foreign country and would also allow the possibility that he had been wrong. Where Mosley had been able to admit (to his son Micky): 'I've had an extraordinary life and I've blown it all,' Diana could not admit this for him. With his death, the hardening carapace of blind loyalty closed over her head. Against this imperative, nothing else counted. 'I once thought I would have to cut Jim Lees-Milne out of my life because he wrote something rude about Kit in a book,' she told Robert Skidelsky. 'Then typically Kit simply didn't mind in the very least, so I wasn't deprived of a lifelong friend.'

Nicholas's book had also brought Diana face to face for the first time with the acuteness of the pain she had caused Cimmie. 'Until this moment I never realised how much she suffered,' she had written to Nicholas on 10 March 1982. 'I am not pretending it would have made me give him up, but in small yet important ways it would have made me behave differently.'

The shift in her attitude was dramatic and irreversible. Against what she saw as Nicholas's treachery neither their years of mutual devotion nor his financial and emotional generosity counted in the slightest. The high priestess had taken over and terrible was her wrath. Nicholas was transformed into the 'duplicitous bastard' or 'my vile stepson', blasphemer against a secular deity whose godlike qualities and oracular pronouncements were not to be challenged.

The reviewers looked at it differently. Nicholas's book was a love story, a family story, a son's story, said one. 'A brilliant two-volume memoir,' said another. 'Lucky the father who has such a son to plead his cause,' said the *Sunday Telegraph*. *The Times* was more guarded. 'One cannot help feeling that, for all his good intentions, Mr Mosley may have dropped another bucketful [of mud] over his father's head.'

Before the second volume, *Beyond the Pale*, was published, Nicholas again sent the typescript to Diana. Again she reacted furiously, accusing him of dishonour in including some of Mosley's letters to her in prison.

'As you well know, I gave them to you at a time when I wanted to help you in every way to write the book that you pretended to me you wanted to write. As soon as I read your first degraded volume I asked for them back.' She did not want to go to law, she said, 'but please make it clear that I strongly object, and that being associated in any way with your book is repugnant to me'.

Although Nicholas included a statement to that effect, nothing softened her attitude. On 30 June 1983 she wrote to Robert Skidelsky. 'I have read Nicky's ghastly effort. It is marginally less spiteful than Volume I . . . I am sorry to say he is the lowest and most dishonourable of men, in addition to being insanely jealous.' The volte-face was complete. She made this clear publicly, in an interview in the *Daily Express* on 4 December 1983. 'I shall never speak to him again, though I used to be fond of him.'

She had dedicated her life to Mosley, now she became the keeper of the shrine.

Chapter XXXVI

To the end of her life, Diana remained faithful to Mosley's beliefs. She likened him to the prophet crying vainly in the wilderness. 'Of course Kit was always a lonely voice,' she wrote to Robert Skidelsky in 1987. 'Think of the blackamoors here in their millions. I suppose a few lunatics like Huddlestone and Nicky are pleased but a referendum would be 99 per cent against. Kit was ahead always.'

In a sense, she *had* to continue believing in him. What she had done she had done for him. If Mosley was the one true voice, this gave meaning to the sacrifices she had made. The alternative was unthinkable: that it was for a flawed god she had suffered the rupture of family relationships, the humiliation of social isolation and, above all, the anguish of losing her children's babyhood while she was detained in Holloway. Thus she could say (and mean it) only years before she died: 'I'm so proud to have been in prison, to have done however little it was to try and stop that ghastly war, which I thought was unnecessary and intensely wicked. All that has happened was foreseen by my husband. So I don't feel prison as any stain on me – quite the opposite. I'm very proud that we did what we could. Of course it wasn't much, unfortunately.'

She remained convinced that the war had been unnecessary and that the threat of invasion in 1940 was nonsense. In a letter to James Lees-Milne (in December 1989) she wrote:

> I *knew* there would be no invasion of England, despite our helplessness when all weapons had been thrown away in France. The reason was so obvious if one knew Hitler as I did. Either the invasion would have failed, because of our navy, or it would have succeeded. Failure a disaster for the Germans, success equally a disaster. Hitler's dream rested upon Anglo-German friendship, even alliance. Instead he would have had a sulky, furious humiliated population on his hands and an

> English government in Canada or US, hotting up the New World to war against Europe. Equally, utterly out of the question that I for example would have collaborated. He would deeply have despised one for anything of the kind. What bad, what hopelessly bad psychologists our rulers were! I don't think anything of this occurred to them.

To questions about what would have happened to the Jews in Germany and German-dominated countries if war had been avoided and a settlement negotiated (as some revisionist historians now argue would have been possible), she would reply that there would of course have been 'a solution'. 'What Hitler wanted was to get rid of them. They could have had a national home. There are enough unoccupied places in the world for such a solution, as long as other countries would co-operate. Then they would have been told, "Either go, or stay at your own risk."' It remains extraordinary that a woman of such high intelligence could talk such heartless nonsense.

Her views on Hitler and Germany never changed. She had gone to prison because her husband was the British fascist leader, but she had remained there because of her own defiant boast of friendship with Hitler in front of the Advisory Committee. Many sophisticated and clever people had been duped by Hitler before the war. Many others had, often despairingly, believed or hoped until the last possible moment that peace could be preserved through appeasement or negotiation. But virtually all had condemned without hesitation the monstrous horrors of the concentration camps and the Final Solution that emerged in the war's aftermath through the irrefutable evidence of film, newsprint and gradually revealed documentation.

Not so Diana. For a long time she refused to believe in the reality of the Holocaust. She would quote *Falsehood in Wartime* by Lord Ponsonby of Shulbrede (published in 1928), a book largely devoted to nailing hearsay and apocryphal reports of German atrocities in the First World War – many, such as the stories of Belgian children with their hands cut off or British prisoners forcibly tattooed with the German Eagle, already refuted in the British press. 'Both sides are as bad as the other,' she would say. Until the end of her life she argued that the number of six million killed in the death camps was far too high an estimate. 'Oh darling,' she would say to the friend who asked her, 'awful, awful things happened on both sides in the war and we don't really know the truth.'

She would point out that Stalin was responsible for more deaths than the concentration camps. She would ask why the Jews did not leave Germany, ignoring the fact that most Jews, despoiled of property and livelihoods, simply could not afford to leave, and that millions were rounded up and sent to the gas chambers from countries overrun by the Nazis. Only finally, if pushed into a corner, would she condemn, usually with the words, 'Murder is always wicked. Whether Hitler murdered one person or six million, it is equally wicked.' Even close friends learned to avoid certain subjects. Those who loved Diana thought the reason she stuck so stubbornly to her views was because she, who hated cruelty, could not bear to look at the reality of what she believed in.

Few things roused stronger feelings than the interview with her on *Desert Island Discs* on 26 November 1989. It had to be postponed three times – from 1 and 8 October and 19 November – because of protests from the Jewish community. The hostility towards her was such that even her choice of 'A Whiter Shade of Pale' as the only non-classical piece of music she would take to a desert island 'because it is based on a piece by Handel' was interpreted as being racially motivated.

Diana did not help herself. Typically, she seized the opportunity to defend her husband, though with a breathtaking disingenuousness. Speaking of the pre-war fascist uniform she said, 'The Black Shirt cost one shilling, so the unemployed could afford it. It was such a success that an Act of Parliament had to be passed to prevent them wearing it.' 'But surely,' replied the interviewer, Sue Lawley, 'the Public Order Act was passed for a different reason – the violence in inner cities.'

When asked if Mosley was anti-semitic, Diana's response was: 'He did not know a Jew from a gentile but was attacked so much by Jews both in the newspapers and physically on marches in places like Manchester that he picked up the challenge. Then a great number of his followers who really were anti-semitic joined him because they thought they could fight their old enemy.' To the charge that Mosley had, after war had been declared, advocated a negotiated peace with 'a system of government and a cause that had become this country's greatest enemy', Diana responded, 'Well, you see, it was the phoney war.'

She was, as usual, devastatingly frank. It had served her ill in front of the Advisory Committee more than fifty years earlier; now, it was to damage her again. When asked if she regretted her friendship with Hitler, she replied:

> I can't regret it. It was so interesting and fascinating. Of all the well known and famous people I have known in my life I couldn't regret having known him because it gives one something by which to measure the nonsense that gets written. I admired him very much. He had extremely mesmeric blue eyes and also he had so much to say. He was so interesting and fascinating, and perfectly willing to talk. You don't get to where he was just by being what people think he was.

What many saw as a wilful refusal to believe established evidence clouded even her condemnation of the Holocaust. 'Having been in Germany a good bit it was years before I could really believe such things happened.'

'And do you believe it now?' asked Miss Lawley.

'I don't really, I'm afraid, believe that six million people were killed,' said Diana, her languid Mitford drawl giving the words a chilling overtone. 'I think this is just not conceivable. It's too many. But whether it's six or whether it's one makes no difference morally. It's completely wrong. I think it was a dreadfully wicked thing.'

After the broadcast, there was a strong reaction from the public. All of it was unfavourable.

Diana cared little. Beside the one great cataclysmic loss she had suffered, the opinion of the world – which she had never much regarded – meant nothing. She had the constant, loving presence of Alexander and Charlotte, who would drive down from Paris to the Temple for lunch most weekends. Her other sons and grandchildren came to stay and there were visits to England, where she stayed with Pam or Debo. Once again, the relationship between the sisters had shifted: it was now Debo, the least political of the family, to whom she was closest. The ten-year age gap had shrivelled to nothing, and letters and telephone calls passed constantly between them. Sometimes younger friends such as Jane Lady Abdy came to stay, or there were summer visits to Lord Lambton and Harold Acton.

Old friends were a diminishing band, often cherished for their feelings for Mosley. Derek Jackson had died in 1982. 'I am desperately sad,' wrote Diana. 'The loyalty and love he had for Kit was adorable and he was such a truly brilliant man.' Gaston Palewski's death in 1984 had meant the loss not only of a neighbour but of a friend. 'His great charm was that, like Kit, he loved life,' she had written to Robert Skidelsky. 'I must go over and see his widow, it is the custom of the country. They rushed over to me when darling Kit died.'

Distressingly, however, the loss of Gaston was followed by a rupture with his widow Violette. Selina Hastings's biography of Nancy, published in 1985, naturally gave full weight to Nancy's passion for Col – and described her wartime love affair with him. Violette Palewski realised that Selina Hastings must have learned much of this open secret from Diana and was outraged. Diana's regular Sundays at the beautiful Le Marais came to an abrupt end.

The routine of Diana's days seldom varied. Sleep had become difficult. She woke at any time from three to five-thirty, often with the brooding half-remembered shadow of some dreadful dream. Her first thoughts were of Mosley and their life together. She would go downstairs to the kitchen, where her breakfast tray was waiting, make coffee or chocolate and toast, to be eaten with Emmy's jam, and take the tray back up the narrow, twisting staircase to her bedroom. She would remain in bed till ten-thirty, listening to the BBC World Service, writing letters and chatting on the telephone. Her most regular friends, such as Walter Lees, rang between six and seven.

At ten-thirty she would dress and, unless the weather made it impossible, go into the garden, in summer often swimming in the long narrow pool. Meals were invariably formal, served by Jerry in white jacket, though she ate little. When she was on her own, lunch might be only a piece of melon and a thin slice of Parma ham. Evenings were usually spent in her own sitting room with its wood fire and were devoted mainly to reading – old favourites including Saint-Simon, Goethe, Proust and Balzac, or books that had been sent for review – interrupted only by dinner at eight.

Her deafness steadily increased. Dinner parties of any size became impossible ('all I hear is roaring, like lions'), though she went every year until 1997 to the dinner in a London pub organised by the Friends of Oswald Mosley on his birthday, 16 November. This band of followers, for whom fascism and OM – as they called the Leader – were virtually a religion, were fellow acolytes whose loyalty ('We will be true to OM till we die') was nectar to Diana. '110 people, each one more adorable than the last,' she wrote in 1987. Every speech at these dinners expressed in some form the view that 'OM was the greatest man this country has ever known or will ever know', as well as echoing Mosley's tenets on Europe – and on race.

Mentally, Diana remained as acute and energetic as ever, writing the occasional article, reviewing books in her elegant, laconic and astringent style for *Books and Bookmen* and the *Evening Standard* and committing historical dates to memory as a mental exercise. 'People are leaving schools after ten years unable to read,' she commented to Robert Skidelsky, to whom she now wrote almost as often as she had once corresponded with Nicholas. 'What do they do with all those thousands of years, I wonder? Sex, I suppose, but you no more have to learn that than how to eat a Mars Bar.'

Mosley's absence was a constant, aching void. 'I long each day for OM to be here,' she wrote to her prison friend Louise Irvine in 1990. Other links were snapping steadily. In July 1992, her first husband Bryan, now Lord Moyne, died. At his memorial service on 15 October Lord Longford gave the address; when he ended with the words, 'Bryan had the sweetest nature of any man I have ever known,' there was a rustle of agreement through the church.

A few days later, there was a loss nearer home: the death of Emmy. For Diana as well as Jerry, it was a bitter sorrow.

At the end of the year Nicholas sent her an extract from the manuscript of his autobiography, *Effort at Truth*, asking if he could quote from her letters to him. He did not expect her permission but it was, he felt, a chance of breaking the log jam of their estrangement. When she refused, he replied that he understood, and would paraphrase them. There was no reply from Diana.

A month after her elder sister Pam's sudden death after a minor illness, in April 1994 ('I always thought,' said Diana sadly, 'that I would go first') she wrote to Nicholas again, prompted by the discovery of a letter she felt belonged to him. 'Nicky,' her letter began baldly:

> Enclosed turned up in Kit's dreaded 'papers', which I am packing up for Birmingham. I've been in England for Woman's funeral, so sad. It made me realise I haven't got long myself, & I want to say if I've been unfair I am truly sorry. I ought to have read your Vol. II [of Nicholas's biography of his father]. I felt physical revulsion and gave way to it.
>
> But however you look at it, a book which induced from a reviewer the exclamation 'What a shit he was!' about an exceptionally brilliant & deeply loved man *can't* have got it right. Can it? Don't bother to answer. D.

Nicholas's reply to this note elicited the nearest she would get to a reconciliation – a postcard written on her eighty-fourth birthday. 'We both love him in our mysterious ways perhaps.'

But within three months the enmity was back. Nicholas had been approached about a television film based on his book about his father and the film-makers were anxious to talk to Diana too. 'No, I couldn't have anything to do with a television drama about your father based on your book as I'm sure you know already,' she told Nicholas. 'I'm sure the book is well-written but that's irrelevant. The telly drama is bound to be absurd & embarrassing but I've really stopped minding such things & am so happy not to have a television set here. It is such an oasis. Love, Diana.'

A month later, just after a small melanoma had been removed from the side of her nose, she was sharper still.

> Dear Nicky, lying bruised and battered in bed with a nose that looks like one of Francis Bacon's Cardinals' noses, I couldn't help laughing about your letter. The idea of you leading the benighted television men to the 'truth' about Kit at the Shaven Crown or anywhere else, is somehow so lovely. Don't you remember that my whole quarrel with you years ago, when his death was still such agony to me, was precisely over that? What I read of your book convinced me that you have no conception of truth. It's something that has been left out of your make up.

When Nicholas replied that the main issue she had taken with his book was the printing of his parents' love letters, this drew another stinging rejoinder.

> Oh, Nicky, how *can* you pretend that the only thing I hated in your book was the vulgar insensitivity of printing those very private little notes. I hoped you might delete them but short of scrapping the whole book & writing it again with a decent regard for truth, which was obviously not on the cards, there was little to be done. You tried to make Kit a trivial rather caddish 'playboy'. He was a philanderer and naturally I don't mind you saying so, but it was a tiny part of a life given up to ideas & work – ideas which are every day proving themselves to have been excellent & possible (the art of the possible) even after the disastrous war which half destroyed our wonderful Europe.
>
> The entire book (or as much as I read of it) is a lie. Your untruthfulness is all-embracing.
>
> I am grateful for a few details you removed. But it is the book and its unfairness that I minded. I mind much less now, it seemed so terrible to launch your hateful attack so very soon after he died . . .
>
> About health. I'm very tired and very hideous and stone deaf nearly, but the beauty of my surroundings and the kindness of many saints keeps me going, just at the moment. Golden days.
>
> P.S. I admit, about your book, I'm going more by reviews than from careful reading of the text.

Only in the signature, 'Love, Diana', was there a hint of softening.

Nor was there any real regret when, after a brief illness, her sister Jessica died on 23 July 1996, though she was greatly upset at being 'quoted' in a newspaper to which she had never spoken as saying, 'I am completely indifferent.' She had never responded to any of Jessica's public attacks on her; to appear to be so callous immediately after her sister's death would, she felt, be deeply distressing both to Jessica's family and to Debo.

Only towards the end of Diana's life did she go some way towards admitting the iniquities of the man and the regime she had so warmly supported. 'You asked about the war,' she wrote on 11 March 1997.

> To tell you the truth the horrible atrocities of all the belligerents make me miserable if I allow myself to think about them. I hate cruelty.
>
> The greatest atrocity of all is war itself. Paradoxically, nuclear weapons have given us 50 years of peace. I know there are horrors in Africa and in Palestine, and I know men are wicked. I know, too, that we are incredibly lucky to live in France or England and not in Rwanda, Harlem or Mao's China. I don't like the prevalent xenophobia in England – or at least in the English newspapers – it's dangerous as well as stupid.

It was her last word on the subject.

Epilogue

Late in 1997 Debo brought one of Diana's oldest friends, Jim Lees-Milne, over to the Temple. They had one happy day and then Jim became so ill that Debo had to take him back. He went straight into hospital, where he remained until his death a few months later, leaving a large gap in both sisters' lives.

Less than a year later, Diana herself decided to leave the Temple. Though family and friends came to stay constantly and her son Aly and his wife Charlotte drove down to lunch or for the day most weekends, living there was growing more difficult. Everyone who knew the house worried that one day Diana would slip and fall down the spiral wooden staircase with its narrow, polished steps which led up from the ground floor to her sitting room and bedroom, and which she would descend every morning to fetch her breakfast tray and return with it to her bedroom.

For Diana herself there was the knowledge that the faithful Jerry, although well past retiring age, would never leave her while she lived at the Temple – how would she manage without him? – and that, if she wished him to enjoy the cottage in Ireland that she and Mosley had given him, she would have to do something about it herself.

This meant selling the Temple. It was bought for £1 million by an American property dealer; and in February 1999 Diana left the house, with its secluded garden where in the summer Mosley would stroll round wearing nothing but a hat and a pair of shoes, and moved to Paris.

Her new home was an elegant first-floor apartment in the rue de l'Université, a few blocks away from Aly and Charlotte. It was entered through a courtyard and its large windows overlooked the chestnuts of a small square opposite. For anyone else the noise of the traffic might have proved a deterrent, but for Diana, isolated in her cocoon of silence, it was immaterial.

Her new home brought her three happy last years. Almost at once, an efficient, smiling Filipino maid was found; like all Diana's servants, she proved devoted. Best of all, there was a new worshipper at the shrine, a young half-Vietnamese writer and biographer, Jean-Noel Liaut, who soon became an intimate friend. With him, she would sortie forth to exhibitions, galleries and museums. When the intense press interest aroused by the reissue of her memoir, *A Life of Contrasts*, brought numerous interview requests, he was able to penetrate her now almost total deafness to get across the questions asked by journalists.

Diana continued her flow of letters, augmented by faxes – often sent within minutes to those perceived as attacking the views of her late husband. Her conviction that Churchill was 'wicked' to have refused to negotiate with Hitler after he (Churchill) had achieved power in May 1940 never altered. 'I think his politics were completely fatal for England,' she would say.

In 2000 she celebrated her ninetieth birthday with a party at the Crillon for more than forty of her descendants – although the breach with Nicholas was never healed. She survived to become the only person alive who had known both Churchill and Hitler, and writers of all kinds made their pilgrimages to her flat, to be entertained with a delicious luncheon, to succumb to the legendary charm and then, quite often, to go home to sharpen their pens.

At the beginning of August 2003 she suffered a mild stroke. A week later, on 11 August, she died peacefully in her Paris apartment. It was the end of a long life that had dazzled, outraged, horrified, charmed, shocked, beguiled and appalled. Through it all, two things remained constant: her love for the man to whom she was married for almost forty-five years – and her affection for the bloodstained dictator who almost extinguished the lights of civilisation over the Europe she loved.

Select Bibliography

Adolf Hitler, Colin Cross. Hodder and Stoughton, 1973
Adolf Hitler, John Toland. Doubleday, 1976
The Alternative, Oswald Mosley. Mosley Publications, 1947
Another Self, James Lees-Milne. Faber and Faber, 1970

Beaverbrook, A Life, Anne Chisholm and Michael Davie. Hutchinson, 1992
Beyond the Pale, Sir Oswald Mosley, 1933–80, Nicholas Mosley. Secker and Warburg, 1983
Blood and Banquets, Bella Fromm. Bles, 1943
Brian Howard, Portrait of a Failure, Marie-Jaqueline Lancaster. Anthony Blond, 1968
The Brown Book of the Hitler Terror, prepared by the Committee for the Victims of German Fascism. Victor Gollancz, 1933

The Candid Review, conducted by Thomas Gibson Bowles
Carrington, Letters and Extracts from her Diaries, ed. David Garnett. Cape 1970
Cousin Randolph, The Life of Randolph Churchill, Anita Leslie. Hutchinson, 1985

Dairy Not Kept, Bryan Guinness. The Compton Press, 1975
Diary of Beatrice Webb, 1924–43, ed. Norman and Jeanne Mackenzie. Virago, in association with the London School of Economics and Political Science, 1985
Duff Cooper, John Charnley. Weidenfeld and Nicolson, 1986

Eva Braun, Hitler's Mistress, Nerin Gun. Frewin, 1969

Evelyn Waugh, Diaries ed. Michael Davie, Weidenfeld and Nicolson, 1976
Evelyn Waugh, Letters ed. Mark Amory, Weidenfeld and Nicolson, 1980

Fascism in Britain, Philip Rees. Harvester, 1979
The Fascists in Britain, Colin Cross. Barrie and Rockliff, 1961
Friends Apart, Philip Toynbee. Sidgwick and Jackson, 1954
From the Shadows of Exile, unpublished MS by Nellie Driver

George VI, Sarah Bradford. Weidenfeld and Nicolson, 1989
The Golden Age of Wireless, Asa Briggs. Vol II, OUP, 1965
Goebbels, Ralf Georg Reuth. Constable, 1993
The Goebbels Diaries, ed. Hugh Trevor-Roper. Secker and Warburg, 1978
Granta 51. Penguin 1995
Greater Britain, Oswald Mosley. 1932

Harold Nicolson, James Lees-Milne. Hamish Hamilton, 1988
Harold Nicolson's Diaries and Letters, 1930–9, ed. Nigel Nicolson. Collins, 1966
The History of Broadcasting in the United Kingdom, Asa Briggs. Vol. II, The Golden Age of Wireless. OUP, 1965
Hitler, The Missing Years, Ernst (Putzi) Hanfstaengl. Eyre and Spottiswode, 1957
The House of Mitford, Jonathan Guinness with Catherine Guinness. Hutchinson, 1984
Hugh Dalton, Memoirs, 1939–45, The Fateful Years. Frederick Muller, 1957

I Fight to Live, Robert Boothby. Victor Gollancz, 1947
I Know these Dictators, Ward Price
In General and Particular, Sir Maurice Bowra. Weidenfeld and Nicolson, 1964
In Many Rhythms, Irene Ravensdale. Weidenfeld and Nicolson
In the Highest Degree Odious, Brian Simpson. Clarendon Press, Oxford, 1992

Knights Cross, A Life of Field Marshal Erwin Rommel, David Fraser. HarperCollins, 1993

The Letters of Nancy Mitford, ed. Charlotte Mosley. Hodder and Stoughton, 1993
A Life of Contrasts, Diana Mosley. Hamish Hamilton, 1977
The Life of Randolph Churchill, Anita Leslie. Hutchinson, 1985
London at War, Philip Ziegler. Sinclair-Stevenson, 1995
Loved Ones, Diana Mosley. Sidgwick and Jackson, 1985
Lytton Strachey, Michael Holroyd. Chatto, 1994

Memoirs, Israel Sieff. Weidenfeld and Nicolson, 1970
Memories, C.M. Bowra. Weidenfeld and Nicolson, 1966
My Answer, Oswald Mosley. Mosley Publications, 1946
My Life, Oswald Mosley. Nelson, 1970

Nancy Mitford, A Biography, Selina Hastings. Hamish Hamilton, 1985

Of a Certain Age, Naim Atallah. Quartet Books, 1992
Oswald Mosley, Robert Skidelsky. Macmillan, 1975
Out of Bounds, Esmond Romilly. Hamish Hamilton, 1935

The Past Masters, Harold Macmillan. Macmillan, 1975
The Pebbled Shore, Elizabeth Longford. Weidenfeld and Nicolson, 1986
Personal Patchwork, Bryan Guinness. The Cygnet Press, 1986
Potpourri from the Thirties, Bryan Guinness. The Cygnet Press, 1982
The Power Behind the Microphone, P.P. Eckersley. Jonathan Cape, 1941
The Prince and the Lily, James Brough. Hodder and Stoughton, 1975
The Pursuit of Love, Nancy Mitford. Hamish Hamilton, 1945

The River Watcher, Hugo Charteris. Collins, 1965
Rules of the Game, Sir Oswald and Lady Cynthia Mosley, 1896–1933, Nicholas Mosley. Secker and Warburg, 1982

Sacheverall Sitwell, Splendours and Miseries, Sarah Bradford. Sinclair-Stevenson, 1993
Shoe, the Odyssey of a Sixties Survivor, Jonathan Guinness. Hutchinson, 1989
Some Account of Life in Holloway Prison, Pat Collins, Roger Page, Kathleen Yardsley Leyland

Unity Mitford, a Quest, David Pryce-Jones. Weidenfeld and Nicolson, 1976

Vile Bodies, Evelyn Waugh. Jonathan Cape, 1930

Water Beetle, Nancy Mitford. Hamish Hamilton, 1982
Whitehall Diary, Vol II, 1926–30, Thomas Jones, ed. Keith Middleton. OUP, 1969
Who Lie in Gaol, Joan Henry. Gollancz, 1952

Author's Notes on Sources

The vast bulk of my information about Diana Mosley's life came directly from her, in hundreds of taped interviews, taken when I saw her or stayed with her in France.

More came from her sisters – Pamela Jackson, the Duchess of Devonshire and Jessica Mitford – and Diana's sons, stepdaughter and stepsons.

There is valuable background information in Jonathan Guinness's *The House of Mitford*, as in other books written by and about the Mitfords, such as Diana's own autobiography *A Life of Contrasts*. (See Select Bibliography.) Several people whom I interviewed were illuminating about the Mitford and Mosley families.

Diana Mosley generously gave me access to her private papers, letters and diaries.

Chapter One

Interviews with Diana Mosley, Jessica Mitford, the Duchess of Devonshire and Pamela Jackson; Derek Hill; Viola Erskine. *Water Beetle* by Nancy Mitford. *Life of Contrasts*. Interviews with James Lees-Milne.

Chapter Two

Diana told me about her time in Paris. She documented many of her adventures there in letters to James Lees-Milne, held at Yale University. Pam described Swinbrook very fully, including the bread handed out after matins at Swinbrook church. Details about the Redesdale pews can be found in the church. Diana described her feelings for Randolph Churchill in a letter to Robert Skidelsky. Information about Diana's dance is in correspondence from Sydney. A letter from Diana Churchill talks about 'the Prof'.

Chapter Three

Bryan's feelings for Diana are made very plain in his letters to her and his own memoir, *Dairy not kept*, is a useful source. Jonathan Guinness gave me a vivid description of his father. Lord Longford, who was Captain of Bryan Guinness's house at Eton, then a contemporary at Oxford and later in the House of Lords, also provided me with much material. Frances Partridge talked to me about Bryan, and Lady Longford paid tribute to his skill as a waltzer. Bryan included much family information in his long letters to Diana.

Interviews with Diana, Pam, Jessica; Diana's great friend Lady Celia McKenna (Cela Keppel); Lady Longford and Lord Longford. Bryan Guinness includes the story of the girl he would not kiss in *Dairy not Kept*. Bryan Guinness's sonnet *Sunrise in Belgrave Square* is included in his book *Potpourri from the Thirties*. The sonnet was written in 1932 and appeared in its earlier form in *Under the Eyelid* (Heinemann), then recast before appearing in the *Collected Poems*, 1927–55 (Heinemann). Bryan described events and his feelings after kissing Diana and after her acceptance of his proposal in letters written immediately afterwards.

Chapter Four

Life with the Malcolms of Portalloch is amply described in *The Prince and the Lily*. Much of the story of the courtship of Bryan and Diana is told in the passionate and adoring letters he wrote to her at the time. He wrote her a long description of his interview with her father. The passage from Bryan Guinness's diary appears in his book *Dairy Not Kept*. I also drew on interviews with Diana herself.

Chapter Five

The story of the Pocket Adonis is from *Dairy Not Kept*. The Tropical Party is described in the 24 July 1929 issue of the *Bystander*. Lady Celia McKenna, who spent much of her time with Diana and Bryan after the death of her mother, gave me a very full description of the life at Buckingham Street, as did Diana in conversations with me. Diana described early married life and how Bryan's father settled on the Bar for him. There is more about Diana and Bryan's life together in *Potpourri from the Thirties*. Letters written to Diana by her friends fill in many details of this picture.

Chapter Six

The parties are described, thinly disguised, in *Vile Bodies*. The Bruno Hat exhibition is mentioned in many volumes of the time; there is a good description of it in M-J. Lancaster's biography of Brian Howard. Diana also told me about it. The Duchess of Devonshire described Biddesden; there is another description of it in *Potpourri from the Thirties*. Elizabeth Lady Moyne told me much about its history. Diana told me how she had furnished it; she and her sister Pam Jackson told me about its haunting. My father-in-law, General Sir Clement Armitage, knew the then Major Clark, who told him of his experiences. Derek Hill described the rich cerulean blue invented by Diana for the buildings at Biddesden. Frances Partridge gave me her impressions of Lytton Strachey, the Strachey voice ('at one time the whole of Cambridge caught it'), Diana's effect on Lytton and Carrington, and how they would visit Biddesden for walks, picnics or riding. She also fully described Diana's visit to Ham Spray, at which she was present. The *Daily Express* of 2 April 1930 describes the April Fool played by Evelyn Waugh on Diana, and the godparents at the christening of her son Jonathan. Lytton's letter about staying at Knockmaroon is quoted in Michael Holroyd's biography of him. James Lees-Milne told me what it was like staying at Biddesden. Diana described the arrival of Desmond in a letter to Lytton Strachey (in the British Library). The verses from the poem *Love's Isolation* are taken from *Potpourri from the Thirties*. The poem is also in *On a Ledge* (Lilliput Press). Earlier versions appeared in the *Collected Poems* and in *Twenty-Three Poems* (Duckworth).

Chapter Seven

Frances Partridge's memories, related to me, contributed to the story of Carrington's death, well charted in various biographies. Carrington's letters are in the British Library; her feelings after Lytton's death are recorded in her Commonplace Book. Diana told me in detail about Bryan's loaning of his gun, and his feelings when he learned of Carrington's death.

Chapter Eight

Many people, including his son Micky, told me about the physical and emotional impression made by Mosley – one of the most vivid these descriptions was by the painter Derek Hill, another was from James Lees-Milne, whose uncle had married Mosley's youngest aunt, and who canvassed for him during the 1931 general election. *The Past Masters* gives a good description of Mosley's resignation over the Mosley Memorandum. Tom Jones's *Whitehall Diary* contains details of the Mosley Memorandum.

Chapter Nine

Harold Nicolson's diaries, both published and unpublished, talk of the New Party in great detail and charts the swing to fascism, as well as Nicolson's initial enthusiasm for Mosley and the New Party and his gradual disillusionment with both. Irene Ravensdale's diaries describe Cliveden weekends. Lord Longford told me about Mosley at Cliveden and the effect of his certainty at a time when no one knew what to do about the state the country was in. Issues of *Action* show how the New Party gradually swung towards fascism. Georgia Sitwell's diary talks about Mosley and his beliefs. Robert Boothby's autobiography, *I Fight to Live*, is illuminating about his great friend Tom Mosley, as is Colin Cross's *The Fascists in Britain*; so were interviews with Nicholas Mosley, and the book he wrote about his father. The resignation statements of John Strachey and Bill Allen were widely quoted in contemporary newspapers in July 1931. Mosley's account of his first sight of Diana is in his book, *My Life*.

Chapter Ten

Frank Cakebread kindly lent me his privately printed *History and Description of Savehay Farm*. Georgia Sitwell's diary talks of her dinners with Mosley and her emotions; the *Tatler*, in the issue of 18 July 1932, talks of Mosley as having John Barrymore's good looks. Nicholas Mosley described his father's many affairs; both Nicholas and Diana described the bachelor flat in Ebury Street. Lady Abdy talked of Mosley and Diana, and Mosley's attitude to women. Irene Ravensdale wrote an essay on her childhood (in *Little Innocents*). Details of the Curzon balls and feuds over money are all in the Curzon papers in the Oriental section of the British Library.

Lord Longford was illuminating on the contrast between Mosley and Bryan. Lady Celia McKenna described Diana falling in love with Mosley and how this separated her from her family and many of her friends. Lady Butler described how Diana dressed her boys in girls' clothes and gave them Aubusson carpets to crawl on. Lady Celia, who was living at Cheyne Walk, gave a vivid description of the housewarming ball (also described in Diana's own memoir, *A Life of Contrasts*).

Chapter Eleven

Diana told me about events in Venice and at the Lido in September 1932. Nicholas Mosley confirmed the story of Mosley telling Boothby that he would need his room; a letter from Diana to Lytton Strachey adds more detail. Diana described the fight at the Lido.

Pam Jackson, who spent most evenings at Biddesden, was a spectator of the disintegration of the marriage of Bryan and Diana. Lady Celia, who was virtually living with the Guinnesses, confirmed that Diana's love affair with Mosley was unmistakable by the time of the Biddesden ball, which she described fully. Lady Pansy Pakenham recalled that Diana told everyone how wonderful Mosley was. Diana told me that Mosley encouraged her to leave Bryan. Lady Celia, who was not only Diana's confidante but also a great friend of Bryan, confirmed that Diana's leaving Bryan 'nearly killed him'. Bryan's son Jonathan spoke movingly of the effect of the divorce on his father.

The extract from Nancy Mitford's letter is printed in her *Letters*, edited by Charlotte Mosley. Pam Jackson confirmed the Redesdales' horror and fury when Diana left Bryan for Mosley; the Duchess of Devonshire confirmed that her parents viewed it as unbearably dishonest. Diana gives a description of her feelings to her stepson Nicholas Mosley in a letter dated 24 April 1981. In another letter, written while Nicholas Mosley was writing his book, she tells of her conviction, so strong that she called it 'knowledge', that she and Mosley were made for each other.

Chapter Twelve

Lady Celia described Diana's stay with her family in Murren, as did Diana. Lady Boothby confirmed the story, originally told in Nicholas Mosley's life of his father, that Lord Boothby had asked Mosley if he had told Cimmie the names of all the women he had been to bed with since his marriage.

There are accounts of Cimmie's death in Nicholas Mosley's book, Irene Ravensdale's diaries and Dr Kirkwood's account in the *Daily Telegraph*. Elsie Corrigan, who was working as a housemaid at Savehay, described the situation there immediately after Cimmie's death. Diana confirmed that she went to Swinbrook rarely, and only as a duty. Nicholas Mosley described his father tapping on the window of the Eatonry with his stick. Hamish St Clair Erskine confided in Diana that he was going to pretend to be engaged to someone else, as he thought it was the only way to get out of his relationship with Nancy. Diana confirmed that Bryan kept hoping she would go back to him until Elizabeth came on the scene; and that she was immediately crossed off the list of those who gave house parties and balls, except for a few 'faithfuls' including Lady Cunard, Lord Berners and Mrs Ronnie Greville. Jonathan Guinness told me how the world saw Diana's leaving his father.

Chapter Thirteen

The lyrics of the Blackshirt songs were written by E.D. Randell. Diana's behaviour at the Boycott German Goods rally in Hyde Park in 1934 is fully described in HO 144/21995/17. Diana gave her views on anti-semitism in a letter to Robert Skidelsky. Harold Nicolson's unpublished diary confirms that anti-Jewish atrocities were even worse than the newspaper stories. His published diary talks of Cimmie's memorial service and Mosley's reaction to her death.

The Brown Book of the Hitler Terror gives authenticated information about the growth of anti-semitism, Hitler's ordering of it to be worked up to a frenzy, and the persecution of the Jews from Hitler's earliest moments of power.

Chapter Fourteen

Lady Butler was among those who knew Peter Rodd and described him. Diana herself told me of her pregnancy and abortion.

Details of Mosley's November 1933 meeting at Oxford, with its descriptions of rough BUF behaviour, can be found in HO144/20141/159 at the Public Records Office. The setting up of fascist groups in public schools and universities is fully documented in HO144/20140/286 and 144/20142/118. The *Daily Mail* of 15 January 1934 prints his belief that the Blackshirts were the party of youth, bursting with patriotic pride. Home Office concern about the increase of support for left-wing activity after Mosley's 7 June 1934 meeting at Olympia can be found in HO144/20142/155, including a police operational order detailing the number and rank of officers on duty for the meeting. Philip Toynbee gives his account in *Friends Apart*. Allegations of violent treatment to members of the public are in HO 144/20140/29–34, 58 and 144/20141/366–7. A full rebuttal by Lord Rothermere of Mosley's fascist and anti-semitic ideas was printed in the *Daily Mail* of 19 July 1934. Papers HO144/20146/136–8 tell of Mosley's stressing of the revolutionary aspect of the BUF to the Left and its authoritarian stance to the Right.

Chapter Fifteen

Diana gave me her impressions of Eva Braun. Neri Gun's biography of Eva Braun, taken from interviews with her family, remains the best. Derek Hill told me how Unity first saw Hitler, after he telephoned her. There are descriptions of the Parteirtag in Ward Price's dispatches, in other newspapers and in Irene Ravensdale's diary.

Copious love letters to Hitler were found in 1946 in an archive in Hitler's former Chancellery in what was then the Russian sector of Berlin. A selection was printed in *Granta*. It is unlikely that Hitler ever saw the most passionate and personal of these.

Papers HO 144/20143/71–80 and 144/20145/14–17 contain reports of the Albert Hall meeting of October 1934, police deployment for it and Mosley's swing to anti-semitism. The full text of Mosley's speech at the October 1934 meeting shows his adoption of anti-semitism as a platform.

Chapter Sixteen

There is much about Hitler's mesmeric effect on those around him in *Knight's Cross*, the biography of Rommel by David Fraser and in many of the biographies of leading Nazis and other figures of the era.

Chapter Seventeen

Mosley's response to Hitler's telegram was published in *Die Sturmer* on 10 May 1935. HO 144/21060/555 holds an MI5 agent's report on Hitler's view that the BUF is a fine organisation and Mosley a fine leader. I have included Winston Churchill's reference to Hitler because, although *Great Contemporaries* was published in 1937, Churchill actually wrote it in 1935.

Nicholas Mosley remembers Diana's early arrival at Posilippo and the furore it caused. There are several references to it in Irene Ravensdale's diary.

Chapter Eighteen

Mosley described his view of Mussolini in *My Life*. MI5's belief that Mosley was receiving funds from Mussolini is recorded in HO/45/25385/38–49. Details of Mosley's funding by Mussolini were discovered by David Irving in the Italian archives.

There is a long description in the *Daily Mail* 23 March 1936, of a demo against a BUF meeting at the Albert Hall. Lady Longford gave me a vivid account of the Mosley meeting at Oxford in 1936, as did Lord Longford; and there is an account of it by Maurice Bowra. Nellie Driver's unpublished MS, *From the Shadows of Exile*, describes Mosley's marches and meetings, the type of people the BU attracted and the place of women in the BU. Lord Monckton of Brenchley described to me its effect on him as a young man.

Full details of the Battle of Cable Street are contained in HO 144/21060/316 and 144/21061/92–101 as well as in many newspaper reports, and Metropolitan Police Report DAC3, signed by the Chief Constable. Speech by E.C. Clarke at Chester Street, so inflammatory that it resulted in 85 arrests and numerous injuries. Police summary of the fascist anniversary rally, i.e. battle of Cable Street, is in a memorandum by Sir Philip Game, Commissioner of Police, on his dealings with Mosley over Battle of Cable Street. 24 June interview in the *Frankische Tageszeitung* on Mosley's 'recognition of Jewish danger'.

Chapter Nineteen

The prices of the Wootton furniture are in a Swindon auction catalogue of Monday 25 May, shown to me by Diana. Ward Price gives a description of a Hitler dinner party. Nicholas Mosley told me about the growth of his extremely affectionate relationship with Diana; his sister Vivien Forbes-Adam also described the life at Wootton. Diana gave me the details of Hitler's flat in Munich. Putzi Hanfstaengl describes the Nazi leaders in *Hitler, the Missing Years*. Irene describes the Parteirtag of 1936 in her diary. There is a long description of the 1936 Olympic Games in Goebbels' diary; entries in it from September and October talk of Diana's wedding. Mosley's reasoning behind the secrecy of his marriage to Diana is in HO283/13/116. Ribbentrop demanded the return to Germany of her colonies at the annual banquet of the Anglo-German Fellowship on 15 December 1936.

Chapter Twenty

Goebbels' diary entries for 5 December, 1936 confirm that money was given to the BU. A report from MI5, in HO45/25776 describes Dorothy Eckersley's career, giving a statement by her, and tells of her imprisonment after sentencing at the Central Criminal Court on 10 December 1945. There are entries throughout 1937 in Goebbels' diary on the question of money for the BU and the radio concession.

Sir Frederick Lawton, who drafted the contracts for the radio concessions, gave me a long and very detailed account of his involvement in the negotiations.

Mosley's dismissal of many of the BU's paid staff, the position of commercial radio advertising in Britain and the recruitment of Peter Eckersley are recorded in *History Today*, Vol 40, March 1990. Lord Trenchard had wished to outlaw fascism (HO 14420158/107); the Commissioner of Police wanted to suppress fascist anti-semitism, though not fascism (HO 144/20159/18); the eventual result was the Public Order Act. The *Spectator* of 24 February 1939 contained an informative article on the state of German broadcasting; there is more on broadcasting in *The Golden Age of Wireless* and *The Power Behind the Microphone*. There are details of Mosley's prospective radio concessions in HO 283/13/110 and of the foreign funding referred to by Goebbels in HO 283/16/49 and 283/10/49.

Chapter Twenty-One

There is much about Esmond Romilly in the memoir about him by his friend Philip Toynbee. A number of very long letters from Lady Redesdale to Diana described Jessica's marriage.

Sir Frederick Lawton gave me a very full account of the long-drawn-out negotiations for the radio concessions and the contracts involved. The news of Diana's marriage broke in the *Daily Telegraph* and *News Chronicle* of 28 November 1938.

Chapter Twenty-Two

Both Nicholas Mosley and his sister Vivien confirmed that they learned of their father's marriage only through newspaper reports. Michael Mosley confirmed that his aunt Baba, of whom he saw a lot, never forgave Mosley for marrying Diana. Irene Ravensdale's diaries describe bringing her niece Vivien out. A brief article about Vivien Mosley as a debutante was printed in the *Sunday Express* of 4 June 1939; in the issue of 6 August 1939 she is canvassed as one of the four contenders for Deb of the Year.

Mosley wrote in 1938 that it was Hitler's habit to convey to him his view of events through Diana. Nancy Mitford's letters cover much of what happened to the Mitfords at the beginning of the war. Irene Ravensdale's diaries describe taking the children up to Wootton at the beginning of the war. The Duchess of Devonshire in an interview with me confirmed that her mother was extremely fond of Mosley and agreed with him politically.

Mosley's message to British Union members, due to be published in *Action* on 9 September but censored, is in the Public Record Office. Details of the Peace Rally at Earls Court in July 1939 are in HO 144/21429/18, including notes on the question of loyalty to fascism as opposed to loyalty to Britain. HO 45 2374/860502 holds a memo to Sir Alexander Maxwell, 1 November 1939, describing the situation over the financial side of the radio concession project and the row between Mosley and Allen (a report from Special Branch, made on 8 December 1939, gives many details of the financial set-up and the personalities involved in Mosley's radio projects).

Unity shooting herself, her spell in the nursing home and her arrival home are described in *Unity Mitford, A Quest*, in long letters from Sydney, and by Diana and the Duchess of Devonshire. Diana told me of her father's visit to Oliver Stanley and its outcome; and also described the Dolphin Square flats.

Chapter Twenty-Three

A long memo by the Chief Constable discusses the pros and cons of detaining Mosley, the importance of freedom and the danger of allowing Mosley to pose as a martyr in the course of a subversive campaign. Irene Ravensdale's diary records her conversation with Sir John Anderson. Hugh Dalton's diaries tell how Churchill believed that Mosley would be put in power by Hitler. Professor Simpson also points out that Mosley and the BU were the nearest thing in Britain to a Führer and the Nazi party.

On 24 May *The Times* reported 167 expulsion orders against IRA sympathisers. Diana gave me a long description of Mosley's arrest and later of her own; they are also detailed in *The House of Mitford*. Richard Stokes's relaying of Nazi peace feelers through Lord Tavistock is mentioned by Liddell Hart, in the Liddell Hart Centre for Military Archives, King's College, London. Nellie Driver describes Home Office interest once the BU had been proscribed. John Beckett's unpublished memoir describes his arrest and the induction procedure at Brixton. HO 283/16/28 holds BU policy notes on how it would silence all opposition if it ever came to power, and decide whether Jews were patriotic Englishmen or not; those who failed would have been expelled.

Chapter Twenty-Four

Diana's order of detention is in Home Office files (Box no 500), signed by one of the principal Secretaries of State, J. S. Hale, and dated 26 June 1940. Many details of the number of people imprisoned under 18b can be found in Professor Simpson's book *In the Highest Degree Odious*. Lord Gladwyn confirmed to me by letter Nancy Mitford's attempt to get Diana imprisoned. Lord Moyne forwarded Desmond's governess's long report on Diana's activities, dates of her visits to Germany and reactions to news bulletins to the Home Office, who in turn sent it to Sir Alexander Maxwell. It also is dated 26 June 1940.

Chapter Twenty-Five

77 CAB. 66/20 WP 279 (41) p 1 details the conditions applied to those detained under 18b. Minutes of the Home Defence (Security Executive, 31 October, 6 November 1940) show how the Home Office – i.e. the Advisory Committee – defeated MI5, who wanted many more interned and much more stringent measures. Louise Irvine, who was in prison with Diana, contributed many memories, as well as lending me a long typewritten memoir. Erica Betts also remembered her time there. Nellie Driver's unpublished MS was full of detail. Jonathan Guinness gave reports of visiting his mother in prison.

John Beckett's unpublished MS contains details of life at Brixton and the attitude of warders and prisoners to Mosley. Nicholas Mosley and his sister Vivien both spoke about their father's handing-over of their guardianship to their aunt Lady Ravensdale.

Sydney's letters give a picture of the situation between herself and David.

Mosley's replies to his questioning by the Advisory Committee are in the Public Record Office, as are Diana's.

Chapter Twenty-Six

HO45/25767 gives details of the proposal to send internees overseas; it also contains a record of the meeting held on 22 July 1940 to decide whether BU detainees should be sent overseas and the Home Secretary's memorandum of 20 July 1940 setting out the reasons why they should not be. HO282/22 is a memorandum from Norman Birkett showing how Sir John Anderson and Herbert Morrison accepted 400 of the 455 recommendations by the Advisory Committee for release by February 1941. Michael Mosley described the incident at St Ronan's. Many prison memories were contributed by Louise Irvine, Erica Betts and Nellie Driver.

Churchill's comments showing his scepticism about detention are in Hansard records of 5 November 1940. Richard Stokes's remark on detention is in Hansard records of December 1940. Family letters give a picture of what is happening to the rest of Diana's family. The Duchess of Devonshire described visiting Diana in Holloway. Francis Beckett contributed recollections of his father's account of life in Brixton, some of which was in his MS. Letters from Norman Birkett are in Home Office files concerning the Advisory Committee.

Chapter Twenty-Seven

HO 45/25733 holds a letter from Mosley to the Prison Governor of Holloway about his and Diana's health. There is an account of Mosley's and Diana's libel action against the *Daily Mirror* in *The Times* of 8 November 1940. There are a great many factual details of the Blitz in Philip Ziegler's book *London at War*. Mosley's letters to Diana are printed in his son Nicholas Mosley's book, *Beyond the Pale*. Baba Metcalfe's diary gives her account of dining with the PM. Dr Rougvie's report on Diana is in PCON 9/880. There is a moving letter from Mosley in *Loved Ones*. Nellie Driver gives a vivid description of the Blitz and its effect on the inhabitants of Holloway. Correspondence over Diana's complaints is in the Public Record Office. Many details of the internees' life in prison, including timetables, letters, requests, etc, are in HO45/25753. HO 144/22495 contains details of the plan and dates of the Mosleys moving in together. CAB 66/20 WF 279 41 p3 explains that Mosley was not to be sent to the Isle of Man because it was thought that he would have an undesirable effect on the other detainees there, i.e., hardening their attitudes.

Chapter Twenty-Eight

HO 144/224/965 details the use of prison labour, and Mosley's fury at the allegation in the *Daily Express* of early November 1943 that the Mosleys had a personal maid. *Loved Ones* tells the story of the Catholic priest coming to visit them. There are more descriptions of the prison in *Beyond the Pale*. Diana's memories of the success of the Post House garden are reinforced by a letter from a fellow prisoner, C.M. Shetland, thanking her for a basket of 'delicious vegetables' grown in their patch. Micky Mosley recalled stories of his father sunbathing in front of the female prisoners' windows; Nellie Driver also recalls this. Numerous letters from the children, from Nannie, from Nannie Blor, from Sydney, from Debo, from Pam, give a picture of life outside.

The discussion about whether wives of detainees who were not subject to detention themselves should be allowed to visit their husbands is in HO 144/224/495, as well as the Home Secretary's letter confirming the decision that all interned couples should have the same right to association with each other. HO 45/25753 gives details of the heating in the prison and the decision that unlike the rest of the internees or prisoners the

Mosleys should not have any until mid-January. The file HO 144/22495 contains details of Diana's requests that her Guinness sons should spend a weekend with her, her longing to bath her baby, reports from Miss Barker of the Prison Commission, the Governor of Holloway's support for Diana's chicken-keeping project and the Commission's refusal, and the story of the 'fashion show'. Lady Celia McKenna remembers meeting Unity in a queue in Oxford.

Chapter Twenty-Nine

Baba Metcalfe's diary describes her conversation with Lord Halifax, whose subsequent letter to the Home Secretary, Herbert Morrison, written from the Privy Council Office, is in Home Office files. HO 45/28492/129 holds the recommendation for Mosley's release on health grounds, the statement that internment had reduced his weight from 14st 7lb to 11st 4 lb; and many extremely long and detailed reports and letters from the various doctors concerned. HO 45/28492/252 and 257 contain letters from the TUC and TGWU objecting to Mosley's release. Copies of the orders for the Mosleys' release, and the reactions of the trades unions, are in Home Office files, as is Herbert Morrison's letter and Baba's letter to Winston Churchill, forwarded to the Home Office with a note from Churchill attached. Long reports on Mosley's health are also there, as is a long letter and report on Mosley's health from Lord Dawson of Penn and Dr Geoffrey Evans, and the prison doctor's report. The reaction to Mosley's release is in HO 262/6. Pam gave me a full description of the Mosleys' arrival at Rignall. Harold Nicolson's reaction to their release is in his unpublished diary.

Chapter Thirty

Letters from Sydney and descriptions from Diana give a full picture of life at the Shaven Crown. In one of Diana's files are letters from would-be servants. James Lees-Milne, training near his friend Tom Mitford at the beginning of the war, heard much of Tom's belief in Nazism and love of Germany, largely influenced by Diana, and confirmed this was the reason he volunteered for Burma. Diana explained to me that the reason Mosley would not see Randolph Churchill at the end of the war was because he regarded the Churchills as enemies. Baba Metcalfe confirmed to me that she refused to see Mosley again after the war because it meant seeing Diana. Family letters make plain what was happening within the family. Frances Partridge described the Mosleys' partial social ostracism.

Chapter Thirty-One

Sydney's letters are full of detail about her life on Inch Kenneth, which is also described in *The House of Mitford*. Several old friends of Diana told me how Diana wanted them to ask her and Mosley to dinner but how they felt they could not (often because of a spouse's reaction). Among the descriptions of life at Crowood by family members is a vivid letter from a schoolboy visitor there. Letters from Jonathan and Bryan discuss the reasons for and against his staying on at Eton. Nancy's letters describe the furore when *The Pursuit of Love* was published. Nicholas Mosley described the summers at Crowood and Crux Easton.

Chapter Thirty-Two

Diana talked to me for many hours about her houses and their furnishings, and my numerous visits to the Temple provided much first-hand observation.

Both Diana and her sisters talked of the overwhelming effect on her life of the migraines she suffered. The atmosphere of the Temple, the Mosleys' way of life there, and the reaction of Parisian society were described by those who stayed or lunched there, such as Lady Abdy, Quentin Crewe, Walter Lees and others. Diana gave a list of Mosley's great men friends to Robert Skidelsky in a letter. Many of Mosley's ideas appear in the *European*, which Diana edited between 1953–9. Joy Law told me of Mosley's remark to Jerry.

Chapter Thirty-Three

Sydney's copious letters describe life on Inch Kenneth and her feelings about it. Quentin Crewe, who knew Jeannie Campbell, told me much about her love affair with Mosley – she had found Crewe, then between places to live, a flat, which she used to borrow to meet Mosley in. Diana confirmed that Mosley had told her that he had had affairs within weeks of his first marriage and that he had told Cimmie all about them when Cimmie was so worried about Diana. She also described, in taped interviews with me, her own wretchedness when Mosley had several affairs when they were living in France.

A letter from Diana and the Duchess of Devonshire in the *Times Literary Supplement* of 8 April 1960 refutes many of the claims in Jessica's *Hons and Rebels*. Diana told me about her differences with Mosley over money. Quentin Crewe described Nancy's mockery of others. The correspondence between Diana and Nicholas describes Diana's difficulties over money and the financial details of sending Alexander to Columbia, financed by Nicholas and Micky. A letter from Desmond Stewart shows a sympathetic and intuitive understanding of the problem. Letters from Diana describe canvassing for the March 1966 election. Diana's letters to Nicholas Mosley describe her unhappiness, then, over Alexander and her misery at Mosley's behaviour to him

Chapter Thirty-Four

Lady Abdy talked to me about the Mosleys' relationship and Diana's total devotion. Simon Blow gave me a very good description of Mosley in old age. Diana's engagement diaries contain timetables of her movements and details of the progress of Nancy's illness; her personal diaries disclose her feelings at the time. Shirley Conran gave me some telling vignettes of life at the Temple. Both Pamela Jackson and the Duchess of Devonshire described the strain Diana suffered visiting Nancy when Mosley did not want her to go. In an interview Jessica Mitford spoke to me about the meeting over Nancy's deathbed, confirming both her feelings about Diana and that she had promised there would be no animosity between them.

Diana's letters to Louise Irvine show her continuing devotion to all Mosley's ideas. Many letters to Nicholas Mosley set out Diana's financial position and show her gratitude for his help to his half-brother. Diana described the reconciliation lunch with Baba.

Chapter Thirty-Five

Frank Cakebread described the vandalising of Cimmie's tomb. June Ducas confirmed that social ostracism of the Mosleys still continued. Michael Bloch gave me a good picture of Mosley's last years, and his attitudes. Many people, including her sisters and children, and close friends including Walter Lees, talked of Diana's state of mind after Mosley's death; her own diary describes her anguish. Michael Bloch was one of those who gave a superb description of Mosley's funeral, as did her Mosley sons and daughter-in-law. Diana's letters paint a full picture of her illness, slow recovery and feelings. Her letters to Nicholas Mosley give a picture of Mosley's attitude to his children over money. The full file of her letters to Nicholas Mosley show her changing opinions, first on the first volume of his biography of his father, then on the second, and his sister Vivien Forbes-Adam elaborated on Diana's attitudes.

Chapter Thirty-Six

Jonathan Guinness described how his mother edited all Mosley's books – one reason Mosley didn't want Diana to write. The transcript of *Desert Island Discs* is in the BBC Archive Library. There are interesting interviews by Ian Waller in the *Spectator* of 8 January 1983, Charles Nevin in the *Daily Telegraph* of 6 November 1983, and Peregrine Worsthorne in the *Listener* of 8 December 1983. There are more reproachful letters to Nicholas Mosley when the television series about his father came out. Lord Longford's address, on 15 October 1992, was reported in the newspapers.

Appendix I

Ten Points of Fascist Policy

1. Blackshirts are loyal to King and country but they stand for far-reaching and revolutionary changes in government, in economics and in life itself. Their watchword is 'Britain first', and they intend to build a nation worthy of the patriotic love of every Briton.
2. Blackshirts stand for action in government. Time after time the people of this nation have voted for policies of action and have been betrayed by those they elected. Fascism combines progress with the executive instruments of loyalty, decision and voluntary discipline.
3. The establishment of the Corporate State is the main ideal of Blackshirt endeavour. Industry is to be divided into national corporations, governed by representatives of employers, workers and consumers. The only limits the State will lay down to industry will be limits of national welfare. Private ownership will be permitted and encouraged provided such activity enriches the nation as well as the individual.
4. The Corporate State, with production properly controlled, and operating to the benefit of the state, will automatically eradicate unemployment and the resulting poverty. The functions of the corporations will be to raise wages and salaries over the whole field of industry. Production will keep pace, with a consequent increase in consumption, for a home market will be provided by the higher purchasing power of our own people. Married women will not be compelled to retire from industry and all women will be paid wages on an equitable basis to those of the men, but the higher wages paid to husbands under the corporate system will not necessitate women working to maintain a home.
5. Britain's new motto will be 'Britain buys from those who buy from Britain'.

6. Blackshirts seek to build an entirely self-contained Empire, withdrawn from the international struggle for markets and the economic dislocation and war such struggles mean. Within the Empire we can produce all the goods, foodstuffs and raw materials which we require.
7. Under a scientifically-controlled three year plan fascism would increase production of food supplies in this country from £200 million per annum to £400 million per annum for the total exclusion of foreign goods, and provide the Dominions with a better market than we have today.
8. Under fascism no alien will enter this country and take a Briton's job, and Blackshirts will deal not only with poor aliens but also the great alien financiers of the City of London, who use the financial power of Britain not in the interests of this country but of foreign countries.
9. Blackshirts bind themselves together in a voluntary discipline. They realise nothing can be done without discipline. The wearing of the black shirt is not compulsory but in wearing it our active members break down all barriers of class within our ranks.
10. Fascism is a dictatorship in the modern sense of the word, Government armed by the people with the complete power of action to overcome problems which must be solved if the nation is to live. At the end of the first fascist Government, elected at a general election, another election will be held, on an occupational and not on a geographical franchise.

Appendix II

Notes of a meeting held at the Berystede Hotel, Ascot, on Wednesday 2 October 1940

Lady Mosley was called and examined by the Advisory Committee. She had appealed three months earlier, in July.

Q You had the particulars which the Secretary of State alleges to be the grounds on which it was necessary to exercise control over you, that you have been active in the furtherance of the objects of the organisation known as the British Union? I just want to ask one or two questions about that. You are, of course, the wife of the Leader of the Party, Sir Oswald Mosley?

D Yes.

Q Is it right that you have been active in the furtherance of the objects of the British Union?

D No it is not. Although of course I am a fascist and a member of the movement, I have never taken any part. I have never spoken, for example or taken any active part at all.

Q But you had sympathy with the movement, naturally?

D Naturally. Yes, of course.

Q Let us take the matters which are alleged. When were you married – I think in 1936?

D In October 1936.

Q Forgive my asking this, in order to get the chronology right. When was your divorce?

D Yes, I divorced my former husband in June 1933 and the divorce was absolute in January 1934.

Q That is six years ago. That is the reason I wanted that date. Between 1933 and 1936, did you know Sir Oswald?

D Yes.
Q Were you a fascist then?
D Yes.
Q Have you always been a fascist?
D Always.
Q How did you come to be a fascist? Who led you to that belief?
D Of course I knew him very well and he told me a great deal about it.
Q Sir Oswald did?
D Yes. I thought, that is the thing for me.
Q Forgive my asking, what age are you now?
D Thirty.
Q In 1933, seven years ago, you would be twenty-three?
D Yes I was twenty-three.
Q Was your first knowledge of Sir Oswald Mosley about that time – in 1933?
D Yes, about then.
Q It was from that date that you really date your fascist beliefs?
D Yes. He used to tell me all about it and so on. We had at that time a National Government and I was always very much against the National Government. I remember the election of 1931 – I think it was the first time I really began to take any interest in politics, except that the father of my first husband was Minister of Agriculture – a Tory. I remember thinking then that if only Lloyd George had had a proper party I might have voted for it, but of course he had no candidate where I lived. I thought if he had had I might have voted. I have never voted.
Q Had you taken any active part in politics before you met Sir Oswald Mosley?
D Only with my father-in-law being Minister of Agriculture, that is all.
Q In the years 1933 and 1934, I think you paid several visits to Rome, did you not?
D Very likely, it is a long time ago. Yes, I used to go and be with Lord Berners in Rome.
Q What was the purpose of your visits to Rome?
D Rome is a beautiful city and I liked to go there. There was no particular purpose.
Q Did you take fencing lessons?
D No.
Q You did not go for the purpose of studying fencing?
D No, I have never done fencing in my life.

Q Did you meet many Italian fascists on those visits?

D No. The only Italian fascists that I have met, I have met more or less in London.

Q Did you ever meet Mussolini?

D Never.

Q I think in August 1934 you flew, did you not, from London to Rome?

A Very likely, yes – oh no, I was in the south of France. I may have flown from there. I was not in Rome very much.

Q Where did you stay in Rome on that trip?

D At the Grant Hotel. As a rule I used to stay with Lord Berners, who had a house there but if I went on my own like that, I used to stay at the Grant Hotel.

Q Were those visits to Rome in 1933 and 1934 quite unconnected with Italian fascism or fascism at all?

D Yes.

Q Did you go alone?

D As I told you, both in 1933 and 1934 I went to stay at the house of Lord Berners who is a friend of mine, but in the summer, in August, he was not there. I think I was en route for Rapallo, where I went to stay with some other people.

Q Were they English people you stayed with?

D Yes.

Q We may take it that those visits were just ordinary social visits?

D Yes.

Q Then I come to the next year, 1935. Do you remember a demonstration in Hyde Park on 27 October 1935 under the auspices of the British non-sectarian Anti-Nazi Council, the speakers being Mrs Despard and Mr Attlee? Do you remember that?

D Yes.

Q You were present.

D Yes. I just happened to go.

Q There were various speeches by Mrs Despard and Mr Attlee and, as I understand, a resolution that all business relations with Germany should stop, that there should be a boycott of German goods. Do you remember that resolution being put to the meeting?

D Yes.

Q Were you the only person who voted against it?

D Yes, I believe I was.

Q While they sang the National Anthem at that demonstration, did you give the fascist salute?

D Yes.

Q Would you explain why you did that?

D It was a demonstration against the meeting.

Q Why did you do it?

D I felt very strongly about it; I did not know that there was even going to be such a thing. It was a Sunday, and I happened to be in Hyde Park, and I saw this rabble listening to a lot of speakers.

Q The rabble being the public listening?

D It was the public, but it was a rabble. You would have thought so if you had seen it. They put this vote to boycott German goods, which seemed to me the most ridiculous idea I ever heard of. Then they said 'Any against?' and I was on one of those carts or wagons where you stand, and I put my hand up against it. Then, when they sang 'God Save the King', I gave the fascist salute.

Q That roused a good deal of feeling, did it not?

D Well, there were many jeers.

Q What were you desirous of demonstrating? Your affection for Germany, was it?

D I thought it ridiculous nonsense to have a boycott of German goods.

Q The meeting itself was expressing the view that it was the only method of bringing Germany to its senses. I am only putting this question to you; I am not stating it was a wise or proper resolution, I am only trying to get the facts. That was the purpose of going there to demonstrate, and that was the only method of bringing this recalcitrant nation to its senses. The meeting was very strongly incensed against Germany, was it not?

D A lot of people had come from the East End and looked very foreign.

Q It really was an anti-fascist meeting?

D Well . . . yes, it was.

Q That was in 1935. In 1936, did you have many contacts with the German Embassy in London?

D Do you mean when Ribbentrop was there?

Q Did you know someone called Fitz Randolph?

D Yes.

Q Who was he?

D I think he was one of the secretaries – I'm not sure.

Q Did you lunch with him?
D No, I do not think I ever did.
Q Did you meet him and have discussions with him?
D I saw him from time to time in other people's houses. I do not think he ever came to my house.
Q Did you ever discuss with him, at all, political matters with regard to Spain and Russia?
D I may have done so.
Q On 6 October 1936, you were married in Germany?
D Yes.
Q Why were you married in Germany?
D Because we wished to keep the marriage secret.
Q You were married in the house of Frau Goebbels?
D Yes.
Q Was she a friend of yours?
D Yes.
Q How long had you known her?
D A couple of years.
Q Hitler himself was present, was he not?
D Yes.
Q Why?
D Because he is a friend of mine.
Q How long have you known Hitler?
D Since the beginning of 1935.
Q How many times do you think you have seen him between 1935 and 1940?
D I do not know.
Q Is he still a friend of yours?
D I have not seen him for some time.
Q Absence makes the heart grow fonder. Do you still entertain the same feeling for him?
D As regards private and personal friendship, of course I do.
Q The history of Hitler in recent years has not affected your view about that?
D I do not know what his history has been.
Q It is plain for all to see. Did you hear the bombs last night? That is Mr Hitler, as we suggest. Does that kind of thing make any difference to you – the killing of helpless people?
D It is frightful. That is why we have always been for peace.

Q You realise he is doing it?
D But we are doing it too. That is what you have if you have war. That is why we are against war.
Q Does the record of Hitler – we shall have to deal with it in some little detail – affect at all your attitude?
[blotted out].
Q Do you still feel the same?
D Of course I do, exactly the same.
Q Supposing he came here – if I might put what I suggest is rather fanciful – would you welcome him?
D No.
Q What would you do?
D Personal friendship, private friendship, has nothing to do with one's feelings towards one's country.
Q Supposing a personal friend of yours was the greatest enemy the country has?
D I can never live if England has been conquered.
Q If you had your choice, would you live in England or Germany?
D What do you mean by 'if you had your choice'?
Q Supposing you had your choice now?
D I should live in England. I have never lived in Germany. I could live in Germany now if it were not for the war.
Q I put the question quite seriously in this way. If you had the choice of the fascist system of government such as exists in Germany, which is at present anathema to most people in this country, would you prefer to live under that regime?
D It has absolutely nothing to do with a system of government. What we prefer is to have the same sort of system of government here, as we think it has done well for that country.
Q That is to say, if you had power, you would displace the present form of government in this country with a fascist regime?
D Yes.
Q The marriage on 6 October 1936 was in Germany because Frau Goebbels and Hitler were friends of yours and it was arranged in that way?
D We wanted this marriage kept a secret. I think this is how it arose. I was in Italy, and I said to someone there at the Embassy, 'What happens if English people are married in Italy or France?' The reply was, 'You would go to the Consul about it and it would have to be posted in the

Consulate, and banns, whatever you call it, would be issued for three weeks. So that anybody might quote it in the press, just the same as having it in England.' Then I discovered that in Germany you can be married in the ordinary way, by registrar, and I thought that would be the best plan. I happened to see Hitler a short time before, and told him that we were interested to be married that way and he said, 'If you want it to be kept a secret, I would not advise that, because it will be posted up in the registrar's office'. And he said, 'Had I not better arrange it?' And he very kindly did.

Q That was a very great favour?

D I do not think so. It was quite easy to do.

Q He would not do it for most people?

D I think so.

Q Did you tell him why you desired the marriage to be kept a secret?

D I cannot remember – no, I do not think so.

Q What was the ground of keeping the marriage secret?

D My husband thought that it would be better. It was not my idea; it was his idea. He said that in our movement they might be apt to think, as he was just married and so on, that he would not work so hard for them, would not spend so much time and so on, and he thought it might be better if they did not know anything about it.

Q Yes, he made some explanation of that kind to us, of it being out of consideration for the Party and for his position and his work. Did you tell Hitler that was the real reason, doing it out of consideration for his own movement and nothing else?

D No, I think I simply said that we did not wish this to come out yet, and thought of being married abroad and he said, 'Why not here?'

Q Did you know the other leaders of the Nazi Party?

D Yes.

Q You knew them all?

D Yes.

Q You knew Hess?

D Yes.

Q You knew him very well?

D Not very well. I met him.

Q Himmler?

D Himmler I met once.

Q Goering?

D Yes.

Q The general leaders of the Nazi party – you knew them all?
D Yes.
Q Did you like them?
D Some of them.
Q Did you like Himmler, for example?
D Very much, yes.
Q Have you read at all what it is alleged the Gestapo had done?
D I suppose I had from time to time, read things, yes. But I did not believe them very much.
Q Did you ever speak to Himmler, for example, about the activities alleged against the Gestapo?
D No, I do not think so, not that I can remember. If I saw him, we generally discussed topicalities.
Q In October 1936, the date of your marriage at that particular place, the campaign of the Nazis against the Jews was already in very full flow?
D I think they began at once.
Q Did you ever discuss that at all with the Nazi leaders?
D We mentioned Jews.
Q Do you agree with their policy about Jews?
D Up to a point I do. I am not fond of Jews.
Q Have you read, at all, that it was alleged that very great barbarities and cruelties were inflicted?
D I saw the book called *The Brown Book*, on Hitler terrorism, but I did not pay much attention to it.
Q You did not take the opportunity of asking, if you could, what the real truth was? Himmler would probably know most about it.
D I remember at the time of the Anschluss, with Austria, asking about Ludwig Rothschild and I remember him saying 'I cannot think he is any worse than we have got in Britain.' Except for that one conversation, I cannot remember ever having any conversation about Jews.
Q Between the year 1936 and the outbreak of war in 1939, you did make many visits to Germany?
D Yes.
Q Were they usually by aeroplane?
D Yes.
Q Did you get calls from Germany asking you to go?
D No.

Q Did you never?

D I am sure I never did.

Q Tell me this – how old is your baby now?

D It was eleven weeks when I was arrested, and I have been in prison fourteen weeks, so that it is nearly six months old.

Q Shortly before you were expecting the baby, did you actually fly to Germany?

D This baby? No.

Q When was the baby born?

D In April this year.

Q In April 1940. Did you towards the end of 1939, shortly before the outbreak of war, go to Germany?

D No. The last time I was in Germany was the end of July and beginning of August 1939.

Q There is one particular question we want to ask you about. You remember your husband's connection with the radio advertising business. He was negotiating with the German Government for a contract or licence?

D Yes.

Q Did you go and see Hitler about that?

D Yes, I did.

Q Did you obtain a concession from Hitler?

D Yes.

Q What did you think was happening? What was the concession you thought you were getting?

D I think you probably know already.

Q Yes I know the history.

D The idea was to get a wavelength in order to advertise English goods. Of course they pay tremendous sums for these things and it was simply a business thing. You had to know people in the Government.

Q It is quite plain, is it not, that it was your friendship with Hitler that obtained that?

D No, I do not think it is quite plain.

Q What part did Hitler play for you?

D It was like this. When they give these concessions in France, to other people, so far as one can make out what happens in France, is that you give a Cabinet Minister a fiver and he would do anything for you. It is not like that in Germany. The Germans had a great number of wavelengths, and I thought they might possibly like to use one of them in this way because they would get foreign exchange. And therefore I went to Goering who had just instituted the four year plan, and suggested it to him. He was very much taken with the idea but when it came down to it, after several weeks of talking to and fro, it was turned down on the ground that the post office would not have it, and that they must have all wavelengths. So I thought that was the end of it. But afterwards I thought, I must have one more go at it. I explained it to Hitler himself and he immediately saw the advantage of it and said yes.

Q Was that in one interview you managed that with Hitler?

D Actually, what happened was that I went straight off to see him and asked him about it.

Q Where did you see him?

D In Berlin. He said, 'I cannot tell you offhand, but if you like, I will have the thing gone into.' I said, 'I think you will . . . that it is a very good idea.' Then he had it gone into. It took months and months and eventually I was told there were pros and cons and so I thought that was the answer to that. Then I went again to Hitler and said. 'They tell me this and that. How about it?' And he let me have it.

Q Did he give instructions to other officials to expedite the matter?

D Well I suppose he did. It was out of his hands then. He does not interest himself much in detail.

Q I quite follow that, but he is all-powerful?

D Yes, if he says they ought to do this or that, they do it.

Q That is what he did?

D Yes.

Q That was a very great favour.

D I do not think so at all. The French have done it because they thought it was business and after all the Germans were incredibly poor as far as foreign exchange is concerned, and there is no doubt that it was a way of attracting foreign exchange.

Q What I meant to ask was, it shows that your standing with him is very high?

D I am a friend of his. I think it shows that he immediately grasped the splendid idea that I put forward to him and understood what a good one it was.

Q Between your husband's arrest, and your arrest, how long a time elapsed?

D Five weeks.

Q During that time, were you active on behalf of the British Union?

D Only in the sense that the thing was left in the most appalling mess, as you can imagine, as all the chief men had been taken, and I tried to clear it up a bit, as far as salaries and wages were concerned. Actually, it happened that we were just going to move house, because the headquarters' lease was falling in, and you know what a lot there is to arrange, and there were very few people to do it. I used to come up occasionally but I was really living in the country.

Q Were you then living at Denham?

D Yes.

Q Did you take steps, for example, to see that the newspaper *Action* continued, and matters of that kind?

D I tried to take steps, but it was utterly impossible because every time anybody started printing it, they went to prison. We were told at that time by Sir John Anderson that people were not put in prison for their opinions and we took his word. But yet every time it was done they went to prison.

Q Were you arrested before the movement was banned?

D No.

Q During that time, before you were detained, you did what you could to see that the movement was carried on?

D I did what I could, naturally. All that I was intending to do was, as I say, to try and clear up the wages side of it, to arrange the move, and possibly to see about *Action*. So far as I know, we only got two *Action*s out. I said to my husband, 'It looks like a passport to prison', and he agreed we had better give it up.

Q Had you, just before war broke out, the intention of taking your son to see Hitler in Germany?

D Not to see Hitler, but we were going to visit Germany.

Q Just one or two matters about your general outlook. Did you agree with the policy of annexing Austria?

D Well I think I did in this sense, I know the Austrians were longing for it. They always said so.

Q What about Czechoslovakia?

D The Sudetenland was German country and I suppose they felt they must take the rest in order to safeguard their industries.

Q At the time of what they call the Munich crisis, September 1938, did you feel that Germany was justified in claiming the Sudetenland?

D Yes.

Q And you felt in regard to the subsequent annexation of Bohemia and Moravia – that almost followed inevitably?

D I think it was the practical solution there. They asked him to come, as far as I remember.

Q Did you yourself, in conversation about that time, express the view that the Germans were perfectly justified in taking the Sudetenland?

D I think I probably did.

Q With regard to incidents, in September 1939, when war broke out for example, the sinking of the *Athenia* – what did you think of a thing like that?

D I was at a loss to know what to think. Of course terrible things can happen in war, and it was night, and it would be very hard to see what ship it was. I do not know anything about it. One does not know who did shoot the torpedo, does one? It is a mystery.

Q Take the subsequent events. Take the annexation or over-running of Belgium. What attitude did you take of that?

D I did not take an attitude. I suppose I thought that if you are in a war you have to try and get at your enemies somehow to fight them. You cannot sit endlessly behind fortifications.

Q When the British Army was in that rather dreadful situation, what attitude did you take?

D My husband thought, and it turned out afterwards he was justified in saying, we should not have sent an army on to the continent. He viewed it as the most fearful thing, and a most fearful danger, and it was wonderful when they got away. But they only got away with their bare skins.

Q Did you feel that was rather a matter in which you took some exultation because it rather justified your position?

D No, not in the least.

Q You made no expression of that kind, to anybody?

D Good heavens no.

Q With respect to information in regard to Germany, and the German Air Force, and the German Army, had you any information in your possession? And the Nazi leaders – did you glean any information?

D No, they did not tell me.

Q Take, for example, a thing like the Messerschmidts, which we know about to our cost, did you know anything at all about them?

D No, I did not even know the name until the war came.

Q Did you think England would go to war?

D There were all over London enormous notices saying 'We have got to be prepared.' I think it was an extraordinary way of putting it. Then they started by forming the ATS and all the girls joined it, and the men joined up. They were simply dying for a war.

Q Did you really think we would fight?

D I knew we would.

Q It is said that Ribbentrop advised Hitler that we would not.

D What a lie! I can't imagine he would be so stupid.

Q Did you ever speak to Hitler about it?

D Yes, often.

Q What did you tell him?

D I did not tell him much. He used to tell me. He used to say, 'I am afraid they are determined.'

Q 'They' meaning the government?

D Yes. That Chamberlain was determined.

Q But he said 'they'? Who was he meaning when he said 'they'?

D When he said 'they' he meant the Government. He did not mean the British people – English people.

Q Did Hitler ever tell you about his own ambitions – what his own policy was?

D Yes.

Q What was his attitude towards this country?

D Very friendly – an attitude of great admiration.

Q Was that a consistent view, held through all the years?

D Yes.

Q Then how do you account for the public utterances of Hitler against this country?

D Since the war began?

Q Yes.

D Once war begins, you cannot go on saying that.

Q Do you think he was sincere in those statements he made to you?

D Yes.

Q You think he was?

D Yes.

Q Did you tell him that Britain would fight?

D I always said that it looked to me like this. After Czechoslovakia my own private opinion was that they sort of hustled Chamberlain in the House, and probably Winston Churchill at the head of them, until the old man could not bear it any longer. He had had enough, and he thought he had been taken in by Hitler over Bohemia and Moravia, and they decided that they must have a war.

Q It would be quite right, would it not, if Chamberlain said he had been taken in, because when he came back in September 1938 he was under the impression that the thing had been settled and that the territorial ambitions of Hitler in Europe were satisfied? He would be quite right, would he not, in saying that he had been taken in?

D No, I do not think he would be right, because I think they asked Hitler to go and clear the thing up and stop a revolution.

Q Tell me this; when was the last occasion on which you saw Hitler?

D At Bayreuth, which ended the first few days of August 1939.

Q Did you not tell him then that there was no doubt that war was coming?

D No, he told me. He said, 'I am afraid they are determined on it.' Danzig had to be settled, and he said, 'I am afraid they are determined on it.'

Q Did you say to Hitler that you thought the Danzig question could be settled by negotiation, as our government had told him?

D He tried that with the Poles. In his view, it could all have been arranged quite peaceably had it not been that just at that time Chamberlain decided to give him this ultimatum.

Q Did you have a good many discussions with Hitler about political matters generally?

D Yes.

Q Did you convey to him the idea that you were rather sympathetic towards his point of view?

D Well, yes, I was.

Q That is quite honest. And therefore you conveyed that idea to him?

D Yes, I suppose so.

Q That was rather siding with him against this country?

D No, of course not.

Q It was rather saying 'My country is in the wrong'?

D Not my country. I absolutely differentiate between my Government and my country.

Q Well, the Government of my country?

D Not even that. I think it was just a few of them.

Q Supposing you had been dictator in this country in the year 1939, in the month of August, what would you have said to Hitler?

D If I had considered it was my business at all . . .

W Assume it was your business. Assume you are dictator.

D Even as dictator, I do not think Danzig would have been my business. It would be Hitler's business.

Q You would have had to take an attitude.

D If I took an attitude, I should say: 'Go ahead.'

Q Would you say: 'This is your business, not mine. I shall not interfere at all.'? Supposing you were absolute dictator and he said 'I am now going to take Denmark and Norway.' What would you say then? Would you say: 'Go ahead.'?

D Yes. It is not our business.

Q Supposing he said: 'I will now take Belgium and Holland.'?

D I think that question is unfair, because he did not want to do that.

Q I would not put an unfair question.

D You may not think it is unfair.

Q None of these questions is designed to put you into difficulty.

D I quite appreciate that, but I still think that it is an unfair question because it could not have happened.

Q Unless war had come, you mean?

D Hitler does not want Belgium. If you have ever been to Belgium, you will know it is a horrible place. It is just because they have to have the ports to fight us. They would soon go out again when the war is over.

Q Let me put this question. Do you think Hitler can be trusted?

D We always say, 'It is not a question of whether he could be trusted'. If you ask me personally whether as a man I trust him, of course I do. But we should not put ourselves in a position where we should have to trust. If I were dictator of this country, there would be no question of trusting anybody.

Q Would you agree, for example, that it would be simple to demonstrate that Hitler has broken his word?

D No, I would not agree.

Q 'With regard to my territorial claims in Europe,' he said, 'this is the last claim I have to make.'

D Perhaps it was, when he made it.

Q His pact with Poland, which was a ten-year pact of non-aggression, was rudely broken. How would you regard that?

D I think he was able to fix that with Poland. He said himself they would certainly agree to the motor road and a railway through East Prussia. He went so far as to say to me once in private conversation, and I believe also in a speech, 'No other German could have made a suggestion like this to the German people.'

Q He said this in a public speech.

D Yes, I believe he did. He was so extremely generous to Poland.

Q What you are really saying is this: expediency has governed the actions of Hitler, and therefore, if it is a matter of truth or breaking of pledges, when he says 'this is my last claim in Europe', it was at the time.

D People always talk about the 'last' claim. Nobody could imagine that Danzig was going to go on being ruled by the League of Nations for ever. It is almost too unreal.

Q It has gone on for a very long time.

D Yes it has – but what an extraordinary position.

Q It went on until Hitler was in such a powerful position by force of arms that he could say 'We are going to take it.'

D No, I do not think so. Danzig wanted to go back, did it not?

Q The situation really was this, was it not? Taking it quite broadly, as Hitler grew to power, and his movement became greater, so his policy adapted itself. There was the Rhineland, Austria, Czechoslovakia.

D You do not make any move, if you are a wise statesman, unless your country is strong.

Q In a word, you are a very great admirer of Hitler and his policy?

D I think he has done well for Germany. When it comes to England or the Empire, that is our affair.

Q Have you at any time, either in private conversation or in public, ever condemned anything that Hitler has done?

D I cannot remember anything.

Q With regard to the universal hate which is to be found among the people of this country for Hitler, and all his methods, do you find that difficult to understand?

D Not in the least, because I do not think they know any better. I dare say they have read nothing but the quotes of the *Daily Express* or the *Daily Herald*. If that had been my daily bread for ten years, I should feel the same.

Q You will forgive me putting it like this. You are exceedingly intelligent on these matters.

D You are being sarcastic.

Q I am not sarcastic. I am quite sincere. I put it this way because you do betray, in the real sense of the word, an intelligent view about these things. When you say 'they know no better', would you not agree that up to 1938, the policy of this country had been 'Let us avoid war at all costs.'?

D Of course I would not agree. That was not the policy.

Q The Government, headed by Mr Chamberlain, said, 'Let us avoid war at all costs.'

D But there was always in the background the Duff Coopers and the Winstons.

Q You mean Mr Winston Churchill? He was saying 'Let us be prepared at any rate.'

D I have known Winston Churchill since I was a small child. He is more interested in war than anything else in the world.

Q You have read a good many of his speeches, I suppose? Have you read any of his books? Have you read *Step by Step*?

D I have not read *Step by Step*.

Q You have probably read his speeches in *The Times*. Does not Mr Churchill, in all those speeches, say 'War is a most terrible calamity'.

D He would do. But I do not think he really feels that.

Q It is very difficult to believe that any man would really desire a cataclysm.

D I am sure he does not either. But I think he is this sort of person. He actually said to my husband once and in my presence too, 'When the country is absolutely calm, and things are going on peaceably, people like Baldwin can govern it. But when it gets to times of stress, then they want a man like me.' He feels that. There is something in it. He has a very great personality. There is no question of it. In a way, he welcomes times of stress.

Q That is your view about the matter. Supposing the Home Secretary were to release you and say 'You may go free', your views would remain unaltered? You would not be anxious for this country to win the war?

D I would be very anxious.

Q And Hitler defeated?

D They obviously could not do that.

Q They are going to do so.

D I could not actually ask to do anything, because our movement has been banned and there is nothing I can do. But my wish and hope would be perhaps we might negotiate a peace.

Q Would you like to see Hitler defeated?

D It is not a question of liking.

Q Let me put it in this way. Suppose it were possible for the might of this country, plus probably the might of America, to defeat Hitler and all his schemes, would you rejoice, or would you be disappointed?

D I think I should prefer not to answer that. It is not a question of rejoicing. I should be very glad to see it end, however it might end. Certainly it would be far better from the point of view of the world that we did not have another Versailles, if we could arrange it.

Q You say 'however it might end'. Most people want war to end very quickly and very badly. Supposing the war ended with Hitler being able to dictate the policy of this country, as he dictates the policy of France now, surely that would be a tragedy?

D Yes, that is the whole reason why.

Q The thing cannot be dealt with without the defeat of Hitler, can it?

D Well, in my opinion it could. I think we could stop now and not lose the Empire, although every week the war goes on deepens the tragedy, because already, I suppose, we shall have lost Gibraltar and the Mediterranean.

Q I just want to ask you one word about your attitude to the Italian war and Mussolini. Do you agree with the intervention of Italy in the war?

D I do not agree with it, but I quite expected it.

Q What possible kind of provocation has Italy had?

D The allies of a country naturally go in with that country. We had France and Hitler had Italy.

Q Why did not Mussolini come in at the beginning of the war? Why did he wait until France was on the verge of collapse?

D I do not know. I assume it was due to the blockade.

Q But the blockade had been going on all those months before Italy came into the war?

D There was no war, practically, going on before April. There was just a small bit here and there. There were just a few bombings.

Q Do you not feel a strong sense of indignation that Mussolini should intervene and declare himself an enemy of this country?

D He has always declared himself that, ever since sanctions.

Q On the contrary. What do you suppose Lord Halifax and Mr Chamberlain went to Rome for?

D Heaven knows. I do not.

Q Do you really think that the Italian people, for example, had any real quarrel with the people of this country?

D There is no question of a quarrel between the peoples. Our people had no quarrel with the Germans, or the Germans with us, or the Italians with us. It is always the Government. Ever since sanctions, I think the Italians have felt rather bitter.

Q Would you like to see Mussolini defeated?

D Yes.

Q By conquest?

D Yes. But I think it is a fearful thing. That is the whole reason why we are so against having a war.

Q Have you ever reflected, during your detention, why you have been detained?

D Yes, often.

Q What conclusion did you come to?

D It was because I had married Sir Oswald Mosley.

Q You are probably right. There is something in that. There are very few people in the country holding the views you hold.

D I think there are a huge number, but they are afraid to speak. We do not mind, but others do.

Q You are in a very exceptional position as being a close friend of Hitler.

D Yes, I suppose so.

Q That is the distinction. You are a close friend of Hitler.

D Yes, I suppose so.

Q He had discussed with you the most important aspects of foreign policy.

D Yes, he always did. Because, as I say, he is an authoritarian, and that is what interested me.

Q And you conveyed to him what you thought the real situation here was?

D Always. I never tried to hide that from him, in the least, as far as I knew it. I only knew it from the papers, because I had completely lost touch with anybody.

Q As to the kind of thing you would say to him, you would speak in strong terms of condemnation of Churchill and Duff Cooper?

D Yes. Strong condemnation of Duff Cooper and in [? defence] of Churchill. I remember him asking me about Winston, and I told him what I believed to be the truth.

Q Pretty well what you have told us today?

D I always said he was an extremely clever man and a great patriot according to his own lights, and I always said he was in quite a different category from Duff Cooper or Eden.

Q And I suppose you discussed all the leading personalities like Mr Chamberlain?

D I never knew Mr Chamberlain.

Q You probably gave your view of him?

D No, I did not. I asked him his opinion of him. He had seen him and I had not.

Q Did he tell you all about the meetings?

D Yes.

Q What attitude did he take – that he liked Mr Chamberlain?

D I think he thought he was all right – a typical democrat.

Q Did he think he had got the better of Chamberlain?

D No, I do not think so; I do not think that he wanted to get the better of him. I do not think he was at all pleased. It never struck me that he was pleased. I do not think he saw it as a great victory at all, in the way it was portrayed here.

Q You have never been a member of the British Union, have you?

D Always.

Q You have, of course, told us about your activities in connection with it. Really, it comes to this, does it not? Your views are utterly unchanged about anything.

D Why should they change? They are confirmed, really, by everything that has happened. I am, as I tell you, a friend of Hitler, but quite apart from that, it is not a question of how he governs his country. We felt it would be a great tragedy for England to declare war about such a thing as Danzig. If we were attacked, then it would be time enough to declare war. If only they had done what my husband said! People used to say, 'If you do that, he will come and take England.' But we always thought that was absolute nonsense. Even supposing it true, we should not have lost 35,000 men in France and spent millions of pounds on guns which we had to leave behind. We should have those to defend us, supposing he had attacked us. And with all those supplies, we could deal with Mussolini. Everything my husband said came true.

Q Are you able to communicate with your husband now?

D Yes, I write to him. But I have sent in a petition to be allowed to see him. It is more than three months since I did. My lawyer went to see Mr Peck at the Home Office, and Mr Peck saw no reason why I should not.

Q Where is he interned?
D Brixton. I wonder whether it would be possible.
Q I will deal with any matter of that kind at the end of this morning. For the moment, those are the questions I want to ask you. My colleagues may desire to ask you some questions.

SIR GEORGE CLERK You knew all the Nazi leaders, you say. Did you by any chance know Streicher?
D Yes.
Q Did you agree with his views?
D I do not know how far all those stories about him are true. I do not know if you have ever seen him yourself?
Q No.
D He is a very simple little fellow. Quite uneducated. He was a schoolmaster and speaks what you would call quite uneducated German. He used to amuse me with his talk and so on. I do not know that he is as bad as he is made out to be.
Q Did the Sturmer amuse you?
D Very much, but only from the pornographic point of view. The Sturmer is nonsense from beginning to end.
Q I suppose he had a certain amount of influence?
D Very little.
Q Did you know Himmler?
D No.
Q Or Frank?
D Frank I think I have met once, but I do not know him.
Q What impression did he make on you?
D I have often heard him speak. I think I have once had lunch with him. I cannot remember really.
Q I am asking you these questions because I gather from what you told the Chairman that you thought, having taken the Sudetenland, Hitler had in a sense no option but to go on and absorb Bohemia and Moravia?
D I believe this was inevitable.
Q I was only trying to get at your point of view. I was interested to know if you had had any close contact with the Franks and Henleins and certain elements of that sort who were involved in it.
D No I have not.
Q There is one small point going right the way back to the famous meeting in Hyde Park. You called it a rabble but I think it was mostly an ordinary sort of crowd, wasn't it?

D No, they had come up from the East End for it. It was mostly Jews.

Q When you gave your fascist salute, I do not quite understand why you did it? Was it a way of saluting our National Anthem or was it a protest?

D If they play it right through from beginning to end, that is how we salute on an informal sort of occasion.

Q At a British Union meeting, for instance, supposing the National Anthem was played, you would all stand and give the fascist salute?

D Not at the beginning, only in the last few lines. Not perhaps in a theatre or anything of that kind, but anywhere out of doors.

SIR ARTHUR HAZLERIGG At the end of one of your big meetings perhaps?

D Yes, at the end of one of our big meetings.

SIR GEORGE CLERK Your fascist salute on this occasion was made a good deal of in the press. You did not mean it as a protest? You made it as a good fascist?

D No. There was a little more to it than that. After I voted against the boycott they started growling, and making most sinister sounds.

Q Had you got the badge on you?

D No. I had quite a new coat on. I gave the salute in order more or less to underline the fact that I had voted against the resolution, and then they started running after me. It was terrifying.

Q They did not get you?

D No, they did not get me.

Q Did you run faster?

D Two young men appeared as if from nowhere and took my arms. They whispered to me that they were members of the movement. It was fortunate for me, as I was entirely alone.

Q One last point. On the whole, you have seen a great deal of it, and you approve of the Nazi regime?

D Yes, I suppose so. I think it has done very well.

Q Supposing the British Union and your husband were in power here, you would consider that a really good way to run the country?

D Oh yes, an excellent way.

Q With a Gestapo, and all the rest of it?

D I do not bother about the Gestapo. I do not know anything about that.

Q But you cannot run the Nazi regime, or the Bolshevist regime, or the Italian fascist regime, without something of that sort.

D You would have to have police, as we do, as far as that goes. I do not know anything about concentration camps, I have never been to one and I only know what people have told me.

Q It is rather more than concentration camps. No expression of opinion is safe.

D It is no worse than in England. It seems to me very similar. As a matter of fact, supposing somebody had come to the equivalent of what I have done in Germany, I do not think they would be in prison. I am sure they would never take a woman from a tiny baby. I absolutely know it. They have a respect for motherhood. It is terrible in our prison. The child was not actually taken away from me – I was sure London would be bombed. It was shortly after France gave in and I did not want him in London.

Q You do not think that if you talked about Duff Cooper, assuming he was a member of the Government, in the same way as you still talk of Hitler here, that that would be subversive in a fascist regime?

D Oh no, that's not so. Privately, you mean?

Q Yes.

D Oh no, they can talk privately. They would not talk in a disrespectful way any more than it would occur to anyone in England to talk about the King in a disrespectful way.

Q To talk about Himmler would be very dangerous in Germany.

D It might be, but I suppose people do it. In fact, I have heard all kinds of expressions passed in trains and so on. Every kind of thing I have had said to me. As to Streicher, practically all the leaders are down on him, always taking every opportunity of putting him down. I think it is a great shame.

Q You think it is a great shame, to put him in jail?

D I should think so. They do not all absolutely agree, you know, any more than we do.

Q But your general feeling, taking it all round, is that the Nazi regime is a good one?

D I think it is a marvellous thing for the Germans.

Q Do you think that the Czechs, Poles, Slovaks, Danes and Norwegians, French and Rumanians are going to find it a marvellous regime?

D I cannot answer that in just one word. But obviously, when the war is over, there will be no more question of the occupation of France or Belgium, and I think not even of Holland.

Q Or Rumania?

D Rumania is not in occupation now.

Q Not yet, but it looks very like it. You think all those acts are the inevitable consequences of our folly in embarking on this war?

D Yes, I do, because I do not think those particular things would ever have occurred. We must never let one power have all the Channel ports. I do not think that should ever have happened.

SIR ARTHUR HAZLERIGG I do not know if you think this is an unfair question. Whose fault do you think the war was?

D I think it was the fault of the people now in power in England.

Q I see.

D I know it was.

Q We will leave it at that. You say you know it was. Your husband's policy was always to prepare against any form of attack and to make the country great and safe?

D Yes.

Q You would not have called him a warmonger?

D No.

Q Winston Churchill's whole attitude during the last two or three years has been exactly the same idea of making the country safe and preparing armaments, and I gather from what you said to the Chairman that because he had done that he ought to be called 'warmonger'.

D I was very much against the weakness of total disarmament that we practically found ourselves in. I remember once talking to Hitler about it and something about Winston Churchill. I said, 'He is a patriot and cannot bear to see England weak. And he is hurrying on our armaments as much as he can.' I remember him saying 'I am the German Churchill'. I admired Winston for that tremendously; I thought it was splendid. But I think in his own character he is a person who enjoys war, and always saw himself as a great leader.

Q If you were a friend of his, you might say his best qualities would be brought out in a time of crisis.

D I think he is always looking back to Marlborough and I think it would be his dearest wish to lead an army in the field. But of course he has not been able to do that.

SIR GEORGE CLERK There is a point that occurred to me when you were talking about Winston Churchill earlier on. I think what you really mean may be put in this way: he regards as the highest form of human activity the command of a great army in the field – let us say the career of Alexander the Great. He is almost obsessed by the career of Marlborough but for all that he would not deliberately work for his country to go to war in order to have that position, whether as Prime Minister or leader of the people.

D Even I would not suggest that. I do feel that he was frightfully jealous of Hitler. The last time I saw him, he wanted to hear about him. I think he felt himself 'If only I had a chance like that.' Of course he knew he never could. It was too late to change. I think he thought 'How maddening – I've missed it again.' He was always thinking of Caesar and Alexander.

Q You do not think he, as it were, seized this crisis?

D No of course not. I do not think he is wicked at all in that sense. I think he is a patriot. In a way I think he longed for war but he did not want to bring it about.

SIR ARTHUR HAZLERIGG You have a great contempt for democracy?

D Yes.

Q And so has Hitler?

D Yes.

Q How can you say then that he admires this country, which is a democratic country?

D Because he admires England tremendously for its general characteristics.

Q But I should have thought the characteristics of this country for years were democratic?

D I do not think it was democratic when we got the Empire and so on. We did not go to the negroes and say 'Look here, you vote to have your rulers.' We went and took bits of the world.

Q I see what you mean. But that was the Government of this country. The country may have been democratic, even when they did some undemocratic things.

CHAIRMAN That is a quality that Hitler admires – that goes and gets something when he wants it?

D He admires a country when it is growing and expanding.

SIR ARTHUR HAZLERIGG I suppose he thinks we ought always to go on growing and expanding?

D He thinks all the Europeans should have their own 'living space' which they should be allowed to develop.

Q Supposing he had most of it?

D No. He wanted Europe, and we could have the world, as far as we want. I do not think he wanted more, except the colonies back.

Q And all Europe?

D I think all Europe east of the Rhine. Not to occupy physically, but simply as an economic unit.

Q It is very much the same thing, is it not, in Hitler's idea?

D Sir George Clark knows more about this than I do, obviously. But I believe those countries are very backward, and in a sense they have to be developed by the Germans or the French. Even Czechoslovakia had to have the French to organise their army, the same as the Germans organised the Russian army. You cannot quite call them advanced peoples, can you?

SIR GEORGE CLERK The Czechs, very far from being a backward people, are an advanced people. The organisation of the army was due to the fact that they never had one. They had always been part of the Austrian army or the Russian army.

D Take for instance Rumania. Everybody who has ever been there, and who knows it well, always tells you they practically do not know who governs them.

THE CHAIRMAN The other point I intended to ask you is this: How many sisters have you?

D Five.

Q There is Miss Unity Mitford?

D Yes. They are all shades. I have a social democrat sister, and a communist sister.

Q What is the name of the communist sister?

D Mrs Romilly.

Q You said you knew that London would be bombed so took your child away. Did Hitler ever speak about the bombing of London?

D No.

Q Did you by any chance know Joycc – Lord Haw-Haw?

D I have seen him but I do not know him.

Q You were not attached to the movement when he was a member?

D Yes, I once heard him speak. That is how I saw him.

Q Some of the questions we have been asking, which have ranged over a very wide field are, I think, from many points of view very difficult questions to put to anybody. But those are really the questions we wanted to ask you. I would, however, like you to feel that if you would like to make any observations about anything we have raised, or anything we have not raised, dealing with the general position, to why you should not be detained, we should be very glad to hear you.

D The reason why I think I should not be detained any longer, or indeed at all, is that they have banned the movement, and my husband has said that it is not to carry on any activity until the war is over; I cannot see any reason for keeping us any longer in prison.

Q Do you think there is any probability that people will disregard the order of the Home Secretary and carry on the movement in an underground fashion?

D They would never do what they were told not to do by him.

Q I have never really quite understood why your husband did not issue a manifesto to his people after the Low Countries were invaded and say, 'The whole policy of the Union must be reversed'?

D He did in a sense, in a statement he made at the beginning of the week he told them to obey the law and fight for our country.

Q But when the Low Countries were invaded he did not say to his people 'The moment I told you about, when we should stand in defence of the country, has arrived.' That would have been the finest thing for him to prove his sincerity that I can imagine. I could never imagine why he did not do that.

D I think he was always expecting – I am certain that I did – that France would give in, because France was so riddled with communism and never had her heart in the war. I think that is what he more or less expected.

Q What he did was take steps to see that the organisation was carried on after he was interned. We have the evidence. He told us quite plainly that certain . . . were given a note to say 'This gentleman has the confidence of the Leader' and they were to carry on the organisation if the Leader was interned. That was always the intention. Lots of humbler folk from the British Union – the rank and file, shopkeepers and all sorts of humble folk who come here – say 'We joined the movement because it was a patriotic movement. Britain for the British and if this country were attacked we should defend it to the last.' That being so, I could never understand, as I say, when the Low Countries were invaded and the German forces were coming to the coast, why Sir Oswald did not say: 'That moment has now come.'

D I think he still thought that we could have a very fair peace.

Q But you see, it would have meant he was saying 'The moment has now come when we should defend this country and defeat Hitler'.

D That was not why he did not say it, of course.

Q I am suggesting that to have made such a statement would have involved saying that and therefore he shrank from saying it.

D I do not think so in the least. I was with him at the time and I remember him taking a great interest in the war, as anyone does. He had always recognised that as a possibility. I think probably the Germans themselves were surprised how quickly they had won.

Q Was there anything further, Lady Mosley?

D There is the question about a visit to him.

Q Is there anything further you wish to say on whether you should be released or that you be detained?

D No, absolutely nothing. It does seem to me, however much you recommend it, I shall never be released until you have got rid of Sir John Anderson. It seems to me that he is our enemy in the matter.

Q You quite appreciate we are only an advisory committee?

D I cannot think anyone would want to detain one in prison. It seems almost too mad and crazy. Above all, my husband is the most tremendous patriot and adores his country. I feel it very strongly. Of course I am very bitter about it. At the same time, there you are.

The witness withdrew.

The Committee consisted of Mr Norman Birkett, KC (Chairman); the Rt Hon Sir George Clerk, CGM, GCB; Sir Arthur Hazlerigg, Bart; and Mr G. B. Churchill, CB, Secretary

Index